D1450951

Michael Ryan's Writings on Medical Ethics

Philosophy and Medicine

VOLUME 105

Founding Co-Editor
Stuart F. Spicker

Senior Editor

H. Tristram Engelhardt, Jr., *Department of Philosophy, Rice University,*
and Baylor College of Medicine, Houston, Texas

Associate Editor

Lisa M. Rasmussen, *Department of Philosophy, University of North Carolina*
at Charlotte, Charlotte, North Carolina

CLASSICS OF MEDICAL ETHICS 3

Series Editor

Robert B. Baker, *Department of Philosophy, Union College, Schenectady,*
New York and Center for Bioethics, University of Pennsylvania, Philadelphia,
Pennsylvania

H. Tristram Engelhardt, Jr., *Department of Philosophy, Rice University,*
and Baylor College of Medicine, Houston, Texas

Laurence B. McCullough, *Center for Medical Ethics and Health Policy, Baylor*
College of Medicine, Houston, Texas

For other titles published in this series, go to
www.springer.com/series/6414

Howard Brody · Zahra Meghani
Kimberley Greenwald
Editors

Michael Ryan's Writings on Medical Ethics

 Springer

Editors
H. Brody
Institute for the Medical Humanities
University of Texas Medical Branch
Galveston
USA
habrody@utmb.edu

Z. Meghani
Philosophy Department
University of Rhode Island
Kingston
USA

K. Greenwald
Ada, USA
greekima@trinity-health.org
kgwald@aol.com

ISBN 978-90-481-3048-1   e-ISBN 978-90-481-3049-8
DOI 10.1007/978-90-481-3049-8
Springer Dordrecht Heidelberg London New York

Library of Congress Control Number: 2009929914

Printed on acid-free paper

Springer is part of Springer Science+Business Media (www.springer.com)

Acknowledgements

We began this project as a team of four. Sara Schuman assisted us in our early stages until other responsibilities forced her to withdraw.

During most of the process of preparing this volume we benefited from the facilities provided by the Center for Ethics and Humanities in the Life Sciences, Michigan State University. We are especially grateful to Jan Holmes and the late Beth McPhail for their assistance in scanning the Ryan texts.

We could not have completed this volume without assistance from a great many people, most of whom are listed in the Endnotes to the Introductory Essay in connection with their specific contributions. Our primary research source was the Wellcome Library in London with its magnificent history of medicine collection, and we are grateful to the various staff of the Wellcome who assisted us in numerous ways.

As none of us are historians of medicine by training, we have relied all along on advice from our better-educated colleagues, including Peter Vinten-Johansen, Declan O'Reilly, Harold Vanderpool, Jason Glenn, and Dayle DeLancey. Our editors, Bob Baker and Larry McCullough, provided considerable advice and guidance on matters both historical and philosophical.

We wish especially to acknowledge the steadfast assistance of Mary Hope Griffin, without whom the many perplexing Latin passages in Ryan's text would have refused to divulge their secrets. She soon discovered that her extensive knowledge of classical Latin was of limited help in deciphering what apparently passed for that language among English medical practitioners in the early nineteenth century. On one occasion she explained to us that the only way she could make sense of some of it was to imagine that it was very bad Italian.

Contents

Part Two. Selected Writings by Ryan

Explanatory Note on Selections from Ryan's Works

Part I
Introductory Essay: Michael Ryan and Medical Ethics

Howard Brody, Zahra Meghani, and Kimberley Greenwald

I Medical Ethics in the English and Scottish Enlightenment

His contemporaries remembered Michael Ryan (d. 1840) as an Irishman, a learned and kind-hearted colleague, a prolific writer, editor of the *London Medical and Surgical Journal*, and a colorful disputationist who regularly went toe-to-toe with Thomas Wakley (1795–1862), the acerbic and outspoken founding editor of *The Lancet*. Later historians identified him as one of the founders of modern medical ethics, and the first to link the medical ethics of his own day with the historical tradition of the Hippocratic Oath. Ryan was a seminal figure in the development of medical jurisprudence and an original thinker on research ethics, apparently the first to publicly defend the necessity of obtaining the informed consent of the research subject.

Ryan appears to have been the only person in Great Britain, during the middle years of the nineteenth century, to attempt to produce a systematic account of medical ethics, for use especially by medical students. While relatively little of his thinking was original, Ryan is noteworthy for addressing medical ethics under the broader rubric of medical jurisprudence, identifying some important ethical issues in his other special-interest area, obstetrics, and breaking some new ground with his ethical reflections. He also advocated addressing medical ethics as a critical portion of the curriculum for medical students.

When, around 1830, Michael Ryan began to compile a work on medical ethics, the subject was dominated by two figures from the end of the eighteenth century, John Gregory (1724–1773) and Thomas Percival (1740–1804). These writers, in turn, were heavily influenced by the philosophy of the Enlightenment period, especially as it was formulated in Scotland.[1] The ethics of the Scottish Enlightenment generally, and the work of Gregory and Percival in particular, therefore provide a context for Ryan's role in the history of medical ethics.

H. Brody et al. (eds.), *Michael Ryan's Writings on Medical Ethics*,
Philosophy and Medicine, Vol 105,
© Springer Science+Business Media B.V. 2009

Understanding the ethical writing of the Scottish Enlightenment is very challenging for us at the start of the twenty-first century. It is difficult not to see the Enlightenment through the prism of whatever issues most concern us today. For example, we imagine that writings on ethics can be divided generally into theoretical and practical works, and that practical works can be subdivided into those that are clearly based on or derived from theory, and those that are not. We also imagine that a fundamental feature of medical ethics is the physician-patient (or patient–physician) relationship, and that the extent to which the physician does or does not respect the autonomous wishes and choices of the patient is a major line of demarcation between an "old" ethic, such as is suggested in the Hippocratic Oath, and a "modern" ethic suitable for today's practice.

Adopting this presentist perspective, we might conclude several things about Gregory and Percival. Since Gregory seems to attend more to the physician–patient relationship, while Percival focuses on relationships among physicians, we might conclude that Gregory is a pioneer of "modern" medical ethics while Percival is primarily a conservative figure from the Hippocratic past. We might also see in Gregory more explicit mention of various theoretical strands of ethical thought and conclude that his ethical writings are theoretically grounded in a robust way, while Percival represents a "merely" practical or applied medical ethics.[2]

While it is ultimately impossible to stand free from our present vantage point, we may at least approximate an interpretation that better situates medical ethics within the intellectual life of the late eighteenth century. This latter interpretation might see Gregory and Percival, despite their differences, as allied within a single tradition of thought. Both are transition figures at the intersection between an old, guild-based medicine and modern institutional and professional medicine.[3]

The educational background of Gregory and Percival shaped their thinking on medical ethics. They both did some of their medical training at the University of Edinburgh, which Ryan attended several decades later. Edinburgh was the home of the Scottish Enlightenment culture. The medical school had inherited the mantle previously held by Herman Boerhaave (1668–1738) and the University of Leiden, as the dominant school for teaching modern medicine.[4] The emphasis in both medicine and religion was on tolerance and free rational inquiry, associated with a loathing of sectarian dogmatism. The medical professors called this "Hippocratic teaching," but this was a modernized Hippocratic approach that stressed critical thought, a Hippocrates of the Enlightenment. "Eventually a new Hippocrates emerged who spoke in Baconian cadences and who, not surprisingly, espoused the popular mechanist philosophy of Newton and Boyle adapted by Boerhaave and his Edinburgh disciples."[5] Since English and Scottish thinkers in the eighteenth century were distrustful of grand systems, Hippocrates was held up as a model of the physician who allowed himself to be taught by nature and not by theory.

Edinburgh influenced its graduates in several more concrete ways. The Edinburgh system allowed access to medical school for religious dissenters, while Oxford or Cambridge accepted only those who would subscribe to the Church of England faith. Lectures were given in English rather than in Latin, which remained the language of instruction in the English as well as many continental universities. Professors were not paid on a salary basis, but supplemented their clinical practice incomes with fees collected from students for admission to their courses of lectures. As a result the professors tended to give more practical and appealing lectures, and were actively involved in the life of the university and community.[6]

Gregory and Percival shared an immersion in the ethical thinking and writing of the Scottish Enlightenment, led by figures including Francis Hutcheson (1694–1746) and Gregory's cousin, Thomas Reid (1710–1796). Their ethical thinking flowed seamlessly into a natural-law view of both science and religion. Neither Gregory nor Percival thought it possible to have ethics without a religious basis. Both were apolitical and represented the conservative Enlightenment. They had no wish to overturn the social order that seemed to provide a stable home both for the medical profession and the advancement of medical science, and for civilized life generally.

Today's label, "practical ethics" as opposed to "theoretical ethics," does not track very well the Enlightenment goal of producing a smooth integration of practice and speculation. "To their contemporaries, ... Gregory [and] Percival ... were qualified to write medical ethics only because they were practical men of speculative bent. Not only did they have theoretical expertise in their own fields but they were also practical moralists."[7] "[Both] were recognized by their contemporaries as moral philosophers, a reputation based on their previously published and internationally well-received works on literary, moral and medical topics."[8] Gregory and Percival believed that medical science must follow the recipe for the inductive method laid down by Francis Bacon. In their ethics, they followed another of Bacon's teachings – that there are two types of duties, general duties owed by all persons to all others, and specific duties owed by virtue of one's office or station in life. Bacon argued further that ethics of the special duties are best written about by those with the most intimate practical understanding of that office. Hence, physicians are best situated to write about medical ethics. Bacon, in *The Advancement of Learning*, realized there would be tendency to magnify the importance of one's own office, leading to a danger of distorting one's ethical pronouncements. His proposed corrective was to require that anyone writing on special duties must address the vices of that office as well as its virtues.[9]

Duties peculiar to special offices or roles could be viewed as social contracts, since the natural order included the organization of society and the interrelationships among its parts. To Thomas Reid, the physician–patient contract was blatantly (but not solely) commercial. The patient voluntarily seeks the physician's advice and pays a fee, and the physician agrees to prescribe according to the best of his skill. "This idea of the implied contract which binds physician and patient together in a moral association that is a small constituent of the common good, is the central point in Enlightenment medical ethics."[10]

The importance of duties incumbent upon social roles help to explain why politeness and decorum are so frequently mentioned in the ethical writings of the period. Some have complained that these "ethical" works are primarily about "medical etiquette" and not about ethics at all.[11] Politeness may easily be misunderstood as "merely" etiquette when in fact it was seen as a core virtue of the scientist. The gentleman-scientist who debates rationally, with liberality and candor, was contrasted with the disputatious, pedantic sectarian who did nothing to advance true scientific knowledge. Some of Gregory's and Percival's comments on the decorum of the physician harked back to Cicero, who agreed that mere fashion (in clothing, for instance) is not of ethical interest. Nevertheless, one's decorum in part establishes what role one asserts for oneself and whether one is going about it honestly and devotedly. Hence some discussion of decorum is appropriate as part of the moral analysis of special duties owed by virtue of offices.

IA   John Gregory

Gregory was born to an aristocratic Scottish family, 9 years after the next-to-last Scottish uprising was put down by England. Gregory's philosophical style was engendered by the Scottish national identity crisis. From his involvement in the Scottish Enlightenment philosophical tradition, he developed the ethical underpinnings that would later form his concept of medical ethics.[12] Women in the Gregory family were articulate and intelligent, influencing his future view of women and how they should be treated in medicine and in society at large.[13]

Gregory's father and half-brother were doctors who studied and taught at Edinburgh. Gregory himself attended Edinburgh from 1742 to 1745. He traveled to Leiden for further study at the school that was viewed as Edinburgh's ancestral home. He was then called back to King's College, Aberdeen, where his grandfather had been principal and where he had received his pre-medical education, and for 3 years taught math, natural history, and moral philosophy. He left teaching to start a private medical practice in Aberdeen in 1749. In 1754 Gregory married and moved to London to practice medicine.[14] He spent only 2 years in London before he returned to King's College to take up a post vacated by the death of his physician-brother, James. While in London he met Elizabeth Montagu, "Queen of the Bluestockings" and her circle of women intellectuals.[15] Mrs. Montagu and Gregory became good friends, and she was a confidant to his four daughters after his wife died.[16]

Teaching at King's College for the decade, 1756–1766, Gregory tried and failed to create a modern medical school there, but succeeded in forming the Aberdeen Philosophical Society with the help of his cousin, Thomas Reid. Gregory's most important popular work, *A Comparative View of the State and Faculties of Man with Those of the Animal World* (1765), grew out of discourses delivered to the Society.[17] By this time Gregory had a clear philosophical position, sharing Reid's view on importance of religion in morals and on the methods of the Common Sense philosophy of the Scottish Enlightenment. He relied on Bacon's rules of induction for the empirical foundations of any science – under which heading he included morals – and was skeptical of any hypothetical system not grounded in induction. He also shared Reid's interest in the mind–body connection.[18]

In 1766, Gregory went back to Edinburgh, where he occupied the chair in medicine until his death 7 years later. As a medical lecturer he was overshadowed by his more famous and influential colleague, William Cullen (1710–1790). But he established an excellent local reputation both as a physician to the elite and as a writer-intellectual, characterized by Boswell, "as amiable as he was benevolent."[19] His reputation and his writings therefore made it easy after his death to place Gregory on a pedestal as a paragon of virtue.[20]

The medical reformers, first at Leiden and then at Edinburgh, advocated learning at the bedside of the patient from concrete experience, rather than blindly following the dictates of doctrine.[21] In Gregory's day, the most obvious voices of outmoded doctrine came from the medical corporations, chiefly the Royal College of Physicians of London and its sister organization, The Royal College of Physicians of Edinburgh. The corporations' overriding concern was self-interest and the preservation of their ancient privileges. They issued strong rules against criticizing a colleague or consulting with any physician who practiced without their license, however impressive that physician's training and credentials. In the period we are considering, only a graduate of Oxford or Cambridge could become a Fellow of the Royal College of Physician of London, even though all dispassionate observers regarded the curriculum at Edinburgh as far superior to that of the English universities. Gregory's goal was to improve medicine from the ground up "by relieving its estate from the reactionary, anti-scientific and therefore harmful forces of the medical corporations."[22]

Gregory understood Bacon's method of investigation to consist of observation, or experience. Experience should be obtained in a disciplined manner to act as the appropriate progenitor of new thoughts, thus making it experimental in nature. Gregory agreed with the direction both Bacon and Boerhaave thought medicine should take, beginning with bedside observations of the sick and then proceeding to analytical thought into the cause of disease. He contrasted this approach with the synthetic way medicine was being taught at Universities, relying on rote acceptance of the humoral teachings of Galen and Aristotle. Gregory believed that only the analytic, experimental method would advance scientific concepts within medical practice.

The Scottish Enlightenment philosophers sought to study scientifically what we would call the contents of the mind and what contemporaries called 'ideas' or beliefs. They tried to make intelligible the external world, God, and the self, as a way of understanding the principles of morality. Gregory's morals laid special emphasis on the concept of sympathy, which he (following Hume) viewed in Baconian, experimental terms. Sympathy functions analogously to the long nerves and muscles that, according to the physiologists, connected the mind to the body. On this view, the mind is active, the body (initially) inert. The nerves and muscles become an apparatus whereby the active mind can impart activity to the body. The Scottish Common Sense philosophers applied a similar active/inactive distinction to reason and passion (emotion), with reason seen as inactive and passion as active. Passion was thus a more powerful moral motivator than reason alone. The "experiment" occurs when the physician (e.g.) tries to imagine himself in the place of the patient and feels or seems to feel the patient's suffering. The activity of sympathy can therefore be seen as a moral and social nervous system, naturally binding everyone together. The physician is more powerfully moved to respond in a useful way to the patient's distress through the exercise of sympathy than he ever could via reason alone divorced from passion. For Gregory, sympathy is central to the moral character of the physician.[23]

Gregory himself did not practice a specific religion, but had a strong belief in God and agreed with most of his fellow Scottish Enlightenment philosophers that religion played a key role in the underpinnings of morality. Religion is important as it admits of the Supreme Being who supports the structure of the world by laws that are "steady and uniform" but that are often unknowable to mere humans. A man with sound understanding realizes that extensive knowledge is the truest teacher of humility. If a man has a weak understanding, he is bound to get too caught up with secondary causes and overlook the "first and great cause" (God).[24]

Gregory first lectured to the Edinburgh students on philosophy of medicine and medical ethics. In a later part of the curriculum, students went to the Royal Infirmary of Edinburgh to see their patients.[25] The Royal Infirmaries of Edinburgh and Britain had a structured environment regarded as perfect for study of the "natural history of disease and the effects of medical and surgical interventions."[26] Many of the variables that influenced the disease process could be controlled within the infirmary setting. Infirmary patients ate the same food, kept the same hours of sleeping and waking, breathed the same air and drank the same water. This allowed physicians to observe the effects of the disease process upon the patient in a setting that controlled for many ancillary factors.

If the Infirmary setting had major advantages from the standpoint of Baconian science, it also created a new set of ethical concerns. McCullough argues, "The Royal Infirmaries made a crucial contribution to the social transformation of the *patient-physician* relationship into the *physician-patient* relationship."[27] When the physician visited the patient at home and the patient paid for the physician's services, the power relation was dominated by the patient. Patients in the Royal Infirmaries were charity rather than paying patients, and had left their homes to enter the life of a regimented institution. The new power relation between the physician and patient was dominated by the physician. This shift in power dynamics challenged Gregory to write medical ethics in a scholarly manner not just for paying outpatients, but for non-paying Infirmary patients as well. Gregory's main moral concept was sympathy; and sympathy could serve as an antidote to class and religious differences. Thomas Percival, coming a generation later, more fully developed a medical ethics to grapple with the new institutional settings in which medicine was increasingly practiced.[28]

Gregory approached the duties and office of the physician first by addressing temperament and understanding.[29] He was interested in two applications of temperament – what best suited an individual to medical practice, and what best prepared a person to advance medical science through new discoveries. Each physician must rest on his own judgment and experience alone. Gregory was fully aware that the physicians of his day had a very uneven foundation of scientific knowledge, and that market conditions encouraged them to denigrate their rivals' methods of practice.[30] A virtuous physician would not participate in such collegially derogatory practices, but would instead dispassionately try to establish which method of treatment worked best. One implication of this view was that Gregory tended to see the gentleman of independent means as a more scientifically inclined individual than one who needed to make his living through medical practice.[31]

Gregory listed the virtues appertaining to the role of physician under the heading of "moral qualities." The most important quality is that of humanity, or sympathy. "Sympathy produces an anxious attention to a thousand little circumstances that may tend to relieve the patient; an attention which money can never purchase: hence the inexpressible advantages of having a friend for a physician."[32] In keeping with an Aristotelian model of virtue, Gregory worried about excesses as well as deficiencies of virtuous qualities. Physicians often had to act under emergency conditions, in which too much sympathy for the patient's suffering could impede the decisive action that might save the patient's life. Gregory argued, "Doctors can feel what is amiable in pity without suffering it to enervate or unman them."[33] Similar balancing was required in giving instructions to patients – avoiding both undue flexibility and undue rigidity – and in responding to apparently trifling symptoms, avoiding both negligence and sarcastic ridicule on the one hand, and excessive attention and anxiety on the other. Gregory warned the students of the common falling off of virtue as the physician matures in practice. In his early years, a physician might be solicitous and attendant, but once established, often becomes "haughty, rapacious, careless, and sometimes perfectly brutal in his manners."[34] Physicians should maintain temperance and sobriety at all times. Because a physician knows the private issues of families, it is important that he remain circumspect in his public utterances, especially where women are concerned. A physician should never take advantage of his office to seduce women in his care.

Gregory next addressed physician decorum. A physician should make considerable effort to fit in with the community where he is practicing. Adopting the manners and dress of the area will allow a physician to further his practice and create a comfortable situation for the patient who is receiving medical care. Once again, Gregory was thinking about both the physician's duties, as a practitioner and as a scientist. The physician who dressed in an excessively gaudy or an excessively slovenly fashion would not have a comforting and reassuring effect on the vulnerable patient, and the physician who imagined that ostentation in dress would impress the public more than real scientific attainments failed in his duty to advance the profession.[35] A physician should listen to his patient's suggestions and consider them rather than imagining that his dignity has been assaulted. Gregory agreed that it was sometimes best to withhold the truth of a dangerous prognosis from patients at a time of crisis, when emotional shock could by itself bring about the feared outcome.[36] He opposed concealing the real situation from the relations of the patient, or abandoning the patient at the hour of their death. Gregory agreed with Bacon that bringing comfort to the dying was as much a defining duty of the physician as curing disease, and was also in keeping with his moral ideal of physician-as-sympathetic-friend.[37]

Gregory approached the contentious issue of consultations among practitioners from his usual dual perspective – what best serves the needs of the patient? What best advances medical science? The medical corporations attacked any public dissension among physicians as bringing dishonor upon the professional body. Gregory also opposed open dissension, but for other reasons. It was highly unlikely that either the patient would get better, or that medical science would be advanced, if the physicians gathered in consultation began bickering. To assure that consultations truly served the needs of the patient rather than the false pride of some physicians, Gregory urged that any medical practitioner who possessed a skill relevant to the treatment of the disease be allowed a role: "If a surgeon or apothecary has had the education, and acquired the knowledge of a physician, he is a physician to all intents and purposes, whether he has the degree or not, and ought to be respected and treated accordingly."[38] Gregory here made perhaps his most radical proposal with regard to the typical practices of his own day. He took to heart Bacon's admonition that anyone who would write of the ethics of a profession would have to address the profession's defects as well as its virtues.[39]

IB   Thomas Percival

Ryan thought the work of Percival sufficiently important to reprint most of the core of Percival's *Medical Ethics* nearly verbatim. Percival supplemented that core with copious Appendices that make up about half of the final volume and expand our understanding of Percival's approach to ethics.

Percival was born in Warrington, near Manchester.[40] Orphaned in infancy and raised by his eldest sister, Percival chose a medical career after his uncle Thomas, a physician, died when the young man was 10 and left him an inheritance sufficient for his education. At age 17, Percival enrolled in the new Warrington Academy set up by his former tutor, Rev. John Seddon. Seddon had left the Anglican Church to found a Unitarian congregation. Although non-Anglicans (Dissenters) were still denied many legal rights, the Unitarians were a powerful and wealthy group in the Warrington area, and had considerable influence on the development of Manchester as the first modern industrial city in Europe.

The Warrington Academy welcomed students from any faith, and prepared them for careers in commerce as well as the professions. All its tutors were graduates of Scottish universities and were heavily influenced by (and indeed actively participated in) the Scottish Enlightenment. Haakonssen thinks it important that Warrington was the first academy founded on institutional principles. Previously, a school was more or less a large house, with a chief tutor as head of the family. The

Warrington experience may have later helped Percival realize the need to develop a medical ethics suitable for institutional practice.[41]

In 1761, Percival traveled to Edinburgh to enroll in the medical school. There he met and admired David Hume, and while Hume's atheism challenged some of his beliefs, Percival did not abandon his Christian faith. Rather than graduating at Edinburgh, he went on to Leiden, where he took his M.D. in 1765. He visited Paris and then walked the wards of the great London hospitals before he returned to Warrington, married, and started practice. He moved to Manchester in 1767, set up a new practice and began his writing career. He later served 21 years as president of the Manchester Literary and Philosophical Society, which he had helped to found. Percival also assisted in establishing two local colleges.

Like Gregory, Percival declared his allegiance to the Baconian ideal for medical and scientific advances. He wrote numerous essays, mostly oriented toward public health, and experimented on the uses of cod liver oil; but his biographer-son had to admit that he had little real talent for experimental work. In 1773, Percival and some colleagues completed a study of bills of mortality in Manchester and Salford, which they presented to the Royal Society. The study formed the basis for several reform proposals, the early creation of a board of health, and the opening of a fever hospital. Percival was also involved in a failed project to promote mass inoculation against smallpox.

The Manchester Infirmary and institutions like it were funded by private charity and staffed on a voluntary basis by medical men and clergy. They provided limited, short-term care for the laboring poor who were not covered by the existing Poor Law relief system. They excluded the majority of the sick poor who had chronic diseases and could not readily return to employment. Percival viewed these institutions' relationship with the poor as a social contract. Justice required that since the poor were willing to work to make articles of value to the rich, the rich had a reciprocal duty to assure the poor a decent wage, assist them in sickness, and provide education for their youth so that a general upward movement in society could continue. Unlike some other charities of this era, the hospitals were successful; they never had a serious financial failure or a crisis of public confidence.

With new medical knowledge came expanded spheres of duty for physicians. By 1803 the Manchester Infirmary included a lunatic asylum, an out-patient dispensary, and the fever hospital. By this time the hospital was emerging as an essential laboratory for clinical medicine, both for observation and experimentation. Percival as a public health expert had to admit the downside – admission to the hospital carried with it an increased risk of dying from contagious fevers. Percival favored home births and out-patient management of many disorders to avoid the risks of hospitalization.[42]

According to his biographer-son, Percival epitomized the virtue of candor, defined as a habitual disposition to form fair and impartial judgments and to avoid bigotry and prejudice. He was an expert at conversing rationally and bringing disagreeing parties into concurrence. When the Manchester Infirmary suffered a crisis among its medical staff in 1790, "[I]t is hardly surprising that [his associates] should have asked him to be Manchester Infirmary's Justinian…. He had a scholarly reputation as one of Britain's best-known medical moralists; he had previously written extensively on hospital policy; and his views had been in circulation for many years."[43] The 1790 crisis was precipitated by an old guard of Tory physicians and surgeons whose families had long enjoyed a relative monopoly in Manchester and who opposed the Trustees' plans to expand the medical staff so as to increase services to the poor. In 1792 the institution asked Percival to prepare a set of rules that would prevent future disputes of this nature, and the associated unseemly publicity.[44]

When Percival prepared the Preface to the second (1803) edition of his code of professional conduct for physicians and surgeons, he noted that the code written for his Infirmary colleagues in 1792 formed the first chapter of the new work. He now wished to expand that work into a general system of medical ethics "that the official conduct, and mutual intercourse of the faculty, might be regulated by precise and acknowledged principles of urbanity and rectitude."[45] He proceeded to list a diverse and distinguished set of readers who had reviewed his manuscript, including Thomas Gisborne (1758–1846) and William Paley (1743–1805). Percival saw his work as a supplement to Gregory, Gisborne's chapter "On the Duties of a Physician," and the ethical proclamations of the Royal College of Physicians – works he saw as helpful but "not sufficiently comprehensive for the existing sphere of medical and chirurgical duty."[46]

Percival explained in his Preface that the earlier draft of his code that he had circulated in 1794 was "originally entitled 'Medical Jurisprudence'; but some friends having objected to the term Jurisprudence, it has been changed to Ethics."[47] In this manner, less than 30 years before Ryan wrote on the subject, Percival became the first writer to employ the phrase "medical ethics."[48]

Percival, like Gregory, devoted a good deal of his attention to the virtues and character of the physician. Unlike Gregory, Percival devoted attention to the institutional character of the hospital. Both issues were informed by the Baconian ideal of scientific progress. When Percival spoke of the reputation and dignity of the medicine, he was not defending a professional monopoly. Rather he was tacitly admitting that in previous years medicine's reputation had been based on an inadequate foundation, and greater scientific accomplishment was required to shore up that foundation. The hospital was the laboratory where that scientific advance could best occur. Gregory had assumed that only the elite gentleman-physician could properly move medical science forward. Percival, living among the commercial class in a modern industrial city, looked instead to cooperation among all classes of practitioners to provide the basis for medical progress. According to Haakonssen, Percival "realized that creation of medical knowledge would no longer be the domain of independently wealthy gentlemen but 'interested' ones who, like himself, lived by their profession and nevertheless remained committed to the Baconian ethos of the Royal Society."[49] Percival wished to show that an "interested" professional could still be a Christian gentleman and scholar.[50]

While the ideal physician was a gentleman, Percival saw the need to demarcate medical ethics from the gentleman's code of honor. The uproar at the

Manchester Infirmary in 1792, like most unseemly clashes within the medical fraternity, had occurred precisely because the "gentlemen" of the faculty felt their honor had been threatened and insulted, and because their code of behavior gave them no other framework within which to settle disputes. Percival's ethics can be seen as an effort to reconceptualize the physician from a private gentleman to a public professional, and to separate questions of personal honor from those of professional ethics.[51]

For Percival, as for Gregory, "[T]he ideal physician was, as ever, an attentive, humane, indulgent, confidential, tactful, faithful, punctual, honourable, candid, and, not least in an age when sobriety was not always considered a virtue, a strictly temperate man."[52] Gregory argued that the physician should be the friend of the patient, not as a social equal (especially in the case of infirmary patients), but as the patient's protector. In Percival's day, the increasing numbers of poor patients, and the rise of institutional settings, further complicated the notion of the physician as "friend." Percival wrote an essay, "False Opinions Concerning Friendship," to show that friendship could never be a duty because it was always particular and interested, whereas true duties were disinterested. Justice and probity were universal and disinterested social virtues, and so they always took priority over friendship. As historian Lisbeth Haakonssen characterizes Percival's ethics, the physician owed every person an equal duty of care and consideration; but as people had very different stations in society, this "equal" duty was compatible with unequal treatment, modified according to custom and decorum in how social unequals were to treat each other:[53]

According to Percival's scheme of ethics, the patients' status was inseparable from their position and role in society. To respect a patient, or to treat a patient as a person, was, essentially, to respect that patient's capacity for fulfilling the duties of his or her particular station. In this view, a physician might best demonstrate respect for his patients by assisting them to return to the scene of their active duties, a goal equally compatible with benevolence and the general welfare of society.[54]

If the physician was to be a friend to the patient while acknowledging social inequality, what does this imply for disclosing the truth about a grim prognosis? Percival addressed this problem in one of his Appendices, which is perhaps today the best-remembered part of his *Medical Ethics*. (Since the detailed discussion appears in an appendix and not in the main work, Ryan did not see fit to reprint it, though he did address the issue.) Percival began by surveying the views of the most important moralists of the time on whether one may lie to produce a good end. Gisborne, for instance, argued not, while Hutcheson said yes. Percival then offered his own analysis, based on the idea that the duty to be truthful was a personal duty for the physician, but only a relative duty within the physician-patient dyad. By appeal to the Golden Rule, duties to patients took priority over duties to oneself. In the patient setting, the most important duty was not to cause harm. Percival interpreted the overriding nature of the duty not to cause harm as effectively eliminating any duty of truthfulness to the patient in circumstances of extreme danger.

A duty to tell the truth is then reduced solely to a duty the physician owes to himself, which he would then unselfishly renounce in favor of the greater obligation to serve his patient. The modern reader tends to focus on Percival's eventual rationalization of physician paternalism.[55] This reader may then miss Percival's method of approaching the question, which epitomized the sort of "candid" free and rational inquiry that he argued was the sure way toward medical advancement. In that regard, the essay on deception might have been intended as much as a model of ideal medical reasoning and inquiry as an action guide to behavior.[56]

Having described the duties of the virtuous physician, Percival proceeded to the nature of the hospital as an institution. The notion of ethical duties within and about institutions was relatively novel at that time. Percival included as an appendix to his volume the "Discourse on Hospital Duties" delivered by his since-deceased son, Rev. Thomas B. Percival, to the Liverpool Infirmary in 1791. While the younger Percival saw the hospital as public and social in a way that made it quite different from the traditional family, he was unsure how to conceptualize it except as a large home with its managers acting *in loco parentis*. He argued that it was critical that patients' family members be allowed into the hospital and encouraged to do as much as they could for the sick: "Domestic virtue was the basis for social virtue, and the hospital should be wary of breaking this tie [between patient and family]."[57] The younger Percival claimed that modern physicians were doing Christ's work within an institution that was an extension of the "sacramentally founded family."[58]

The elder Percival realized, however, that the hospital was clearly something other than a large household. He saw that new institutional relationships required a new ethical analysis. The domestic ethic of the family, and the traditional code of conduct of the English gentleman, would no longer serve medicine's needs. One pressing issue was the power balance between the physicians and the hospital's lay managers. On the one hand, physicians needed and generally received considerable discretion to prescribe as they wished for each patient. On the other hand, physicians were of necessity subservient to budgetary decisions as to how much and what kind of medicine and food the hospital would purchase.

Medicine practiced in the hospital, unlike traditional practice in the patient's home, was fully visible to one's peers. This visibility created an opportunity for disputes among practitioners that could undermine public faith in the hospital and dry up the wellspring of philanthropy. Percival was in favor of retaining the three traditional medical orders (physicians, surgeons, and apothecaries), and much of his ethical code was devoted to mechanisms to produce a harmonious collaboration among them while maintaining a useful division of labor.[59] While Percival was not concerned about the physician–patient relationship in the modern sense, he was extremely concerned about patient safety, in a time of high hospital mortality. Hygiene and fever control, argued Percival the public health reformer, were critical *ethical* issues in hospital medicine and required a collaborative orientation among the three orders.

Percival's proposed rules to govern consultations among practitioners maintained both the hierarchy of the three medical orders, and also the preferential treatment of practitioners long settled in town. True to his Edinburgh education, however, he also sought to assure that any practitioner with intelligence and experience would receive a polite hearing from others, whether or not that individual had a degree. He emphasized that the goal was the well-being of the patient, not the honor or pride of any physician. Percival the ideal Baconian scientist might have hoped for a system in which errors were candidly admitted so as to lead the way to new discoveries and improved future practice. Percival the diplomat and peacemaker designed a practical system for people to cooperate and resolve the inevitable disputes. The compromise required that most disputes be kept in-house among the medical staff.

Percival recognized the tension between individual patient benefit and the advance of medical science. He recognized that the public at large had a horror of being subjected to experimentation. He thought that the right sort of education and rhetoric, continued for a long enough period, might convince them that medical experimentation was in the general public interest. He believed that the poorer classes treated in hospitals, who were most likely to be subjected to experiments, were also the most likely to benefit from the experimental results. He favored consultation among the medical staff prior to any experiment, in effect anticipating today's system of committee review of research. As one would expect in a day when there was no concept of patient autonomy in any ethical writing, Percival saw no need for explicit patient consent for participation in research. He did, however, urge that hospitals try whenever possible to accede to the preferences of patients, such as to be admitted under the care of one physician rather than another. He argued that it was important to attend to feelings and emotions of patients as well as to their symptoms.

Percival, in one of his most often quoted passages, wrote: "[Physicians] should study, also, in their deportment, so to unite *tenderness* with *steadiness*, and *condescension* with *authority*, as to inspire the minds of their patients with gratitude, respect, and confidence."[60] Haakonssen urges us to read this passage with an understanding of the intellectual environment within which Percival practiced. True condescension was guided by education and training. The right sort of condescension allowed the physician calmly and without inner turmoil to attend carefully to the patient's situation and needs, and wisely to employ the best means to relieve them. Percival contrasted the "turbulent emotion" of pity with the "calm principle" of condescension.[61] For Percival, "sympathy," "condescension," "fellow feeling," and "benevolence" were all allied concepts that undergirded a physician's legitimate authority over his patients.[62]

Nevertheless, focusing too much on the physician's *authority* risks missing the truly radical aspects of Percival's thought. The authority of the medical practitioners over the sick poor in hospitals and infirmaries was guaranteed by custom and social status. Percival, Gregory, and Gisborne were nearly alone in their day in stressing the physician's *obligations* to their patients. Ryan would later extend this reform movement in ethical thinking.

Gregory and Percival succeeded both in addressing some of the most pressing ethical issues specific to the practice of medicine in their day, and also in situating that ethical discussion within the larger framework that characterized ethical inquiry in eighteenth century England and Scotland. We can now turn our attention to Ryan's life and writings.

II Michael Ryan: A Biographical Summary

IIA Origins

Ryan's birth and early life remain mysterious. Many authors continue to give his dates as 1800–1841, following the *Dictionary of National Biography*. The 1841 date for his death is demonstrably incorrect, as his death certificate shows that Ryan died on 11 December 1840.[63] His death certificate also indicates that he died in his 40th year, but an anonymous obituary, which appears otherwise to be reasonably accurate, states clearly that he died in his 46th year.[64] We believe Ryan to have been admitted as a Member of the Royal College of Surgeons in 1819.[65] The official rules of the college in those days required a minimum age of 22, so that would seem to argue against a birth date of 1800.[66] Adrian Desmond, in his study of radical reform in British medicine, accepts the obituary account and gives Ryan's dates as 1794–1840.[67]

The records of the Royal College of Surgeons and of the University of Edinburgh Medical School agree that Michael Ryan's address in 1818–1819 was Barris O'Leigh (today Burrisoleigh) in County Tipperary, Ireland. Ryan dedicated his M.D. thesis at Edinburgh to a clergyman at Glankeen, which is near Burrisoleigh, and he returned to that locale for at least a year following graduation from Edinburgh. We conclude that Burrisoleigh, Tipperary was most likely the location of Ryan's birth and family home. All contemporary birth records for that parish have been destroyed, so it may be impossible today to trace any further information about Ryan's origins.[68] When Ryan was brought to insolvent debtor's court in London in 1836, he was reported to have previously had income from land in Ireland to the sum of about £150 annually, raising the possibility that his family was of the landowning class.[69] Ryan once referred to the "views of his relative, Dr. O'Ryan, on 'Consumption,' " suggesting that other members of his family may have practiced medicine.[70] Ryan also at some point became sufficiently well versed in both French and Latin so as later to be able to translate volumes written in those languages.[71] These characteristics would be consistent with higher rather than lower class origins.

Other recent authors have assumed from available evidence that Ryan was Roman Catholic.[72] We believe it more likely that Ryan was affiliated with the Church of Ireland (Anglican). Most importantly, he dedicated his 1821 M.D. thesis at Edinburgh to Rev. Richard Boyle Bernard, rector of Glankeen, a Church of Ireland clergyman.[73] Also, if his family were indeed Irish landowners, this status at that time would be more likely to be enjoyed by a family of Protestant affiliations. Finally, if Ryan had indeed been Catholic, he might have had even more trouble establishing himself in the world of London medicine than he later did.[74]

IIB Medical Studies

Ryan wrote much later that he began his career as a medical student in 1815.[75] He also mentioned in passing that he had been clinical assistant in the practice of

Professor Dease at the Charitable Infirmary, Jervis Street, Dublin.[76] The Charitable Infirmary was the first voluntary (non-religious) hospital in the British Isles, opening in 1718, 1 year before the first such hospital in London, the Westminster Hospital. A small medical school was attached to the Infirmary between 1808 and 1833. Richard Dease (1774–1819) held the chairs of Surgery and Anatomy in the medical school of the Royal College of Surgeons in Ireland, besides attending and lecturing at the Infirmary.[77]

Apparently on the strength of his Dublin medical training, Ryan submitted his credentials to the Royal College of Surgeons of England and was admitted Member (MRCS) on 2 April 1819. The College archives list his address at the time as "Burris O'Leigh, Tipperary."[78]

Ryan next sought to enhance his credentials by matriculating at the medical school of Edinburgh University, where John Gregory had taught some 40 years before. The Edinburgh M.D. was not recognized by the Royal College of Physicians of London and gave one no legal entitlement to practice in England. Nevertheless the Scottish university retained the same high prestige that it had enjoyed in Gregory's time, though it would soon be supplanted by Paris as the preferred academic destination of American physicians seeking wider educational and research opportunities in Europe.[79]

Ryan apparently began his studies at Edinburgh in 1818 or 1819.[80] He graduated in 1821, the title of his M.D. thesis being, "De Genere Humano ejusque Varietatibus," which he later translated as, "on Man, and the apparent varieties of the Human Species."[81] The anonymous obituarist in the *Provincial Medical and Surgical Journal* stated, "During his collegiate career he attracted the attention of the then professor of midwifery, Dr. Hamilton, from whom he received many marks of kindness, and it was this intimacy which laid the foundation of that taste for the theory and practice of the obstetric art which Dr. Ryan so much excelled in."[82] In his M.D. dissertation, Ryan included sections on human reproduction and birth, as well as on the division of humanity into discrete racial groups.[83]

IIC   Irish Medical Career

Following completion of his Edinburgh studies, Ryan returned to his native Ireland to practice. Apparently he spent at least 1 year back in Tipperary before relocating to Kilkenny.[84] The 1824 Kilkenny City directory listed Michael Ryan as practicing in Patrick Street.[85] Ryan also began his voluminous literary production. He published papers on topics including the various causes of aphonia, the use of prussic acid in heart and lung conditions, and blindness caused by cataract.[86] He also expanded one article into a 46-page book entitled, *A Treatise on the Most Celebrated Mineral Waters of Ireland containing an Account of the Waters of Ballyspellan, Castleconnel, Ballynahinch, Mallow, Lucan, Swadlinbar, Goldenbridgre, Kilmainham, &c., &c.: and of the spa lately discovered at Brownstown, near Kilkenny, with Plain Directions during the Use of Mineral Waters and an Account of some of those Diseases in which they are Most Useful.*[87]

Mineral Waters bore several features suggesting an ambitious young physician eager to make his name known to a wider audience. He stressed that the book was intended for both professional and lay readers, and made note of the fact that he had been selected by the local authorities to do an analysis of the Brownstown mineral waters.[88] He commented that the water "has been used by an immense number of persons last season, as soon as my opinions were known on its nature..."[89] He also mentioned the attractions of Kilkenny town in an apparent ploy to draw potential patients from a distance: "From its central situation, it is well adapted for the reception of invalids..."[90]

In two ways this relatively slim pamphlet provided a model that Ryan would utilize repeatedly in his subsequent books. Most importantly, Ryan began his discussion of mineral waters with a historical review, citing instances of water used for healing purposes in "Heathen Mythology," the Bible, and the works of Hippocrates and Aristotle.[91] Ryan also defended himself from possible accusations that he had slighted the better-known mineral spa at Ballyspellan in order to sing the praises of his local Kilkenny waters by adopting for himself the motto, "amicus Socrates, amicus Plato, sed magis amica veritas" ("Socrates and Plato are my friends, but truth is a greater friend still").[92] This phrase (minus the word "sed") appeared in the *Medicus-Politicus* of Roderic a Castro, a Renaissance writer on medical ethics from whom Ryan would later borrow extensively.[93] Ryan would repeat this phrase in the preface or introduction to many of his subsequent books, claiming that he personally always wished to avoid controversy, but that the facts of the matter forced him to disagree with the statements of others.

We have no record of the precise "circumstances [that] induced him to seek a larger field for his exertions," as his obituary stated.[94] By July 1827, Ryan was residing in Hatton Garden, London, and serving as a physician to the Central Infirmary in adjacent Greville Street.[95] He announced in 1828 that he was offering lectures on midwifery and on medical jurisprudence at the Infirmary.[96] Since his 1828 *Manual on Midwifery* was characterized as "an analysis of my lectures," it is likely that he had been lecturing for at least a few months prior to that date.[97]

In 1828, Ryan announced his presence within the London medical community by the publication of two books. One, *Remarks on the Supply of Water to the Metropolis...*, was an opportunistic cut-and-paste. His "remarks" on London water – a comment on the polluted state of the Thames, with a recommendation to pipe purer drinking water from a distance, as had been done successfully in Edinburgh – occupied all of three pages. The next portion of the volume reprinted a journal article that he had published, discussing the chemistry of water and listing all known mineral waters in Europe with notes as to their composition.[98] He then inserted the comment, "As the mineral waters of Ireland have been noticed by few writers, and are consequently little known, the author deems it necessary to add a brief account of them, in order to complete this work."[99] This was Ryan's cue to append, verbatim, his entire 1824 pamphlet on the Irish mineral waters.[100] In this way he managed to produce what purported to be a new book while preparing exactly five pages of new material.

IID   New Directions – London, 1828

Ryan's second work of 1828, based on his midwifery lectures, was a more substantial production. Its title page hinted that Ryan was playing with the idea of a unique career direction. The courses of lectures at the London medical schools, like the lectures he had attended at Edinburgh earlier, followed a rigidly prescribed formula: anatomy, physiology, chemistry, botany, surgery, principles and practice of medicine, materia medica (therapeutics), and, bringing up the rear in prestige, Medical Jurisprudence or Forensic Medicine.[101] The mere fact that a professor happened to be interested in a given subject did not justify offering a formal course of lectures in a topic not part of the standard list. But Ryan boldly broke the mold. On the title page of his *Manual on Midwifery*, he characterized himself as "Lecturer on Midwifery, Medical Ethics, and Medical Jurisprudence" (Fig. 1). As Robert Baker has noted, this made Ryan the first person anywhere in the world to claim the status of lecturer on medical ethics.[102] Ryan, it appears, was exploring in his own mind the relationships among medical jurisprudence, medical ethics, and midwifery (or obstetricy, as he preferred to call it).[103] He laid stress on the fact that his manual of midwifery would thoroughly discuss "obstetrico-legal medicine [which] has had no place previously in systematic works on midwifery."[104] He perhaps already envisioned the "obstetrico-legal" portion of the present work as later to form part of a manual specifically devoted to medical jurisprudence. He also, it seems, envisioned that later manual to include a detailed discussion of medical ethics. He promised the reader of the midwifery manual, "The moral conduct, and professional duties of the medical practitioner, are clearly defined, and will be reduced to a few simple precepts."[105]

As he would later make clear in his writings on medical jurisprudence (reprinted in this volume), Ryan regarded it as scandalous that so little attention was paid to medical ethics in an educational program designed to train young men to become physicians. Moreover, he proposed also to discuss as part of his midwifery manual "the moral and religious question – whether one life should be sacrificed to save the other. The opinions of the profession in this country, on the continent of Europe, and in America, are fully detailed."[106] Finally, he advised prospective students: "Dr. Ryan's lectures on Medical Ethics and Forensic Medicine … are intended to Explain the Moral Duties and Obligations of Medical Men in public and private Practice, in Hospitals, Infirmaries, Dispensaries, in the Military and Naval Service, in private life, and also in Courts of Justice, on whose Evidence the lives, liberty, honour, reputation, property, happiness or misery of all classes, so often depend…"[107] The physician's role in offering evidence to the courts, the traditional focus of medical jurisprudence, was therefore but one of the public roles of the professional life that might entail moral duties.

One other aspect of the midwifery manual suggested that Ryan had finally determined his special niche within the medical literature:

> Such are the advantages attempted to be affected in this work, which is intended as a Pocket Companion for students, and the junior members of the profession, who are actively engaged in practice, and whose time will not permit them to consult the larger works upon

A. o. 13

A

MANUAL ON MIDWIFERY;

OR, A

Summary of the Science and Art of

OBSTETRIC MEDICINE;

INCLUDING THE

ANATOMY, PHYSIOLOGY, PATHOLOGY,

AND

THERAPEUTICS, PECULIAR TO FEMALES;

TREATMENT OF

Parturition, Puerperal, and Infantile Diseases;

AND AN EXPOSITION OF

OBSTETRICO-LEGAL MEDICINE.

BY

MICHAEL RYAN, M.D.

Member of the Royal Colleges of Surgeons, in London and Edinburgh, Member
of the Association of Fellows and Licentiates of the Royal College of Physicians
in Dublin, one of the Physicians to the Central Infirmary, and Dispensary,
Greville Street, Lecturer on Midwifery, Medical Ethics, and Medical Jurisprudence.

LONDON:

PRINTED FOR MESSRS. LONGMAN AND CO. PATERNOSTER ROW; AND
J. ANDERSON, MEDICAL BOOKSELLER, SMITHFIELD.

1828.

Fig. 1 Title page of Ryan's *A Manual on Midwifery*, 1828 edition, on which Ryan lists as one of his
credentials "Lecturer on Midwifery, Medical Ethics, and Medical Jurisprudence" (reproduced with
the kind permission of the Wellcome Library, London)

> the subject. To the student it affords great advantages, for he may fairly consider it a review
> of the ancient and most modern works on obstetric medicine.[108]

Ryan would henceforth focus on writing the pocket handbook or medical manual, not the learned, systematic treatise or the original monograph. And to those who would criticize such works for what they necessarily omitted, or for lack of originality, he had a ready excuse:

> In justice to myself, I must allude to the style of this work, and briefly explain how it was
> arranged and sent to press. It has been composed within a few weeks, since the cessation
> of my summer courses of lectures; amidst the daily discharge of official duties, at one of
> the most crowded Dispensaries of this metropolis, and amidst the innumerable interrup-
> tions attendant on private practice. Allowance ought, therefore, to be made for ... want of
> polished style...[109]

Ryan's determination of his niche within the medical literature led directly to another recurring feature of his work – lack of appreciation from reviewers and critics. The reviewer for the *London Medical Gazette* took off sarcastically after Ryan's tone of self-promotion. It saves the reviewer time and trouble, he noted wryly, when the author himself so loudly announces to the world the exalted quality of his own work. The book itself "does contain a great deal of information, ill arranged, and mixed up with much extraneous matter."[110] (As if to prove the reviewer's point, when Ryan quoted from this review in advertisements to later editions of his works, he repeated only the phrase, "contain[s] a great deal of information.")

The reviewer concluded by giving Ryan the author some detailed advice:

> Should Dr. Ryan's work come to a second edition, we would advise him to devote three
> times as long a period to its correction as he represents himself to have done in its original
> composition. Let him not be persuaded, either by 'talented friends,' or impatient pupils, to
> give it prematurely to the world, for the world will wait without complaining till he is
> ready. Let him expunge all passages laudatory of his own productions. Then – and we fear
> not till then – with the industry he has displayed, may we hope for a work entitled to receive
> at the hands of impartial judges a degree of praise equivalent to that which the present edi-
> tion has met with from a less disinterested source [i.e., Ryan himself].[111]

Ryan, unfortunately, gave little indication of ever heeding this advice.[112]

After the publication of his midwifery manual, Ryan moved a few doors down from the Central Infirmary to 61 Hatton Garden; and by October, 1829, the *Lancet's* listing of his lectures suggested that he was giving them at his private residence.[113] The proposed *Manual of Medical Jurisprudence* duly appeared in 1831. By then, Ryan had apparently rethought his titles, if not his career direction. The title page to this new *Manual* listed him, among other distinctions, as "Lecturer on Practice of Medicine, Obstetrics, and Medical Jurisprudence," with no mention now of Medical Ethics.

IIE   The Medico-Botanical Society

The medical societies formed an important part of the social and cultural network that made up the world of medicine in London. A young physician, arriving in the metropolis without formal letters of introduction to people of influence, might hope to establish himself by becoming active in one or more of these societies.[114] Ryan became affiliated particularly with the Medico-Botanical Society, a group organized to study the medical uses and the toxic properties of substances derived from plants, and to support the discoveries of new plants of potential medical use. He would appear to have joined this society fairly soon after his arrival in London, since by 1831 he had already served as a member of the Society's Council.[115]

The President of the Society was Philip Henry, 4th Earl Stanhope (1781–1855), who is best remembered today for his involvement with the case of the Bavarian 'wild boy,' Kaspar Hauser.[116] Ryan dedicated the first, 1831 edition of his *Manual of Medical Jurisprudence* to Earl Stanhope, one of only three formal dedications we have been able to discover in his books.[117] Later Stanhope was a relatively generous contributor to funds aiding Ryan during periods of financial embarrassment (such as in the wake of the Ramadge lawsuit, discussed below). It is possible that Stanhope in some way became a patron of the young Ryan and aided his entry into the world of London medicine. It would otherwise be difficult to explain, for example, how an unknown from Kilkenny could have received an appointment as physician to a London dispensary so soon after arriving in the metropolis.

While apparently an active member of the Society, Ryan seems not to have contributed much to its meetings. A review of 38 reports of Society meetings in *The Lancet* during Ryan's London years shows him mentioned as a speaker only twice.[118]

Ryan appears to have attended meetings of at least one other medical society, the Medical Society of London. He was reported to have made comments following the main presentation at a meeting of this society held 13 April 1829.[119] He wrote a letter to the editor of the *London Medical Gazette* complaining that his reasoning had been inadequately captured by the journal's reporter.[120]

IIF The London Medical and Surgical Journal

When Ryan took over editorship of the monthly *London Medical and Surgical Journal (LMSJ)* in 1829, radicalism and reform were the big issues of the day. Very generally, being "pro-reform" meant several things. One opposed the traditional power of the medical graduates of Cambridge and Oxford, schools which had deteriorated into medieval backwaters of medicine compared to the modern schools at Leiden and Edinburgh, and compared to the schools of Paris where statistical analyses of autopsy series were leading to rapid advancement of knowledge. One opposed the nepotism of the Royal Colleges and the London teaching hospitals, where positions were usually awarded based on patronage and not on merit. And one tended to favor the interests of the new, rapidly expanding group of general practitioners (officially, dually trained surgeon-apothecaries) against the restrictive and hierarchical practices of the College of Physicians.[121]

The *LMSJ* had been founded the previous year by John Davies, John Epps, and Joseph Houlton. In two ways the editors of the *LMSJ* set out to compete with Thomas Wakley's weekly journal, *The Lancet*, which had begun publication in

1823. First, the *LMSJ* attempted to position itself as a pro-reform journal, but without the vituperative personal attacks that characterized the *Lancet's* editorial style. Second, the *LMSJ* sought to undercut the *Lancet* in price and so appeal to a general practitioner readership.[122]

Ryan's work as a medical editor ultimately had a strong influence on all of his writing. For one thing, he must have been extremely busy to have kept up a full schedule of lectures, a reasonable medical practice, and his editorial duties all at once. More specifically, he initially wrote many of his longer works, including his *Manual of Medical Jurisprudence*, as series of brief articles for his journal.[123] The articles were crafted to meet a deadline and to fit into available space, rather than being designed as fully developed, thoughtful scholarship. The journalistic style and its limitations explains many of the less desirable qualities of Ryan's book-length works.

James Fernandez Clarke, MRCS, who published a memoir many decades later, assisted both Ryan and Wakley at different times during his early professional career. Clarke reported:

> Ryan, however, at this time, did valuable service to the cause of the Profession. He was editor of the *London Medical and Surgical Journal*, then published by Renshaw, in the Strand. This journal might be regarded as an *equipoise* between the *Lancet* and the old *Medical Gazette*; it avoided the personalities and virulence of the *Lancet*, and the tameness and milk-and-water contents of the *Gazette*…. I have reason to believe that if the *Medical and Surgical Journal* had been carried on in the manner in which it was originally framed, it would have had a most beneficial influence on the Profession. While it excluded from its pages the disgraceful attacks on personal character which then were a prominent feature of the *Lancet*, it afforded to a large class of contributors to periodical literature a medium of expressing or publishing to the world their experiences in the science and practice of Medicine.[124]

Wakley in the *Lancet* included Ryan, as a rival editor, as fair game for his character attacks, notwithstanding their shared views on medical reform. In April, 1832, when the *LMSJ* had published the amounts paid to members of the Central Board of Health and intimated that their pecuniary interests influenced their views on the contagiousness of the then-prevalent epidemic cholera, Wakley replied with a strong editorial condemnation and referred to the *LMSJ* as a "filthy repository."[125] Ryan in turn gave back as good as he got, arguing when he introduced the transition of *LMSJ* from a monthly to a weekly journal in 1832, "We deliberately believe that the coarse and violent importunity with which medical reform in this country has been insisted on for the last few years, has materially retarded the application of a wise and efficient measure for the removal of all our grievances."[126]

In March, 1832, Ryan had indulged in a bit of self-congratulation, noting that the *LMSJ* was increasing its circulation and that many of the top authorities were writing for it. He promised to continue to try to rise above petty personal attacks, and suggested that he could not completely avoid the appearance of *ad hominem* arguments when momentous issues were discussed.[127] The overall pattern of Ryan's writing shows that his claim to avoid personal attacks unless he was driven to them was disingenuous at best. He was driven to such attacks amazingly often for a man who claimed that he never sought them out. For example, when the *Lancet* indulged in a lengthy and totally negative review of the first edition of the *Manual of Medical Jurisprudence*, Ryan replied with a letter that occupied more than two pages, which

Wakley happily printed, taking the opportunity to add two further pages of anti-Ryan invective.[128] Ryan reprinted in the *LSMJ* the whole of this letter to *The Lancet*, and proceeded to grace the pages of his own journal with a lengthy follow-up letter in his own defense which Wakley did not see fit to print, as well as letters from several sympathizers.[129]

One may excuse Ryan for feeling obliged to leap to the defense of one of his major works when it was attacked by an unfriendly critic. It is somewhat harder to excuse an earlier exchange of letters in the *Lancet* in which Ryan traded accusations with Dr. John Gordon Smith, whom he recognized in his *Manual of Medical Jurisprudence* as one of the few major British authorities on that subject. Smith accused Ryan of publishing material from their private correspondence and said that as a result the "acquaintance between them necessarily terminates."[130] Ryan replied with a long diatribe in which among other things he insisted that Dr. Smith "ought to have discovered that our acquaintance had ceased" at Ryan's own insistence some days before Smith's letter appeared in *The Lancet*.[131]

Perhaps the most striking example of Ryan's penchant to trade invective with any real or supposed critic was provided by his concluding comments in his volume on *Prostitution in London* (1839). Ryan asserted, "My strenuous endeavour, and most anxious wish have been to diminish immorality, and crime, as well as disease, in this and every other country; and I am perfectly regardless of the dishonest, partial, unjust, and venal criticisms of many of the medical reviewers, or ignorant critics of this kingdom…"[132] He went on to claim that as a former editor of *LMSJ*, he wrote his share of negative reviews, though he was always ready to praise works of real merit. Those whom he criticized before, he now suggested, took full advantage of the fact that Ryan was no longer editor of his own journal: "They are now armed, and cowardly attack one unarmed."[133] He added gleefully, "Their unjust censures have not, however, prevented repeated editions of the several standard works of mine, which they so loudly, unsparingly, and unjustly abused and condemned."[134] After arguing that this problem was peculiarly English, and that other European nations were not harassed by a similarly unprincipled pack of reviewers, Ryan turned now to the work that he was in the process of completing: "I enter into these particulars to prepare the reader for the attacks of the critics on this work. He must know, that every volume I produce, however useful or successful, is unsparingly assailed; although the profession and the public induce me to continue my feeble and humble exertions for the promotion of the interests of science and humanity."[135] Ryan thus launched a pre-emptive strike against his future critics before they even had an opportunity to put pen to paper.[136]

IIG   The Ramadge Case

Whatever the benefits of editing a journal, Ryan soon got a full taste of the disadvantages. A Dr. Francis H. Ramadge was known to have frequent social contacts with St. John Long, regarded by regular physicians as a quack.[137] Long was accused in the deaths of two patients and Ramadge wrote a long letter to *The Lancet* defending Long.[138] This letter apparently prompted the London Medical Society to expel Ramadge. Ramadge responded to the motion of expulsion with threats and vituperations against the Society members.[139]

In September, 1831, Ryan's *LMSJ* printed the following article:

> TWEEDIE v. RAMADGE. – Dr. Ramadge was in attendance on a case of typhus; the patient (a young lady) was bled from the arm on a Friday, and eight dozen leeches (96) applied to the head and neck. On Saturday both temporal arteries were opened–the patient fainted, and the apothecary (who was likewise in attendance) left her; the nurse brought her round with wine and water. On the Sunday, another dozen leeches were applied, and immediately she became delirious; when Dr. Tweedie's advice was requested by the relatives.
>
> Dr. Tweedie having spoken apart with Dr. Ramadge, addressed Mrs. Reynolds, the sister of the patient, and said, 'That having attended before, he should be happy now to give his assistance to the young lady, but that Dr. Ramadge's conduct in a late correspondence with John Long had been such, that no medical man of respectability could call him in or consult with him without injuring himself in the eyes of his brethren. That he (Dr. Tweedie) bore no private pique against Dr. Ramadge, – he believed him indeed to be clever; but his character, as respected the above transaction, rendered it imperative for all medical men to decline acting with him; and Mrs. Reynolds must choose which she would entrust. Dr. Ramadge replied in great anger, that he was a gentleman by birth, education, and profession, but that Dr. Tweedie was neither. *** Dr. Tweedie answered him by turning cooly on his heel and walking out of the room. Dr. Tweedie was retained, and cured the patient by exactly opposite treatment. Dr. Ramadge (it is said) frequently is at supper with John Long. – *Lancet.*
>
> Dr. Tweedie has honourably and faithfully discharged his duty to his medical brethren; and we hope every one else will do the same. We are well aware who it is–and a medical man to boot – that makes the trio in these family suppers. Let him be warned in time. He takes upon him to defend this nefarious quack and manslaughterer in the face of the whole profession: let him take warning, or we will not spare him. – *Ed. Med. Surg. J.*

Thus Ryan primarily lifted his account of the Tweedie–Ramadge affair directly from *The Lancet*, but added his own further editorial comment attacking Ramadge, Long, and an unnamed third associate.[140]

Ramadge brought suit for libel individually against both Wakley and Ryan. The trials were held respectively on 25 and 26 June 1832. In the first case the jury found against Wakley and assessed damages of one farthing (1/4 cent), plus costs. In the second case the jury found against Ryan and assessed damages of £400 plus costs. It appears that the court was heavily influenced by the additional editorial comment that Ryan had appended; his defense that he merely passed on to his readers an article already printed in another journal was not accepted. There was also some confusion engendered by the fact that Ryan had at one point apparently offered Ramadge an apology for what he had published, raising questions as to whether he had already tacitly admitted that it was libellous.[141]

Wakley commented at some length editorially on the outcome of the trials, and indicated that despite his editorial disagreements with Ryan, he viewed the jury verdict against his rival as both unduly harsh and illogical in light of the very different verdict in the *Lancet* case. A number of physicians, including some luminaries among the London medical establishment, contributed to a fund to help defray Ryan's judgment.[142]

IIH   Later History of the LMSJ

Ryan's association with the *Journal* – and, perhaps, his tendency to get himself into conflicts with others – continued to bring him financial woe even after the Ramadge case was over. James F. Clarke recalled the later fate of the publication:

> ... [U]nfortunately, a dispute between the editor [Ryan] and the publisher [Renshaw] led to a disruption between them.[143] The consequence was, Renshaw carried on the *Journal*, which was edited by John Foote, then a general practitioner in Tavistock Street, Covent Garden. Foote was a man of undoubted ability, but conceited and pragmatical. Ryan published an opposition Journal, to which he appended the title of "original." Foote was not a match for his learned and cumbrous antagonist.
> ... Unfortunately, the quarrel between Renshaw and Ryan terminated in a collapse of both the journals, each having deteriorated in every way, and Ryan's "original" was badly printed on bad paper.... Ryan had become involved in pecuniary difficulties, and the journal was neglected by him.[144]

According to Desmond, Ryan was actually imprisoned for nonpayment of debts.[145] Ryan was taken to Insolvent Debtors Court in 1836 by his new publisher and printer, Mr. Henderson, for the sum of £1,000 (though Henderson "claimed to be a creditor for £1,431"). Ryan had given up his furniture and medical library as payment to his creditors. However, Henderson took Ryan to court because "the whole of the books were not given up." The court found that no part "of the furniture or books had been unfairly kept back." The court also found that "the work (Ryan's journal) had not been a losing concern, and that the insolvent had laid out a large sum upon it." However, the court came to the conclusion that "There were a few books that must be given up."[146]

III   Ryan the Medical Lecturer

Ryan's career as a lecturer is depicted in Table 1.

Ryan 's affiliations were solely with the smaller, private medical schools rather than with any of the large teaching hospitals. The private schools experienced sinking fortunes and a precarious existence during this decade. The medical schools attached to the large hospitals were in the ascendency, and within another 10 years most of the private schools had closed.[147]

Ryan apparently left the Central Infirmary and began offering his own private lectures down the street.[148] In 1830 he joined the staff of a new school set up in Brewer Street, near the much-better-known Windmill Street school that had been founded by the great eighteenth-century anatomist, John Hunter. He may have gone there in company with a young Edinburgh graduate, John Epps (1805–1869), one of his immediate predecessors as editor of *LMSJ*. A problem with one of the other lecturers led to this school dissolving, and Ryan and Epps then threw in their lot with a lecturer on anatomy and surgery, George Dermott.[149] According to Desmond, the trio became the nucleus of another medical school in Gerrard Street, Soho, which gained a reputation

Table 1 Ryan's lectures and medical school affiliations

Dates[261]	School	Subjects
1828	Central Infirmary, Greville Street, Hatton Garden	Midwifery, medical jurisprudence
1829	Private residence, 61, Hatton Garden	Medicine, midwifery
1830	New Medical School, 34, Brewer Street, Windmill Street, Golden Square; and ? Private residence[262]	Medicine, midwifery, medical jurisprudence
1831[263]	Westminster School and Dispensary, 9 Gerrard Street, Soho; and Private residence, 61, Hatton Garden	Medicine, midwifery, medical jurisprudence
1831–1835	Westminster School and Dispensary, 9 Gerrard Street, Soho	Medicine, midwifery, medical jurisprudence
1836	Hunterian School, 16 Great Windmill Street	Medicine, midwifery
1837–1839	North London School, 20 Charlotte Street, Bloomsbury	Medicine, midwifery
1840	Charlotte Street School of Medicine, 15 Charlotte Street, Bloomsbury	Midwifery, medical jurisprudence

as a center of radical medical reform.[150] George Darby Dermott (1802–1847) was impulsive and unpredictable, but was widely reputed to be one of the best teachers of anatomy in the kingdom. He also lectured on surgery and physiology. Epps was well regarded as a lecturer in botany, chemistry, and materia medica.[151]

Dermott had first set up his own school in 1825 in Little Windmill Street, also trying to borrow some of the reflected glory of the longer-established Hunterian school in Windmill Street. When Ryan joined, the school was situated in Gerrard Street and shared some quarters with the Westminster Dispensary. Dermott continually had problems paying the rent and the school next had to move to Charlotte Street in Bloomsbury. For a brief time, Ryan and Epps broke off and allied themselves with the Old Hunterian School in Great Windmill Street, but Ryan then rejoined Dermott in Charlotte Street.[152] Just before Ryan's death, a rival "Hunterian" school took over the Charlotte Street address, and Dermott and Ryan moved to a new building a few doors down.

James Fernandez Clarke enrolled as a student in Dermott's school in 1833. Since we have located no graphic portrait of Ryan, Clarke's verbal portrait of him gives us the most vivid picture of the man available:

Dr. Michael Ryan ... was a genuine Irishman.[153] His lectures were of a somewhat "rigmarole" kind, or, more properly speaking, compilations of almost every author from the time of Hippocrates downwards. Ryan was totally deficient in originality, and was a very inferior Practitioner; but he was learned, had most agreeable manners, and was strongly impressed with the importance of appearance in a Physician.[154] He dressed in black, wore a white "choker," carried a large gold watch, as big as a turnip, to which was appended a massive chain, with seals and keys. It is true, he did not carry a gold-headed cane, but he had a huge gold snuffbox, the contents of which were frequently applied to his olfactory organ.[155]

Ryan was one of the Physicians of the Western Dispensary, in Charles Street, Westminster, and most of his students at the Gerrard Street School were also his pupils at that institution... . Ryan had a curious way of eking out a lecture. He would select for his theme some case which he had treated in the Dispensary, and if any remarkable case occurred there, he had a fine opportunity of availing himself of this privilege. I recollect, on one occasion, he had treated a case of hypertrophy of the heart "successfully" with the iodide of potassium. He did not think it sufficient to state the fact; he would take from his pocket a list of the names of students who attended his practice at the Dispensary. "Gentlemen," he would say, "I have the honour of giving you the names of those who witnessed this remarkable case;" and he read from the list the names of 150 of those who were, or ought to have been, present. The reading of this list necessarily occupied a considerable portion of his lecture, and he would conclude with the remark – "With such a host of witnesses, gentlemen, I think that I am entitled to assert that my diagnosis was correct, and my treatment most successful." His audience, of course, cheered and laughed.[156]

... Ryan was just above the middle height, and inclined to corpulency; he was of fair complexion and red haired. His face was truly Milesian.[157] He had a large head. He was a good-natured, kind-hearted man; but too facile and with little consistency and firmness. He was a favourite with his pupils.[158]

IIJ   Ryan's Medical Writings

Table 2 summarizes Ryan's book-length publications.

Several of these works are purely of medical interest. *Lectures on Population* is more of a pamphlet (72 pages) and consists of lectures from his midwifery course. The two "vademecum" volumes are handbooks of aphorisms on medicine and obstetrics, respectively, initially compiled by other authors, then edited and expanded by Ryan.[159] The *Practical Formulary* fully achieves Ryan's goal of writing a "pocket guide," since the book measures only 5 by 3.5 in. (and is 1.5 in. thick). *The Medico-chirurgical Pharmacopœia* was, most likely, a reissue of the *Practical Formulary* with only minor modifications. The remaining works have some ethical interest, and we will discuss them individually below.

According to booksellers' advertisements, Ryan was at the time of his death preparing two more volumes for the press – an atlas of midwifery, and a book on auscultation and on internal medicine more generally.[160] Earlier he had given notice of a volume titled, "A Treatise on the Physical Education and Diseases of Infants, from Birth to Puberty, compiled from the best National and Foreign Works, and arranged as a Manual of the Practice of Medicine." Such a work would have been quite innovative for its time, when pediatrics was unknown as a specialty, and when lectures on diseases of children were tacked onto the midwifery course almost as an afterthought. However, even though Ryan stated that this book was "In the Press" in 1836, nothing more was apparently heard of it.[161]

IIK   The Metropolitan Free Hospital

The formation of the Metropolitan Free Hospital was announced at the end of March, 1836. The new institution differed from existing hospitals in opening its doors to people "whose only recommendations are poverty, destitution and disease."

Table 2 Ryan's published books[264]

First British edition	Title	First American edition	Later British editions	Later American editions
1821	Tentamen medica physica inauguralis, de genere humano, ejusque varietatibus (M.D. thesis, Edinburgh)			
1824	A treatise on the most celebrated mineral waters of Ireland containing an account of the waters of Ballyspellan, Castleconnel, Ballynahinch, Mallow, Lucan, Swadlinbar, Goldenbridge, Kilmainham, &c. &c.: and of the spa lately discovered at Brownstown, near Kilkenny, with plain directions during the use of mineral waters and an account of some of those diseases in which they are most useful			
1828	Remarks on the supply of water to the metropolis:with an account of the natural history of water in its simple and combined states, and of the chemical composition and medical uses of all the known mineral waters being a guide to foreign and British watering places			
1828	A manual on midwifery; or a summary of the science and art of obstetric medicine, including the anatomy, physiology, pathology, and therapeutics, peculiar to females; treatment of parturition, puerperal, and infantile diseases; and an exposition of obstetrico-legal medicine	1835(?)	1830, 1831, 1841 (title varies in later ed.)	
1831	A manual of medical jurisprudence, compiled from the best medical and legal works comprising an account of (I) the ethics of the medical profession; (II) the charters and statutes relating to the faculty; and, (III) all medico-legal questions, with the latest decisions: being an analysis of a course of lectures on forensic medicine, annually delivered in London. And intended as a compendium for the use of barristers, solicitors, magistrates, coroners, and medical practitioners	1832	1836	

Year	Title		
1831	Lectures on population, marriage, and divorce, as questions of state medicine, comprising an account of the causes and treatment of impotence and sterility, and of the morbid and curative effects of marriage…	1835	
1833	The physician's vade-mecum, by Robert Hooper/improved by M. Ryan	1837, 1838	
1835	A new practical formulary of hospitals of England, Scotland, Ireland, France, Germany, Italy, Spain, Portugal, Sweden, Russia, and America; of MM. Magendie, Lugol, etc. Or a conspectus of prescriptions in medicine, surgery, and obstetrics. With the doses of all new and ordinary medicines/translated from the new French ed. of Milne Edwards and P. Vavasseur, and considerably augmented		1836, 1839[265]
?	The medico-chirurgical pharmacopoeia; or, a conspectus of the best prescriptions in medicine, surgery, obstetricy, and infantile medicine; with a table of the doses of all medicines in use; the additions in the London Pharmacopoeia, 1836; M. Magendie's formulary; aphorisms on the treatment of poisoning; reduction of dislocations and fractures; and on natural and difficult parturitions, with puerperal diseases[266]		1838
1836	The obstetrician's vademecum; or aphorisms on natural and difficult parturition; the application and use of instruments in preternatural labours; on labours complicated with hemorrhage, convulsions, etc. [by Thomas Denman]. Considerably augmented and arranged according to the present state of obstetricy, by M. Ryan[267]	1848 (title varies in American ed.)	1837(?)

(continued)

Table 2 (continued)

First British edition	Title	First American edition	Later British editions	Later American editions
1838(?)	The philosophy of marriage: in its social, moral, and physical relations; with an account of the diseases of the genito-urinary organs ... with the physiology of generation in the vegetable and animal kingdoms; being part of a course of obstetric lectures delivered at the North London School of Medicine		1839, 1843	1844, 1870, 1886, 1974 (title varies in later ed.)
1839	Prostitution in London, with a comparative view of that of Paris and New York ... with an account of the nature and treatment of the various diseases, caused by the abuses of the reproductive function			

It did not, as was the general practice at the time, require a letter of recommendation from one of the governors of the hospital, thus making its services much more accessible to the larger mass of the poor.[162] Prince Albert was one of its early supporters.[163] The hospital was located at 29 Carey Street, adjacent to Lincoln's Inn and a short distance from Lincoln's Inn Fields where the headquarters of the College of Surgeons was located; it was a relatively brief walk from Hatton Garden and Greville Street, where Ryan had lived and worked when he first arrived in the metropolis. Despite Royal support, the hospital had serious financial difficulties in its early years, having trouble paying its tradesmen's bills, and because of limited funds it was unable to add beds for inpatients until near the end of its first decade of existence.[164]

Ryan attended meetings of the hospital committee during March and April, 1836, when the plans for the institution were being laid down. He ceased to serve on the governing committee after that but along with a Dr. Uwins was appointed as a Physician to the Hospital on 7 April 1836.[165] Ryan listed this affiliation under his name on the title pages of some of his later books.[166] In June, 1839, when Ryan was in the process of revising the final edition of his *Manual of Midwifery*, he mentioned a case of a pregnant woman with a severe prolapse of the uterus "under my care at the Metropolitan Free Hospital."[167]

Ryan no doubt hoped that this small hospital would grow and that eventually his affiliation would lead to more lucrative connections in the London medical world. The hospital, however, remained a small, struggling operation during the few remaining years of Ryan's life.

IIL   Ryan's Death

Ryan died on 11 December 1840 following what his anonymous obituarist called a "lingering illness."[168] Nine days earlier, on 2 December 1840, the General Committee of the Metropolitan Free Hospital had discussed Ryan's lack of attendance at meetings, and the fact that he had sent in as a substitute a Mr. Earles, who was not yet fully qualified. It was decided to send a respectful letter to inquire about the situation.[169] The Committee's reaction suggests that however "lingering" the illness may have been, at least some of Ryan's associates had no inkling of its severity even a short time before his death.

If James Clarke's recollections are to be believed, we know rather more about Ryan's death than we do about many aspects of his life: "He was attended at his house in Charlotte Street, Bedford Square, by Drs. Bright, Roots, and myself. The post-mortem examination was made by me at 7 o'clock on a January morning, the two Physicians being present at that early hour. I never saw so much general visceral disease in any body before or since."[170] Unfortunately there is evidence that Clarke's memory, recalling these events decades later unaided by notes, was faulty.[171] Since Ryan died in early December there is no reason to expect that the autopsy would have been postponed until January. The death certificate lists "John

Daniel Earles, surgeon" as having reported the death and as having been present at the time of death, with no mention of any other physicians. The cause of death according to the certificate was "chronic bronchitis with dilatation of the heart."[172]

The physicians and surgeons of the Metropolitan Free Hospital officially reported to the Committee on 16 December 1840:

> The Medical Officers beg to announce to the Committee and Governors of the Hospital, the melancholy and premature decease of their colleague, D[r]. Ryan. The circumstances attending this sad event are calculated to excite in their minds universal feelings of regret, for they have not only been deprived of an active, intelligent, and kind-hearted colleague, and one which they feel will be difficult to replace, but they have also to deplore the hopeless and destitute condition of his widow and children, who were entirely dependent upon D[r]. Ryan's professional and literary exertions, and who are now thrown upon the wide world without a refuge unless speedy relief be afforded them by the benevolence of the Profession and the public. The long and faithful services rendered by D[r]. Ryan to this Institution constitute, in the opinion of the Medical Officers, a very strong reason for pressing upon the Committee and the Governors at large the expediency of taking such measures as may testify their regret at the loss of their invaluable Physician, and their sympathy in the deplorable state of his widow and family.[173]

Wakley in *The Lancet* saw no reason to mention Ryan's death. A sympathetic obituary appeared in the *Provincial Medical and Surgical Journal*, and a short notice of death in the *Gentleman's Magazine*.[174] The *Lancet* did, however, later open its pages to an announcement of the subscription taken up "for the widow and children of the late Dr. Ryan." As of 10 March 1841, approximately £150 had been collected.[175]

III Ryan's Writings on Medical Ethics

IIIA Manual of Medical Jurisprudence

Ryan's major writing on medical ethics went through three versions. First, it appeared in the form of filler interspersed among other articles in his *LMSJ* in 1829–1830.[176] He then reprinted these articles as the first four chapters of his *Manual of Medical Jurisprudence*, in its first London edition (1831). This volume, as we have seen, was the object of a long, highly critical review in *The Lancet*.[177] The reviewer took Ryan to task for, among other things, plagiarizing and inadequately indicating the differences between his own writing and direct quotations from other works; for his apparently gratuitous diatribe against phrenology; and for his antipathy to medical experiments on patients. The reviewer also accused Ryan of padding the book by reprinting large sections of his earlier-published *Manual on Midwifery*.

While Ryan discounted both the content and the motives of the *Lancet* review, he gave evidence in his second (1836) edition that he had taken at least some of the criticisms to heart. He included in his Introduction an extremely long list of the works on medical jurisprudence that he had consulted in compiling the present book. He also appeared to devote more concern to distinguishing quotations and giving credit to other authors.

The second edition of *Manual of Medical Jurisprudence* differed from the first in a number of other ways. Ryan expanded his discussion at several points. In the first edition he had been content to offer a summary and paraphrase of Percival; in the second edition he reprinted the major portions of Percival's ethics verbatim. He also added two chapters – one on "American Medical Ethics," the second a miscellany directed at the student or new graduate – that had not appeared in the first edition. Therefore the second edition seems to represent Ryan's more carefully constructed and considered work, and we have chosen to use that edition for reprinting in this volume. In the preface to the second edition, Ryan stated that he had "considerably enlarged the first part, on Medical Ethics, by introducing the codes of continental Europe and America, and also those of Dr. John Gregory and Dr. Percival."

Ryan intended, by producing a manual of medical jurisprudence in which the essentials of the topic were compressed into a relatively small space, to meet the needs of a number of audiences – "legislators, barristers, magistrates, coroners, private gentlemen, jurors, and medical practitioners," as he stated somewhat ambitiously on his title page. Nonetheless, in the Preface to his first edition as reprinted in the 1836 edition, Ryan suggested that he had one particular audience in mind when he included the material on medical ethics: "[The author] hopes that the promulgation of Medical Ethics, or the institutes of professional conduct, will contribute in no small degree to maintain and support the honour, dignity, character, and utility of the profession, by impressing the minds of medical students with a sense of the noble and virtuous principles which have always characterized their predecessors, and which ought ever to distinguish the scientific cultivator of medicine."[178] Since his writings on medical jurisprudence grew out of the scattered articles in his *LMSJ*, which in turn were most likely derived from Ryan's notes from his own lectures to medical students, it would be natural for him focus on a student audience.

The American historian, Chester Burns, argued that Ryan's particular contribution to the history of medical ethics lay in his having incorporated ethics into medical jurisprudence:

> Ryan is singularly significant because he attempted to correlate medical ethics, health legislation, and forensic medicine. He realized that any society could incorporate its values about professional behavior into civil statutes and, consequently, impose both moral and legal obligations on professional persons. Moreover, practitioners could not satisfactorily discharge professional obligations without understanding community expectations embodied in laws, and a satisfactory fulfillment of certain community obligations involved a special knowledge of law as well as medicine. Thus, Ryan sustained Percival's emphasis on the moral import of laws as well as his understanding of the forensic responsibilities of practitioners.[179]

Following an Introduction which covered the work as a whole, Ryan then devoted the first six chapters of his *Manual* to medical ethics. We will now follow him through the organization of his material.

The Introduction pays little attention to medical ethics and focuses rather on the larger subject of medical jurisprudence. Ryan discusses various definitions and synonyms of this term and also indicates how its contents may be subdivided. He ends with a long laundry list of the important works written on medical jurisprudence through the ages, tending to confirm Clarke's assessment that whatever he may have lacked in originality, he made up for with comprehensive reading and learning. Scattered through this list of works are a few that we would now classify as pertaining to medical ethics.

Chapter 1 is intended to drive home two important points. The first Ryan selects to begin this section of his work: "All medicine is derived from God, and without his will it cannot exist or be practised. Hence the healing art, if disunited from religion, would be impious or nothing."[180] Ryan's view of the relationship between religion and medical ethics seems to have two components. First, he views all ethics as beginning with humankind's duty to God, and commonly refers to the Bible (often elliptically as the "sacred book") as the repository of God's word. Second, he takes for granted a natural-law view of both ethics and science. To do what is in accord with God's law is to do both what is ethically right, and also what promotes the highest level of human health. So physicians have a dual reason to dedicate their activities to furthering the laws and dictates of God.

Ryan's second major point in Chapter 1 is to claim (based on a selective winnowing of the historical evidence) that physicians were uniquely esteemed in virtually every civilization, and at every historical period, in which medicine was practiced. "The concurrent testimony of historians of all ages proves [medicine] to be the noblest and must useful of human pursuits; and hence the esteem and veneration universally entertained for its cultivators by mankind. The dignity of medicine arises from the nobleness of its subjects and its end; its subject is the human body, which excels all other material bodies; its end is health, which is the greatest temporal blessing of man."[181] There seems to be implicit in this discussion a version of a social contract view – since society expresses such high esteem for physicians, physicians ought to feel obligated to adhere to the highest ethical standards so as to be worthy of this esteem:

> That this confidence and esteem should be fairly merited, the father of medicine, and all his eminent successors, to the present period, required an oath of their disciples, the principal obligations of which were, the cultivation of every virtue that adorns the human character. A code of professional duties, or ethics, was arranged, which all were obliged to obey, and which still governs those who have been properly educated.[182]

Given the importance of this implicit social contract, Ryan thought it scandalous that medical schools of his day spent so little time teaching medical ethics:

> But there never was a period in medical history, in which ethics was so neglected and violated as in this "age of intellect," nor the dignity of the science so degraded and disregarded. It is therefore necessary to inform the rising members of the profession, of those virtuous and noble principles which regulated the professional conduct of their predecessors, and procured that unbounded confidence and universal esteem, which was bestowed on them by society in every age and country. I shall, therefore, describe the ethics of the founder and father of medicine, and those of his successors to the present time, with the hope of exciting my readers to imitate their example, and practise those precepts which have always characterized the erudite and scientific portion of the profession.[183]

Baker and McCullough argue that by electing to organize his "medical ethics" within a historical framework, Ryan was making one of his few original contributions to this field. The very term "medical ethics" dated back only to Percival's use of that title in 1803. Ryan was thus the first writer to offer a history of medical ethics. Ryan, moreover, claimed for the first time that what he (and Gregory and Percival before him) were addressing was not a set of issues or concerns that had recently arisen due to the particular circumstances of medicine in England and Scotland in the Enlightenment era. Rather, Ryan was making the much grander claim that he and his immediate predecessors were participants in a very long historical tradition that extended back to the earliest days of medicine on Earth, and certainly extended as far back as Hippocrates. (By contrast, Gregory and Percival had not referred explicitly to the Hippocratic Oath in their writings, nor did they regard that oath as in any way constituting an authority.)

Ryan thus became the first in a long line of physician writers on ethics to wrap himself in the mantle of venerated tradition, and to imply that his own work gained authority and credence based on that tradition.[184] As we have seen, while this may have been a new departure for the nascent "field" of medical ethics, it was, for Ryan, merely standard operating procedure. He did here for ethics little more than he had done previously when writing about mineral healing waters, obstetrics, and caesarian section. Ryan tended to begin a discussion of any medical topic with the implicit assumption of continuous historical progression.[185]

Ryan ostensibly devotes Chapter 2 to the ethics of Hippocrates, but finds ways to supplement this discussion with further observations of a religious nature. He summarizes the Hippocratic Oath, and notes with approval that his own medical school, Edinburgh, was the only school in Britain to require that its graduates take the Oath. Ryan supplemented the Oath with a list of physicianly virtues derived from other Hippocratic writings, addressing how the good physician treats both patients and colleagues. He included here the injunction, "Some, on account of friendship or acquaintance, expect attendance gratuitously, but these deserve neglect."[186] This passage, in the first edition, had brought down on his head the wrath of the *Lancet* reviewer, so Ryan tries to make more clear here that he is merely citing Hippocrates.

Ryan has no problem creating a segue from Hippocrates to the Bible, insisting as he will again later that in the pages of "the sacred volume ... is laid the foundation of all ethics, 'do unto others as you would they should do unto you.' "[187] Ryan adds, "Hoffman was right when he said, a physician ought to be a Christian."[188] He then lists distinguished physicians who have written learnedly on philosophy and morals, including the philosopher, John Locke.

Despite having been chastised by *The Lancet*, Ryan cannot resist repeating here his diatribe against phrenology: "Vain attempts have been lately made by materialists, among whom are certain phrenologists, to prove that the mind is a mere function of the brain, and consequently dies with it; that the embryo has no mind or soul either in the uterus or at birth, but that its mind is built up by the five external senses, and is annihilated by death."[189] Ryan here stuck to his guns and dared the anonymous reviewer to do his worst: "I shall repeat my denunciation of this false philosophy as published in the former edition of this work, and advise certain commentators not again to misquote and misrepresent my language, as on a former occasion."[190] Ryan's ire was apparently aroused by what he took to be the claim of phrenologists that the mind was but a secretion of the brain, or in some way was totally dependent upon the physical function of the brain, and so ceased to exist when the body died. Ryan interpreted this as a denial of the immortality of the soul. By denying the immortality of soul, materialists undermine all religion and threaten the foundations of morality, which alone can restrain the passions. Therefore, Ryan argued, if a person who cared about his fellow beings should by misfortune become a materialist, he would be obliged never to divulge that he held that opinion: "Let the patrons of the revived and long refuted philosophy persuade their wives that their souls die with their bodies – let them instil the same doctrine into the minds of their children – let the doctrine become generally received – unfaithful wives, unchaste daughters, rebellious sons, and general confusion and anarchy will be the blessed fruits of their philosophy."[191]

Chapter III is called "Medical Ethics of the Middle Ages," and consists of a discursive listing of aphorisms loosely organized around a list of the virtues displayed by the good physician. The reader would naturally wonder what the Middle Ages have to do with all this, as one would imagine that here we are reading Ryan's own views on the subject. It is only at the start of the next chapter that Ryan states that he has adapted virtually the whole of Chapter 3 from the *Medicus-Politicus* of Roderic a Castro (Rodrigo de Castro, 1546–1627), published at Hamburg in 1614.[192] Winfried Schleiner has argued that Castro's book is "a milestone in the history of medical ethics" and that Castro and some of his contemporaries deserve credit for "creating something that can be called medical ethics, rather than a general (and usually Christian) ethics applied to medical practice."[193] Unfortunately, Ryan's third chapter as a whole is rather confused, since he periodically inserts comments that pertain to nineteenth century practice and that could not possibly had appeared in Castro's original work.

In keeping with most other ethical writings of both the seventeenth and the nineteenth centuries, little distinction is drawn here among duties owed to the patient, duties owed to other physicians, and prudential advice that will make the physician more successful in attracting fees. The behavior of the ideal physician is discussed in the form of both virtues to cultivate and vices to avoid. Ryan addresses one perennial ethical question – whether the physician is justified in lying to a dangerously ill patient about his condition – by appealing to his favorite authority. After acknowledging that truthfulness is a general moral duty, Ryan gives a number of examples from the Bible in which famous persons lied to get themselves out of difficulties. Ryan agrees with the assessment, common in the heyday of humoral medicine, that the mind powerfully influences the body: "a tone, a word, a look, will destroy life in delicate and dangerous cases."[194] He then concludes, "Official mendacity is less pernicious than that which is malicious."[195] The physician who lies to spare the patient a potentially fatal shock is lying in his "official" role and intends no harm; therefore his behavior does not fall under the usual injunction against telling falsehoods.

Ryan devotes the last several pages of this chapter to the subject of fees. He quotes a number of classical sources to illustrate the perennial problem of the patient who would give his fortune to be cured while in the throes of illness, but once recovered begrudges the physician the most reasonable fees. (Ryan seems here to have forgotten his arguments in Chapter I, about how all societies hold physicians in the highest possible esteem.) He argues against the British practice of the standard "guinea fee" in favor of the Continental habit of varying fees according to the experience of the practitioner. This is fair both to the patient and also to the less established medical practitioners; Ryan argues that so long as the most experienced physicians are willing to accept low fees, the less experienced ones are likely never to earn a living. Finally, Ryan briefly addresses the profession's role in policing malpractice among practitioners, and argues that the tarnishing of the individual's own reputation through his misdeeds is sufficient punishment.

Chapter IV, "Ethics of the Present Age," is Ryan's longest. The bulk of the chapter is taken up with a precis of Gregory's writings on the duties of the physician, and a mostly verbatim reprinting of Percival's *Medical Ethics* (minus the extensive Appendices). Between them Ryan adds verbatim transcripts of the Edinburgh University oath; the "promise" and Statutes of Morality of the Royal College of Physicians; and the oath of the Royal College of Surgeons.[196]

Ryan gives as his excuse for merely reprinting extracts from his predecessors: "Want of leisure precludes me from making sufficient research for the compilation of a complete system of ethics."[197] Additional commentary is nevertheless needed for several reasons – neither Gregory's nor Percival's work was regarded as authoritative;

medical students seldom studied them; and Gregory's work was now half a century out of date and so did not reflect the modern conditions of practice.[198]

Ryan has few critical comments on any of Gregory's injunctions. On the physicians' duty of secrecy and discretion, Ryan takes for granted that civil law overrides medical ethics, and that the physician is helpless when the revelation of patients' private confidences is demanded by a court. This point is very important as it appears to be the first acknowledgment in modern medical ethics that the duty of confidentiality is constrained by law.[199] But Ryan does note that some French authorities are of the opinion that physicians ought to risk civil punishment in order to remain true to the traditional Hippocratic promise to maintain confidentiality.[200]

On one important issue, Ryan disagrees with both Gregory and Percival. Both of the older authors, full of admiration for Baconian scientific advance, spoke indulgently of experimentation on patients. Percival's relevant passage is as follows:

> XII. Whenever cases occur, attended with circumstances not heretofore observed, or in which the ordinary modes of practice have been attempted without success, it is for the public good, and in an especial degree advantageous to the poor (who, being the most numerous class of society, are the greatest beneficiaries of the healing art) that *new remedies* and *new methods of chirurgical treatment* should be devised. But in the accomplishment of this salutary purpose, the gentlemen of the faculty should be scrupulously and conscientiously governed by sound reason, just analogy, or well authenticated facts. And no such trials should be instituted, without a previous consultation of the physicians or surgeons, according to the nature of the case.[201]

In one way, Percival's admonition parallels the modern view of the ethics of research on human subjects, by calling for a special sort of peer review. Percival clearly realized that the same moral foundations that justified therapeutic interventions by physicians failed to extend so far as to justify research aimed at scientific advancement. This led Henry Beecher, the American anesthesiologist and advocate for increased scrutiny of the ethics of research during the 1960s, to credit Percival as a pioneer.[202]

Ryan's own argument goes like this:

> The duty of *caution* in practice, means 'care not to expose the sick to any unnecessary danger.' The best rule of conduct on this important point, is the simple and comprehensive, religious and moral precept, 'Do unto others as you would they should do unto you.' Whatever the practitioner does or advises to be done for the good of his patient, and what he would do in his own case, or in the case of those who are dearest to him – if he or they were in the same situation – is not only justifiable on his part, but it is his indispensable duty to do. The patient should have the chance, whether it be a hundred to one, or only one in a hundred in his favour. Whatever may be the result, the practitioner has the greatest of all consolation–the consciousness of rectitude – "mens conscia recti;" this will be his solace, should the case terminate unfavourably, when the vulgar, the ignorant, the envious, the malicious, and the interested, will not fail to blame him for the death of his patient. But if he administers a dangerous medicine, merely to gratify his own curiosity, or zeal for science – to ascertain the comparative advantage or disadvantage of some new remedy, either proposed by himself or suggested by others – he is held guilty of a breach of ethics, and of a high misdemeanour, and a great breach of trust towards his patient; and if the patient dies, he might be severely punished.
>
> Medical men have tried the most dangerous experiments upon themselves, from their zeal for science, and even sacrificed their lives; but patients, in general, have no such zeal for science – no ambitions for such a crown of martyrdom – and generally employ and pay

their medical attendants for the very opposite purpose. It must be admitted, that men who would try experiments on themselves, would be very apt to try experiments on their patients. It is a melancholy truth, but cannot be denied. The profession, however, has always reprobated such conduct; and the medical phrase of reproach and contempt for it, "corio humano ludere," to play with the human hide, abundantly testifies in what abomination it has been held by the faculty. It is unnecessary to dwell on this point in this age, because all experiments are made upon the inferior animals ... But every man of common understanding well knows, that neither physic nor surgery can be practised without some danger to the sick. It is universally known, that many surgical operations are dangerous to life, and that our most powerful remedies are highly dangerous; and more especially when improperly employed, or when they cannot be borne. A safe medicine is often extremely dangerous, from the peculiarity of constitution: and the great and urgent danger, in many diseases, requires the immediate use of dangerous remedies. It is admitted, by the best practitioners, that many remedies are still wanted for the cure of disease, and this want leads us most justifiably, and almost inevitably, to try new remedies on many occasions; and such experiments are not blamable, for they are necessary: – *sic enim medicina arta; subinde aliorum salute, aliorum interritu perniciosa discernans a salutaribus*. From these causes there results much inevitable danger in the practice of physic. From this acknowledged danger, results the important duty of caution in a physician, or care to make the danger as little as possible.[203]

Aside from his historical frame of reference, this passage is perhaps Ryan's most important *original* contribution to medical ethics. He acknowledges that all medical therapy has an inherently experimental nature, because patients have idiosyncrasies and one can never predict with full confidence how any individual will react to any treatment. But he distinguishes this dangerousness due to medicine's inherent uncertainties, from added risk brought about from the physician's curiosity or scientific zeal. This foreshadows the important ethical distinction between *research* and *experimentation* made by the national commission in the U.S. that authored the Belmont Report on the ethics of research in the 1970s.[204] Ryan even implicitly admits that his favorite fount of ethical wisdom, the Golden Rule, might fail here. Physicians in their zeal for knowledge often try dangerous experiments on themselves; so the injunction, "do unto others" might erroneously suggest that this is equally acceptable for their patients. Ryan appeals to an implicit form of contract theory once again: patients do not pay physicians to put their lives at risk by experiments that are designed rather to advance scientific knowledge than to bring about any real hope for cure. Ryan discerns, in a way that goes beyond even the patient-centered ethics of Gregory and Percival, that a fundamental duty to the patient prohibits at least some forms of medical research.[205] Why he imagined that experiments on animals had made this question obsolete in his day is never adequately explained.[206]

Ryan dissents from Percival on one other, minor point. Gregory, Percival, and Ryan all concurred that consultations among practitioners were a prime flash point for disputes that generally served none of the patients' interests and that therefore needed strong regulation to keep them amicable. Percival had addressed this need by stipulating in great detail the order in which all involved parties would give their opinions. In the process he suggested that seniority should be determined by how long one had practiced in the local community, not how many years of practice experience one had. He apparently had somewhat exaggerated fears of physicians who had no intention to settle permanently in the area coming into town, disrupting local relationships, and then moving on. Ryan very reasonably offers the counter-argument: If Sir Astley Cooper (one of the most prominent London lecturers on surgery) were to move to Manchester, and were called into a consultation, it made no sense to give him the same ranking as a surgeon just 6 months out of training.[207]

Ryan devotes Chapter 5 to "American medical ethics." One would have imagined that he would here spend time discussing the writings of Benjamin Rush (1745–1813), who had attended Gregory's lectures in Edinburgh, corresponded with Percival, and written *Observations on the Duties of a Physician* (1789), which was widely reprinted in England.[208] Instead the entire chapter is a verbatim reprint from a lecture by Dr. John D. Godman of Rutgers Medical College, New Jersey. Ryan notes that the volume of Godman's collected lectures had been "politely sent to me."[209] Apparently Ryan chose this work to "place American Medical Ethics before British readers" simply as a matter of convenience.[210]

The lecture Ryan chooses to reprint is intended for medical students and is on the subject of "professional reputation." Godman is mainly concerned here to distinguish between a reputation built up slowly over many years from virtuous behavior and solid accomplishments, and a fame based on a few lucky chances that is destined to fade as quickly as it came. He urges students to bear poverty and privation in their early years of practice with fortitude, arguing that almost all successful practitioners had to pass the test of going through such a period in their own lives.

Godman also argues forcefully that a classical education, including languages, is an essential preparation for any physician who would wish for true professional success. Among the virtues of the ideal practitioner, Godman lays special emphasis on truthfulness; neither he nor Ryan comment on the apparent contradiction between the moral duty to be truthful and the oft-cited injunction against truthful disclosure of prognosis to the seriously ill patient.

Chapter 6, dealing further with medical education, is also a new feature of the second edition, but appears to be a hodgepodge of miscellaneous topics, including long sections on physical diagnosis and the "art of prescribing" that seem especially out of place to today's reader. Ryan takes up the topic that Godman had addressed, the professional reputation of the young physician, and inserts a further (rather repetitious) quotation from the American author. Indulging in a bit of self-interest, perhaps, as a journal editor, Ryan advises the student that buying medical periodicals is cheaper than buying books, and by that means one can assemble a very complete medical library.

Ryan next addresses the medical reform proposals then before Parliament, and notes the general testimony to the effect that the required period of medical education in Britain is too short, so that too many illiterate and unqualified persons have managed to earn diplomas. "It is now generally known, that almost all medical appointments in England, Scotland, and Ireland, are filled by jobbing and bribery. The majority of the physicians and surgeons selected, do not possess any, or sufficient, or superior merit; and obtain their situations by intrigue, family influence, or purchase. Such is the case as regards our court, hospital, army, navy, and all other public medical appointments."[211] Ryan much prefers the French system by which professional juries decide who gets medical appointments. After some additional swipes at the privileges and self-interested behavior of the medical corporations in England, Ryan notes with regret that almost all new scientific advances were then coming from the Continent; and he hopes that with the anticipated Parliamentary reforms, Britain would once again take its place in the vanguard of medicine.[212]

The general public is the next target of Ryan's criticism: "If the public do not derive the usual advantages from medical practitioners, they should blame themselves for encouraging illiterate pretenders to medical knowledge, nominal doctors, and empirics. Such imposters are very numerous in all countries; and if the public do not take the trouble of distinguishing the learned from the ignorant, the fault is their own."[213] He fails to ask whether the profession's own traditional secrecy (e.g., concealing the contents of prescriptions by the use of Latin) might itself further public confusion about what counts as good medicine. Indirectly, also, the public is to blame for some of the non-virtuous behavior among physicians: "When a thirst for gold is the only object of professional reputation, it leads to meanness and disreputable behaviour. We see empirics, illiterate and professional, amassing great wealth at the expense of every virtue which adorns the true medical character. The public is the cause of this, as it awards reputation by caprice, and this is the reason that charlatans share, in common with educated physicians, reputation and renown."[214]

Ryan continues here to address some of the dilemmas of character that Godman had discussed – how properly to balance desire for virtue, fame, and money. Desire for fame, if it could be directed into the right channels, should be encouraged, since the behaviors it promotes among physicians contributes to the betterment of society at large. Ryan admits that some distinguished physicians had also amassed great wealth, and again cites Astley Cooper as an example, but he is generally disapproving of wealth-seeking and thinks that the physician of moderate means is more likely to be a person of virtue and professional attainment.

The young physician is instructed on various ways to achieve public acknowledgment without stepping over the line into charlatanism – how, for instance, to spend enough time in society to develop pleasing and easy manners, without wasting time in aimless amusements and detracting too much from scientific study. Ryan's strong interest in obstetric practice comes through in his admonition to the young physician to marry, as a married man is much more likely to be trusted when care is needed for a female patient.

A set of guidelines for the medical writer comes next:

- Don't write at all unless you have something really new to say.
- Don't use hurry as an excuse for error or poor style; the public would have happily waited for a better quality work.
- Praise others and be very sparing with vituperation, but also be modest and ready to reveal errors and untruths. "When criticism is unjustly severe, it often becomes pointless, and only serves the circulation of a work."[215]
- Don't write or try to get mentioned in the press merely as a means of self-advertising.

Perhaps Ryan, in a moment of self-awareness, was offering advice that he wished he could have followed. Sadly, his impetuous and disputatious nature prevented him from adhering to these guidelines very often.

Ryan admonishes the physician to speak well always of his colleagues, something he perhaps neglected to do in the Gordon Smith and Ramadge case. He also praises the use of mental influences as a mode of healing, thereby reminding his readers that the attitude and manner of the physician are of therapeutic importance. While the latter portion of this chapter, on the methods of prescribing, seem quite out of place to us in a volume supposed to be about medical ethics, a few passages appear in which Ryan addresses the extent to which different types of information ought to be shared with patients, and to which patients ought to be allowed to make their own decisions. Incongruously, Ryan appends at the very end a couple of paragraphs about the care of the dying patient.

IIIB   Manual on Midwifery

Ryan included in his *Manual on Midwifery* (1828 and later editions) passages that by today's standards would be viewed as having to do with ethical issues that arise in obstetric practice. He offered no explanation of why he omitted these same passages when he set out to write about medical ethics in his *Manual of Medical Jurisprudence*.[216] He noted in the latter work that he had failed to find sufficient time to write up a complete "system" of medical ethics; and perhaps he imagined that in a more comprehensive work, these ethical issues related to midwifery would be included. More likely, he was simply dependent on his distinguished predecessors, Gregory and Percival, for his sense of what belonged or did not belong in a work on medical ethics. Since midwifery-related issues had not been discussed (or much discussed) in their earlier works, Ryan defined the contents of these passages as "midwifery" and not "ethics." We have elected to reprint two threads from the *Manual of Midwifery* in this volume because regardless of how Ryan categorized them, the passages do in fact show something of how he reasoned about ethical matters.

The first set of passages comes both from the *Manual on Midwifery* of 1828, and the later (1841) edition. The "Professional Approach of the Obstetrician" describes the physician's manner of dealing with the patient in active labor. The long passage from 1828 is heavily dependent on Gregory, in some instances following the Scottish physician almost word for word. Ryan mixes a discussion of the attitude and demeanor with which the obstetrician should approach the patient, with technical details as to how to conduct the examination and prepare the chamber. "The weakness and peculiar delicacy of the female constitution, call forth our greatest tenderness and compassion," Ryan advises, "and on no occasion so powerfully as in the agonies of child-bed, when she is stretched upon the rack on which she is laid by nature."[217] He would here seem to have accepted fully the paternalistic attitudes common in his day, regarding labor as one of the medical crises described by Percival, during which one should never bluntly tell the patient her true situation if it is in any way frightening, but do all to spare her any emotional shock.

The continuation of the equivalent passage from the 1841 edition raises at least the possibility that Ryan was gradually modifying these attitudes. Consistent with the generally accepted formula, Ryan argues that *the family and friends* of the laboring woman have a right to know if any danger threatens, so that they may seek a second opinion or other aid. But he now adds that no instrumental operation should be performed without advising the family of its necessity, or discussing the matter with "[the woman] herself, when she possesses a strong mind."[218] He then adds that it is often desirable to ease the woman's fears of instruments such as the forceps, by showing them openly to her and demonstrating how they are used, rather than trying to conceal them from her or keep her in the dark about their application (a tip that he credits to his predecessor, Dr. Denman, whose book of aphorisms he had edited).

These passages regarding more openness and truth in the obstetrician's relationship with the laboring woman can certainly be interpreted in a solely instrumental fashion. Ryan may simply have been worried about the woman moving at the wrong time and injuring herself when forceps were used, and have determined that these techniques assured a patient who would reliably lie still when needed. He may have had no idea whatever that the woman as patient had any *right* to this information. But it is harder similarly to dismiss his statement about telling a woman of the impending need for an operation. By distinguishing between a woman of "strong mind" and other women, Ryan seems to imply that at least some women, even in the throes of labor, lack the characteristic that seemed to justify withholding an unwelcome truth from the sick person. Absent the "weak mind," with its attendant risk of emotional over-reaction to grim news, on what grounds could a physician who *generally* was devoted to truthfulness recommend concealing vitally important information from the patient? It was not, then, too much of a stretch from the patient's representatives' undoubted *right* to know the truth, so that they could seek other aid if necessary, to the patient's own "right" to know. We might wonder what further thoughts Ryan would have had on this subject had he lived to write yet later editions of this work.

The next long passage we have reprinted, this time only from the later edition, addresses the relative values of preserving the life of the newborn or the mother when there is a risk that both cannot be saved. Ryan contemplated this question in relation to four obstetrical procedures, three of which he considered appropriate options:

1. Gastro-hysterotomy, or caesarian section: An incision is made in the abdominal wall and carried through the wall of the uterus, to deliver the infant via the incision, and then to suture the edges. We must recall that in Ryan's day, such a procedure would necessarily be done without either anesthesia or antiseptic precautions.
2. Embryotomy, craniotomy, or crochet: The fetus's head is perforated, skull contents removed, and the head (and sometimes the body) crushed, so as to allow passage through a pelvic opening too narrow to admit the intact fetus.
3. Induction of premature labor: Uterine contractions and cervical dilatation are stimulated before their natural onset, to allow the passage through the pelvis of a smaller fetus, when the full-term fetus would presumably be too big to pass.
4. Symphyseotomy: The cartilage between the two halves of the pubic bone is incised surgically to allow greater separation of the pubic bones and hence a wider pelvic outlet for passage of the fetus. This procedure was favored by some

French obstetricians but was not recommended as an option for English practice. Ryan alludes to it only in passing.

Ryan's sense of Divine law leads him then to a number of conclusions about the use of these procedures. Induction of premature labor, if used as a method of abortion before the fetus is viable, would be absolutely prohibited, as is abortion in any form. (In a passage we do not reprint, Ryan describes the technical aspects of this procedure in Latin, as an extra precaution against uneducated persons adopting the method as an abortifacient.) Used in hopes of delivering a living infant, when there is good reason to believe that the maternal pelvis would never allow a full-term infant to pass, it should be considered an acceptable option.

English obstetricians routinely recommended embryotomy on a still-living fetus in cases where delivery appeared impossible, arguing that it was better to lose the infant's life than the mother's – by an adage Ryan wryly quoted, "the tree should be preferred to the fruit."[219] Ryan forthrightly rejected this position. As a matter of principle, no one has the authority to claim that the life of one human being is worth less than the life of any other human being.

Ryan adopts an "out" in cases where the fetus cannot descend through the pelvis in the natural way – he notes that the forces of attempted expulsion will probably lead to fetal demise long before the mother's life is truly in danger. There is no moral objection to embryotomy performed to remove a dead fetus. Ryan also believes that with the examinations available, the life or death of the fetus can be accurately determined.

In cases where the mother's life is endangered while the fetus is still alive, or when the pelvis is so contracted that one can predict with certainty that even embryotomy would fail, Ryan advocates caesarian section, even while admitting its high mortality rate. On the one hand, it is such a dangerous operation that Ryan strongly criticizes the physician who would employ it when any less radical means is available. On the other hand, Ryan suggests that if physicians were willing to perform it when truly indicated, without waiting until the mother had been weakened by a prolonged labor, its mortality rate would probably be a good deal less.

Obstetricians had encountered cases where a woman with a severely contracted pelvis, who previously had been delivered only by means of caesarian section, repeatedly became pregnant. A few of Ryan's obstetric colleagues advocated in such cases either dividing the fallopian tubes at the time of caesarian section, to prevent future pregnancy, or operating on the woman in early pregnancy effectively to abort the fetus. We will see below that Ryan strongly opposed contraception as contrary to the Divine plan. Ryan is therefore equally condemnatory of any attempt to interfere with natural fertility even in this extreme instance. He proposes sardonically that if the goal is to prevent fertility, why do not obstetricians recommend sterilizing the woman's husband, since he is at least equally responsible if not more so?

Ryan supplements appeals to Divine law with consequentialist reasoning. For example, he mentions that a consequence of the English practice of performing embryotomy on the still-living infant, to save the life of the mother, is that occasionally infants are born partly dismembered yet still alive. He suggests that the suffering thereby inflicted on such an infant becomes yet another reason not to perform the procedure while life remains. A further argument against embryotomy is that the woman, as well as the fetus, may be injured by the sharp instruments. He also strongly condemns the use of invasive techniques to deliver an infant simply so that it can be baptized. He argues that on the one hand, scriptural requirements can be met by other modes of baptism, such a using a syringe to inject water against the amniotic membranes; and on the other hand, the risks and damage caused to the mother can never justify an operation such as caesarian section, when the infant very likely would die soon no matter what was done. Ryan also occasionally appeals to authority, such as a decision of the theological faculty of the Sorbonne in 1648, that it would be better for both mother and baby to die without assistance, than that the physician should be the direct cause of death of the infant.[220]

IIIC   Prostitution in London and Philosophy of Marriage

Prostitution in London and *Philosophy of Marriage* (in passages not reprinted in this volume) provide further insight into Ryan's moral thought. *Prostitution in London* focuses on what he considered to be the two primary social causes of venereal diseases – poorly conceived legislation contributing to increase in prostitution, and immoderate overuse of the reproductive organs (masturbation and sexual excess). *Philosophy of Marriage* provides in addition a fairly comprehensive account of cross-cultural reproductive norms. Since he assembled the volume from his medical student lectures, *Philosophy of Marriage* also repeats many of Ryan's observations on the physician's approach to the patient in labor.

The main lessons we learn from these volumes about Ryan's ethics are, first, that he attributed to physicians a moral responsibility to reveal the Divine laws of nature to both their own students and educated lay persons, and to act as social critics where current practices deviated from those laws; and second, that he advocated compassionate treatment of some groups of patients whose behavior deviated from those Divine laws and who, as a consequence, were commonly mistreated by physicians as well as by society.

The Physician as Social Critic. The diversity of topics covered in *Philosophy of Marriage* is indicative of what Ryan considered to be the scope of his moral responsibility:

> [It is] the [ethical] duty of every lecturer of on physiology, on obstetric and on legal medicine, to describe ... [the function of reproduction] according to ancient and modern conclusions; because it materially influences population, morals, public health, disease, mortality; as well as personal reputation, property, legitimacy, and even life, together with a vast number of other questions ..."[221]

Physicians should reveal the laws of nature to the masses and to exhort them to act in accordance with them: "the sole object of the medical inquirer is to display nature in her true character, to defend her laws, and to expose the errors and follies of mankind in her violation."[222]

Ryan particularly believed that sexual overindulgence was inconsistent with the natural purpose of the reproductive organs, compromising the ability to produce healthy offspring. That in turn would increase crime and disease, undermining the well-being of any empire:[223]

> Nothing is more certain that this, that animals and plants shorten their existence by multiplied sexual enjoyments. It was to secure vigour of the mind and body that the founders of certain religious sects prescribed chastity and celibacy to their ministers. This rule is in some degree accordant with physiology; for it is well known that our moral and physical powers are diminished by coition, because we impart a portion of our endowments to our offspring, and diminish them in ourselves.
>
> A most important train of consequences to society and government follow from this prodigious influence on the population of empires.[224]

Ryan urged his students to encourage their patients to practice sexual moderation in conformity with the Divine edict. He also vehemently opposed the use of birth-control measures advocated by so-called Malthusians:

> The most zealous disciples of Malthus were said to be the Westminister political economists, including Bentham, Ricardo, Place, (James) Mill, Tooke, Brougham, Miss Martineau, and others of minor note. A number of grossly immoral men followed their example, and in 1822 distributed the most infamous handbills throughout the large manufacturing districts in England, which purported to contain 'the important information for the working classes, how to regulate the number of a family.'[225] Various abominable means were proposed, which few if any one would follow, for all were contrary to the dictates of nature, to the precepts of revealed religion, to morals, to the divine and primitive command – 'go forth and multiply.'
>
> It must be scarcely necessary to observe, that the doctrine of limiting population is based upon a most irreligious doubt in the conservative power of the Divine Creator; which regulates, preserves, and reproduces the illimitable number of organized beings in the animal and vegetable kingdoms.
>
> It was, however, most erroneously contended by the advocates of this cold-hearted and immoral doctrine, the consequences of controlling the faculty of reproduction would be moral, civilizing, would prevent much crime and unhappiness, that they would improve the manners and moral feelings, alleviate the burdens of the poor, diminish the cares of the rich, and lastly, they would enable parents more comfortably to provide for, and educate, their off-spring. But to these conclusions it may be unanswerably replied, that the limitation of offspring is based upon principles severely condemned and reprobated in the sacred volume, which are subversive of every virtue, and holding out inducements and facilities for the degradation of our daughters, sisters, and wives.[226]

From Ryan's vantage point, the neo-Malthusian proposals to limit fertility could not count as true social reform, since they deviated from Divine law as related in the "sacred volume" (the Bible). Because all biological laws had to be in conformity with God's design, it therefore followed logically that the beneficial results promised by the Malthusians *could not* be produced by the means they suggested. Virtually by definition, only human misery could result from disobeying the natural law.[227]

Ryan was not, however, unresponsive to Malthus's contention that population growth surpassing food supply promised misery. Speaking of the multitudes of the poor in Ireland, Ryan argued that late marriage was the solution to the dilemma of over-population. Since sex within marriage, the institution of marriage itself, and the practice of the sexual moderation and self-restraint necessary to postpone marriage were all seen as in accord with Divine law, Ryan saw no conflict between the advice to marry late and the Biblical injunction to "be fruitful and multiply." Noting that those with secure earnings tended to marry late and have fewer children than those with little or no income, he contended that alleviating poverty would result in late marriages, leading to a reduction in births:

> The evidence proves that the sons of farmers, and even the few of the labourers who are in better circumstances than the generality of their class, are much more cautious in contracting marriage than their poor brethren, and that they wait until they meet with a woman who has some little property or other means of assisting to maintain a family. In short, it appears quite certain that in every part of Ireland the more destitute the labouring population are, the more recklessly, nay, the more eagerly, and at a much earlier age, do they marry.... In this county of Clare, M'Mahon, a labourer, stated that, "It is always the poorer man that marries first, because he knows he cannot be worse off by it; it is better for him to marry early than to seduce the girls, who are so poor and wretched that this would often happen. Besides, we poor people have a strange idea that it is a good thing to have children as soon as possible, to help support us when we both begin to grow old."[228]

Ryan espoused late marriage for reasons beyond the problem of over-population. He held that men ought to make a careful, considered decision about whom to marry because it determined, at least in part, the quality of their progeny.[229] Citing John Gregory, he lamented the fact that while men invested considerable effort to improve the quality of their livestock by carefully breeding them, they were all too thoughtless with respect to their own offspring.

> Dr. [James] Gregory maintains, in his Conspectus Medicinae Theoreticae, that children often resemble their parents, and are certainly like them, not only in features, but in forms, mind, virtues, and vices.[230] Dr. John Gregory, the father of the last-named distinguished author, makes the following remarks in his Comparative View of the State and Faculties of Man with those of the Animal World.
>
>   Thus by a proper attention we can preserve and improve the breed of horses, dogs, cattle, and indeed of all other animals. Yet it is amazing that this observation was never transferred to the human species, which were it would be equally applicable. It is certain, that notwithstanding our promiscuous marriages, many families are distinguished by peculiar circumstances in their character. This family character, like a family face, will often be lost in one generation and appear again in the succeeding. Without doubt, education, habit, and emulation, may contribute greatly in many cases to preserve it; but it will be generally found that, independent of these, nature has stamped an original impression on certain minds, which education may greatly alter or efface, but seldom so entirely as to prevent its traces from being seen by an accurate observer. How a certain character or constitution of mind can be transmitted from a parent to a child, is a question of more difficulty than importance. It is, indeed, equally difficult to account for the external resemblance of features, or for bodily diseases, being transmitted from a parent to a child. But we never dream of a difficulty in explaining any appearance of nature, which is exhibited to us every day. A proper attention to this subject would enable us to improve not only our constitutions, but the character of our posterity. Yet, we every day see very sensible people, who are anxiously attentive to preserve or improve the breed of their horses, tainting the blood of their children, and entailing on them, not only the most loathsome diseases of the body, but madness, folly, and the most unworthy dispositions; and this too when they cannot plead being stimulated by necessity, or impelled by passion ...[231]

The inclusion of these sentiments in both *Philosophy of Marriage* and *Manual of Midwifery* suggests that Ryan considered it a part of his social reform agenda to argue that only the physically and mentally healthy should reproduce.[232] His commitment to the idea that physical, mental, and moral traits were hereditary may also be found in his account of Wrolick's and Velpeau's thesis. Foreshadowing Darwin's theory of evolution, Wrolick and Velpeau contended that the various human races were in a hierarchical relationship with each other and non-human primates. Ryan summarized their views:

> There is a vast variety in the pelvic structure of animals, which as we gradually ascend the zoological scale, becomes insensibly more perfect. Thus, the pelvis of the monkey and ourang outang have a strong resemblance to that of the human species; and we can trace the shades between these and the Boshiemans, who are considered by some to be the connecting link between monkeys and man, the Ethiopians, Negroes, Malays, Japanese, and Caucassians, the last of whom are most different from the other mammiferae; and thus, we perceive, that the parturition is more painful as the species is more perfect, and *vice versa*; a most admirable and singular provision of nature, that the dangers are increased and most accumulated, according to perfection and degree of intelligence of the animal.[233]

Given Ryan's commitment to the ideas of racial superiority and inferiority, it is surprising that in *Philosophy of Marriage* he extols the benefits of interracial alliances:[234]

> Nature also condemns [marriage between close relatives]; she wishes marriages between families and nations, because these intermarriages, or crossing of the races, are the true means of improving and invigorating the species. Some of the ancient legislators framed their laws on this ground (Plutarch, Quaest. Roman. St. Augustin, City of God, &c. Vandermonde, Essai sur le Perfect. De l'Esp. Hum., Buffon, &c.). Daily examples confirm the validity of the opinion. Pallas adduces the fact that the intermarriages of the Mogul Tartars with the Russians produce very fine individuals; while Humboldt has observed, that the offspring of a negro and European is more robust and active than that is a white with an American; because the best mode of effacing hereditary diseases, gout, scrofula, phthisis, mania, epilepsy, &c., is by the commixture of the species in intermarriages, as this corrects the defects of one individual, by the soundness of the another. The Jews, who refuse to intermarry with other nations, transmit many hereditary diseases to each other; and they preserve, by this custom, their Hebrew cast of countenance in all countries (For other proofs of this position, the reader may consult *Mr. Combe's Essay on the Constitution of Man, considered in Relation to External Objects*, 1835).[235]

Ryan addressed topics such as the abuse of the sexual organs and prostitution in his works although he was well aware that he would be condemned by those who believed that such matters were inappropriate for discussion. He described such critics as "... possessed of a weak, erotic, or prurient imagination ... or those who pretend to excessive modesty and rigid chastity," and made it clear that their censure would not sway him from attending to his responsibilities as a medical teacher.[236] He argued, for example, that strategies used by the corrupt to coerce female children and adolescents into prostitution should be made public so that parents would be forewarned:

> [H]ow ... could parents, ignorant of the artifices employed and the extent of the evil referred to, give the needed warning [to their children]? ... I most fully agree with Mr. Talbot [of the London Society for the Protection of Young Females and Prevention of Prostitution], and therefore place the information with which he has kindly supplied me

before the reader, convinced as I am, that every moral and religious individual will approve my motive, which I am sure cannot fail to be productive of a vast deal of good, by assisting in urging the legislature of this empire to correct or diminish, the horrible state of immorality amongst us.[237]

Ryan's conviction that it was his responsibility as a physician to bring social ills to light and to argue for legislative reform is particularly evident in *Prostitution in London*, modeled on Dr. Parent-Duchatelet's *On Prostitution in Paris* (1836).[238] Parent-Duchatalet reported on the prevalence and causes of prostitution in Paris, which he attributed in large part to poorly-conceived legislation. Agreeing with his French colleague, Ryan intended that his work be read by legislators, magistrates, and philanthropists as well as by physicians and surgeons.[239]

Prostitution in London is a scathing critique of laws that allowed men to seduce and impregnate lower-class women and then abandon them.[240] The 1834 Poor Law Bill ended the old system by which poor unwed mothers could receive financial support from the parish while remaining in their own homes. Hoping to decrease moral delinquency, the framers of the new law demanded that such women now be admitted to houses of correction. But rather than be incarcerated in an institution, a number of these unwed mothers resorted to prostitution.

> Here is legislation with a vengeance. All the expenses, the odium, and the ruin, are placed on the mother, who is, in ninety-nine cases in a hundred, the victim of seduction, by some heartless, vicious, and unprincipled scoundrel. It is proved by the host of evidence on the state of prostitution in this and other countries, in the work already quoted, that for the most part girls and women are decoyed and ruined by men, and not by reason of inherent licentiousness … It is, therefore, in my opinion, a most glaring defect in our legislation, to exonerate the male sex from all responsibility and punishment for seduction and bastardy, because this sex is by far the most guilty and vicious; for all physiologists admit, that the amorous impulse is stronger in the males than in the females of all grades of the mammiferae, from the lowest to the human species, because the male imparts vitality and perpetuates the species.[241]

Ryan had no doubt that the male's stronger sexual impulses were biologically determined, but did not see this fact as a moral excuse for men who victimized and abandoned women. Properly designed legislation (as well as religion), he thought, could serve as a check against men's biological impulses.

Ryan considered it his moral duty as a physician who was familiar with the plight of unwed mothers to argue for legislative reform that would give these women better options than the house of correction or prostitution.[242] He noted that the trades open to females did not pay a living wage. For instance, he argued that seamstresses ought to be paid enough so that they could support themselves and their children, without having to resort to prostitution.[243] Ryan argued that legislation on the subject should be formulated in consultation with medical practitioners acquainted with the social causes and the pathology of venereal diseases.[244] Finally, Ryan wrote, "… I must take leave to observe, that there can never be sound legislation unless in strict accordance with physiology or nature, and the sacred scriptures," which of course for him amounted to the same thing.[245]

Ryan admitted one wrinkle in his generally rosy view, that violations of Divine law always went hand in hand with unnatural practices and deleterious consequences, while adherence to God's will always ensured good health. He admitted that in one instance, society was forced to choose the lesser of two evils. The issue was what he delicately called 'profligately licentious' behavior in sex-segregated penal colonies for males. Ryan considered the presence of female prostitutes to be a potential solution to the problem of sex between males, which was a clear violation of the true purpose of generative organs, resulting in disease. Faced with a forced choice between homosexual acts and prostitution, Ryan favored the latter:

> If notwithstanding laws, punishments, public contempt, even gross brutality, and frightful disease, the inevitable consequences of prostitution, prostitutes still exist, is not this an evident proof that they cannot be put down, and that they are unavoidable in society. Remove them from among mankind if you can, and all will become profligately licentious ... The truth and force of these conclusions are, alas! too apparent in all communities where women are not allowed, or only a few admitted; as in penal colonies, barracks, ships, prisons, modern workhouses, schools, colleges, &c. &c. It must be acknowledged by everyone conversant with physiology, or the history of nature, that all legislation which enforces such conditions, is contrary to nature, reason, and morality; and is productive of the most horrible crimes ... It is horrible to contemplate the consequences of compelling the most depraved men, in the prime of life, in perfect health, and too often of most vicious and wicked minds, to be congregated in prisons, barracks, ships, workhouses, and even chained together in penal settlements- wholly deprived of the other sex.[246]

Still, as a more satisfactory longer-term solution, Ryan hoped that convicts would not be banished to sex-segregated penal colonies and that existing settlements would be well provided with clergy.[247]

Compassionate Treatment of Stigmatized Groups. Ryan, as we have noted, had sympathy for prostitutes, the majority of whom he thought had been driven into their plight by "seduction, by some heartless, vicious, and unprincipled scoundrel."[248] His own work in charity hospitals had enlarged his experience among classes of society shunned by the average citizen. *Prostitution in London* and *Philosophy of Marriage* are notable for Ryan's clear statements that certain stigmatized groups of patients deserve more compassionate treatment than they often received.

Ryan, for example, generally followed Parent-Duchatelet's ideas in *Prostitution in London*, but categorically rejected the Frenchman's argument that prostitutes had no claim to civil rights because they had violated the law of the land by virtue of their profession. Besides the sheer impossibility of designing laws that would protect the rights of all women but that would specifically exclude prostitutes, Ryan argued that it would be immoral to deny prostitutes civil and legal protection:

> [H]umanity dictates, that such persons must be protected by the laws, as well as others. It would be very difficult to enact laws to deprive this class of persons of civic or social rights, to allow them, as unfortunate human beings, to be maltreated or murdered .... Be it remembered, that our laws most wisely protect the most abandoned prostitute as well as the most virtuous woman. The violation of either would be equally punished.[249]

Ryan also advocated the creation of asylums for female prostitutes as an exercise of Christian charity:

> Such institutions exist in every civilized country. It is to be recollected, that their unfortunate inmates are members of the human family, and that they have erred like all their species; but, 'to err is human, to forgive, divine.' If these unfortunates have contributed to demoralize society, they have been most severely punished by an unfeeling, wicked, and depraved world; and if some few good Samaritans pour oil and wine into their bleeding wounds, they but imitate a Christian and humane example.[250]

Another group of patients that Ryan viewed as deserving more compassionate treatment than they often received were the sufferers from venereal diseases. At one level his concern was the respectful treatment of these patients as fellow human beings: "For my own part, I cannot but consider that such patients require great commiseration as well as relief, and that their distressing disorders deserve as much attention from medical practitioners as any other class of human infirmities."[251] In a similar vein he expressed respect for the privacy of these patients: "I might easily introduce several cases which fell under my own care, in exemplification of the statements made in the preceding volumes; but I could not do so, without, in some instances, giving pain to many of my patients now living."[252] But a more practical level of concern was that venereal sufferers, fearing disrespectful treatment, would be driven away from the physicians who could truly help them and into the hands of quacks:

> Every individual who entertains a doubt as to his capabilities for generation, is anxious to obtain medical advice on his condition; and it is much to be regretted that it is too often the practice of the profession to treat the matter with levity or derision. Hence few of the faculty are consulted, an unreserved disclosure of the symptoms is seldom given, and the inquirer is often fearful that his condition may be made known to his acquaintances. Every duly educated physician is bound to secrecy, in all delicate matters, by an oath, and so far from treating his patients with levity or carelessness, should consider his case as attentively as any other that may come before him. Were this line of conduct generally adopted by the medical profession, an immense number of the public would not be driven to seek advice from the low, ignorant, and unprincipled empirics, who not only defraud them of immense sums of money, but also destroy, what is far more important, their health.[253]

IIID   Ryan's Later Influence

Ryan spent all of his short life following one bit of his ethical advice, strenuously seeking fame and public acclaim through his accomplishments as a writer, lecturer, and practitioner. Leisure was something with which he probably had little familiarity. It is a shame that he lacked the "leisure," as he put it, to conduct "sufficient research for the compilation of a complete system of ethics." He apparently recognized that what he had written fell short of a "system." One would like to have known what form he imagined a system of medical ethics would take if fully developed, since he implied that neither Gregory's nor Percival's work rose to that level of organization.

The next "system" of ethics to appear on the scene in Anglo-American medicine after Ryan was the Code of Medical Ethics of the American Medical Association (1847). Based primarily on Percival, the Code organized its content under the headings: duties of physicians to patients; obligations of patients to physicians; duties of physicians to each other and to the profession; duties of the profession to the public.[254] The next British author to attempt to write a code of medical ethics, Jukes Styrap (1815–1899), also relied heavily on Percival and adopted a set of internal categories much like that of the AMA code.[255]

Burns notes that Ryan's *Manual of Medical Jurisprudence* was widely reprinted in the U.S., and therefore, along with Gregory's and Percival's work, was in a position to influence the development of medical ethics in America.[256] Nonetheless, Ryan's writings seem to have had little influence in his own adopted country, Great Britain. Ryan's six chapters on medical ethics may have been the most extensive writing on this subject in Britain during the long drought between Percival's 1803 edition and Styrap's code of 1878. But even the few writers who attended to the topic during the middle years of the century appeared to pay little attention to Ryan's work.[257]

IIIE   Ryan's Place in the Development of Modern Medical Ethics

A final way to assess Ryan's legacy is to evaluate Robert M. Veatch's important recent work on the history of Anglo-American medical ethics between 1770 and 1980.[258] Veatch argues, following previous writers such as McCullough, that medical ethics as practiced by Gregory and Percival was characterized by open dialogue between physicians and the moral philosophers and other humanities scholars of the day. Soon after Percival died, this dialogue was broken off, and to the extent that physicians addressed ethical issues in the intervening years, they did so in near-total isolation from the ideas of contemporary humanists. Due to a variety of factors that emerged during the 1960s, this dialogue was rather abruptly re-established around 1970 and has characterized the new field of bioethics since then.

Veatch proposes two grand traditions in medical ethics, that he calls the Gregory and Monro views. The Gregory view entails the close connection between work in medical ethics and work in moral philosophy, whether religious or secular or both. The Monro view, by contrast (named for the three generations of Alexander Monros who taught anatomy at Edinburgh before and after Gregory's day), calls for physicians to spend all their time and energy mastering the increasing intricacies of medical science. As physicians disengaged from the philosophical and humanist dialogues of their own time, they needed something to retain a sense that there was a moral core or commitment to medical practice, and the Monros and their followers found that a simplistic, mostly ceremonial attachment to the Hippocratic Oath would serve. In this they contrasted with Gregory and Percival, who were fully immersed in the intellectual discourse of the Scottish Enlightenment and accordingly saw no need to draw any ethical inspiration from the teachings of a two-millenia-old Greek mystery cult.

Veatch assesses briefly the work of Ryan and treats him as a typical nineteenth-century devotee of the Monro view.[259] It is easy to see how Veatch could come to view Ryan in this way. Clearly Ryan was a derivative, not an original thinker; while he quoted copiously from Gregory and Percival, he added little of substance. He spoke fulsomely of Hippocratic ethics and implied, at least, that the words of Hippocrates could be pertinent in his own time. And Ryan was ignorant of most of the work being done in his own day in moral philosophy and theology. But a strong case can be made that this dismissal of Ryan as a *mere* Monro disciple misses the mark – despite all the weaknesses in Ryan that we have catalogued above. In four ways, Ryan could lay more claim than Veatch acknowledges to being a disciple of Gregory and Percival.

First, none of the Monros or their progeny seemed concerned about the neglect of medical ethics in the curriculum of the nineteenth- or early twentieth-century medical school. So long as one could consult the less-than-one-page Hippocratic Oath, apparently, no sustained study was required. Ryan, by contrast, strongly condemned the neglect of this topic. While he agreed that the works of Gregory and Percival had become somewhat out of date by the 1830s, Ryan nevertheless thought it scandalous that medical students of his day knew so little of these leading thinkers.

Second, it is shortsighted to see in moral philosophy or theology all that is worth studying in the "humanities." True, Ryan did not examine the moral-philosophy literature of his day to determine how best to do ethics. But, as we have seen, he is given credit for the first thorough-going application of a different humanities discipline – history – to the study of medical ethics. His claim that medical ethics forms a historical continuum from pre-Hippocratic times to his own day may be, from our point of view, a poorly-supported and amateurish hypothesis. Ryan was, after all, a physician and medical journalist, not a trained historian – though it is important to recall that in his day, no one became a historian by getting a doctorate of philosophy in history, and that in some sense, all historians of that time were amateurs. But nevertheless it is a highly original way of looking at medical ethics for his time, and it is, arguably, a humanities-oriented approach.

Third, Veatch assumes that physicians would naturally look to work on moral philosophy for guidance in ethics, because the material provided by medicine itself is insufficient to develop a satisfactory ethical framework. This assumption does not apply to Ryan in any meaningful way because of his deep religiosity. For Ryan, it was quite clear that physicians needed to look outside medicine for moral guidance, and it was just as clear where that guidance was to be found – in the Bible, and more specifically, in his favorite Golden Rule. It would have been foolish and indeed incoherent for someone of Ryan's religious views to consult the works of philosophers such as Hume, Reid, Adam Smith, or Bentham, as if he was going to find something in those works that provided *superior* ethical wisdom to his preferred Biblical sources. To accuse Ryan of attending inadequately to moral theological discussions of his time would be more to the point, but that would in turn assume that the moral theological scholars of his day were doing work of a sort that was directly helpful to medical–ethical issues. Like Gregory and Percival, Ryan *did* look outside medicine to find ethical guidance. He found it in a different place, which, according to his lights, was much sounder. His disagreement with Gregory and Percival over where to look for moral wisdom did not blind Ryan to the value of their contributions to medical ethics, which is all the more to Ryan's credit.[260]

Fourth, the Monro disciple retreated from any engagement or dialogue with the humanities mostly because the task of keeping up with a burgeoning medical-scientific literature was becoming all-consuming. For the same reason, that physician would usually retreat from any broader social agenda save for the care of individual patients, with the possible exception of anything that could clearly be labeled "public health." Ryan, as we have seen, was of the opposite opinion. He repeatedly called upon his colleagues to remain engaged in wider social movements, to influence jurisprudence and legislation, and to share their understanding of Nature's laws with the populace toward the development of enlightened policies.

The most accurate way to characterize Ryan's role in the unfolding of Anglo-American medical ethics from the later eighteenth century to the present day, therefore, is as a minor disciple of Gregory and Percival. To see him as falling under the Monro tradition highlights his limitations while ignoring his relatively few real strengths.

Notes

1. Our treatment here follows very closely Lisbeth Haakonssen, *Medicine and Morals in the Enlightenment: John Gregory, Thomas Percival, and Benjamin Rush*. Atlanta: Rodopi, 1997:1–35. Haakonssen surveys a number of historical and philosophical interpretations of the works of John Gregory and Thomas Percival, and proposes a largely persuasive account that situates these writings within the intellectual world of the Scottish Enlightenment. We are grateful to Robert Baker and Laurence McCullough for a better understanding of the limits of Haakonssen's analysis.

2. Examples of such interpretations can be found in Chauncey D. Leake, Preface to Percival's *Medical Ethics*, Baltimore, MD: Williams & Wilkins, 1927; Jeffrey Berlant, *Profession and Monopoly: A Study of Medicine in the United States and Great Britain*, Berkeley, CA: University of California Press, 1975; and Laurence B. McCullough, "Virtues, Etiquette, and Anglo-American Medical Ethics in the Eighteenth and Nineteenth Centuries," in *Virtues and Medicine: Explorations in the Character of Medicine*, ed. Earl E. Shelp, Boston, MA: D. Reidel, 1985. The purported contrast between Gregory and Percival is inaccurate, as both addressed physician-patient relationships as well as interactions among practitioners.

3. Gregory and Percival rejected the "traditional guild ethos of the Royal College of Physicians [of London] in favour of one which, they believed, better reflected the social and moral goals of their society as well as those of the new inductive science." (Haakonssen, *Medicine and Morals*: 17) In their day, Haakonssen asserts, "professional" was a tainted term, seen as allied with "interested." This compared unfavorably to the "disinterested" amateur gentleman of science who pursued truth for its own sake and not because he was tied to the interests of a sect or guild. Gregory, Percival, and Ryan all tended to show their disdain for the physician who merely wished to make money and had no devotion to the cause of advancing medical science.

4. Early in the eighteenth century Leiden (in the Netherlands) assumed prominence, thanks to Boerhaave, who became known as the medical instructor of Europe. Boerhaave's goal was to cultivate good clinicians and his first lecture was a call to return to the study of Hippocrates. Leiden had a 12-bed charity ward that was set aside for clinical teaching, and this ward in effect became the world's first teaching hospital. (Generally, hospitals were prohibited from placing patients in a teaching environment because of their religious ethos.) Boerhaave's principal innovation in medical was to stress the importance of the students seeing real patients, both as a source of experience and as illustrations for material covered in lectures. Other teaching centers began to mimic Leiden's manner of teaching medicine. Edinburgh became the first such school in Britain; its medical school dated from 1726 when Leiden-trained Alexander Monro (the first) was appointed as professor of anatomy. Edinburgh was not known for its bedside teaching approach even though it did pioneer infirmary-based teachings. Edinburgh's strength in medical education lay in its more modern, practical orientation, compared to the classical, highly theoretical curriculum then taught at Oxford and Cambridge. After 3 years at Edinburgh, a medical student who had been trained in medicine and surgery was ready to practice medicine in the capacity of general practitioner. Roy Porter, *The Greatest Benefit to Mankind*, New York: W.W. Norton, 1997: 290–291.

5. Haakonssen, *Medicine and Morals*: 14.

6. Gregory and Percival defended the Edinburgh system against critics who viewed it as "a free-for-all and the school a degree-mill" (Haakonssen, *Medicine and Morals*: 16). To some extent their lectures on ethics were a defense of the Edinburgh approach of free inquiry and egalitarianism, as contrasted with monopolies enjoyed by Royal College of Physicians of London, which they viewed as opposed to best interests of patients and the profession. Ryan, as an advocate of radical reform in the 1830s, strongly concurred with these sentiments.

7. Haakonssen, *Medicine and Morals*: 21

8. Haakonssen, *Medicine and Morals*: 22. Moral exhortation was a well known and popular genre of writing; and Gregory and Percival both viewed their writings as appealing to a wider

audience beyond physicians and medical students. Both had written in this genre elsewhere, notably Gregory on the duties of women (*A Father's Legacy to his Daughters*, published posthumously, 1774, 1779).

9. There was indeed an important distinction between moral theory and moral practice in Scottish Enlightenment thought; but it did not track the modern distinction. Moral theory was seen as branch of "theory of mind, or pneumatology: what, in modern terms, is generally referred to as moral psychology or moral epistemology" (Haakonssen, *Medicine and Morals*: 25). Moral theory was contrasted with practical principles of action, often organized as virtues viewed as interchangeable with duties (again defying twentieth century categories). Since little was known empirically at the time about workings of the mind, the moral theory of the day was highly speculative and necessarily, by today's standards, extremely muddled. Thus it made sense to say that moral theory appeared to be quite unnecessary for the practical moralist. A precise theory of moral faculties and moral psychology "did not matter, as long as it was understood that the active power was cognitive, educable, and potentially authoritative over the passions" (Haakonssen, *Medicine and Morals*: 26).

10. Haakonssen, *Medicine and Morals*: 31.

11. Leake, Preface to Percival's *Medical Ethics* (1927).

12. Laurence B. McCullough, *John Gregory and the Invention of Professional Medical Ethics and the Profession of Medicine*, Hingham, MA: Kluwer, 1998: Introduction. An early mentor of Gregory was his cousin, Thomas Reid, a prominent philosopher in the Scottish Enlightenment; Haakonssen, *Medicine and Morals*: 46–48.

13. McCullough, *John Gregory*.

14. Haakonssen, *Medicine and Morals:* 48–54.

15. McCullough, *John Gregory*: 57.

16. "She became his exemplar of a woman of learning and virtue, a concept that he championed and made central to his medical ethics." McCullough, *John Gregory*: 69. Haakonssen places somewhat less stress on Gregory's association with Montagu; *Medicine and Morals*: 48–54. Gregory held relatively advanced views for his day on the status and education of women, but nothing in his medical ethics suggests views that would today be termed feminist. In his posthumously published *A Father' Legacy to his Daughters*, he suggested that women be passive and amiable and hide from men any scholarly inclination they might have; Haakonssen, *Medicine and Morals*: 50–54.

17. Haakonssen characterizes Gregory's *Comparative View* as very popular in both Britain and America, considered "ingenious and elegant" (*Medicine and Morals*: 51). Gregory recommended demystifying medicine and disseminating medical knowledge to laymen. (Gregory would later assist William Buchan in preparing one of the first works that tried to make medicine accessible to the general public, his *Domestic Medicine*, 1769 (65–70). *Comparative View* "shares with the writings of Gregory's fellow literati, particularly the moderate clergy, the ethical preoccupations, literary style, and scientific methodology of the Scottish Enlightenment. Like many physicians and educators, Gregory adopted the polemical strategies of the more reform-minded clergy and adapted them to his own campaign to reform medicine. He attacked the character of the professional physicians as 'tyrannical expert men,' just as the clerics assailed 'zealous orthodox divines.' In *Comparative View* he laid the groundwork for his creation of the ideal physician – 'the man of science and art ... [with] a liberal, knowing or disinterested spirit' – who was the medical equivalent of the reforming moderate clergy." (52)

18. Haakonssen, *Medicine and Morals:* 48–54.

19. Quoted in Haakonssen, *Medicine and Morals:* 54.

20. Gregory's writings comprise *A Comparative View of the State and Faculties of Man with those of the Animal World* (1765, 1772); *Observations on the Duties and Offices of a Physician* (1770); *On the Method of Prosecuting Enquiries in Philosophy* (1770); *Elements of the Practice of Physic: For the Use of Students* (1772); and *Lectures on the Duties and Qualifications of a Physician* (1772). *A Father's Legacy to his Daughters* (1774, 1779) was published posthumously and also became immensely popular. His son, James Gregory, edited

a revised edition of *Lectures on the Duties and Qualifications of a Physician* in 1805, and subsequent editions of this work appeared in 1817 and 1820. McCullough, *John Gregory*: 44.

21. Haakonssen, *Medicine and Morals:* 14–16.
22. McCullough, *John Gregory*: 187.
23. Laurence B. McCullough, ed., *John Gregory's Writings on Medical Ethics and Philosophy of Medicine*, Boston, MD: Kluwer, 1998:112. There is some dispute between McCullough and Haakonssen as to whether Gregory's concept of sympathy followed David Hume or Adam Smith. In Hume's conception of sympathy, A sympathizes with B and that leads A to have same sentiments as B. In Smith's conception of sympathy, A sympathizes with B and imaginatively puts self in same situation as B, but this may or may not lead A to have same sentiments; see also Haakonssen, *Medicine and Morals*: 70–74. Fortunately we need not resolve this dispute in order to understand Gregory's medical ethics as one source used by Ryan. Haakonssen suggests further that Gregory seems to view sympathy for the suffering patient as "feeling for" that individual's distress, not feeling the distress itself. This makes his sense of sympathy equivalent to benevolence, a standard view found among all Scottish moralists of his time. Haakonssen further suggests that Hume was "moral anatomist," concerned with theoretical underpinnings of morality in a theory of mind. Gregory by contrast was a "moral painter" (73) more interested in practical morality, trying to bring out good behavior and concrete actions by exhortation and example. A final reason for Gregory to have resisted too close an adherence to Hume's teaching was Hume's openly atheistic posture; Gregory "loved Mr. Hume personally as a worthy agreeable man" (50) but "detested Hume's philosophy" for its atheism (50).
24. McCullough, *Gregory's Writings:* 127.
25. McCullough, *Gregory's Writings:* 4.
26. McCullough, *John Gregory:* 133.
27. McCullough, *John Gregory:* 134.
28. See for example Haakonssen, *Medicine and Morals*: 101–102.
29. Gregory called his Edinburgh lectures "The Duties and Offices of a Physician" in reference to the idea from Cicero and the Stoics, of the implied contract embodied in a social role or office. This idea was taken up by Scottish natural law philosophy, so that each social role had associated virtues and duties seen as intrinsic to it. Haakonssen, *Medicine and Morals*: 56–59.
30. Haakonssen, *Medicine and Morals:* 59–62.
31. Haakonssen, *Medicine and Morals:* 54–56.
32. McCullough, *Gregory's Writings:* 102.
33. McCullough, *Gregory's Writings:* 103.
34. McCullough, *Gregory's Writings:* 104.
35. McCullough, *Gregory's Writings:* 107; Haakonssen, *Medicine and Morals:* 76–79.
36. We are grateful to Robert Baker and Laurence McCullough for suggesting that Gregory's position here could not properly be characterized as paternalistic. They note that Gregory treated telling the patient unpleasant prognostic news in these circumstances as an unpleasant duty. According to his sense of role responsibilities, it was a duty that family members, not the physician, ought to discharge. Shielding the patient for his own good (paternalism) was not properly part of the analysis.
37. Haakonssen, *Medicine and Morals:* 76–79.
38. Haakonssen, *Medicine and Morals:* 81; quoting Gregory, *Lectures*, 49–50; McCullough, *Gregory's Writings*: 69.
39. Haakonssen, *Medicine and Morals:* 81–82.
40. Again we rely especially on Haakonssen, *Medicine and Morals*: 94–173. Her primary source in turn was Percival's biography written by his son Edward.
41. Haakonssen, *Medicine and Morals*: 100–102. Percival may also have owed to the Academy the original choice of "medical jurisprudence" as the title for his work. One of his tutors, John Aikin Sr., was a student of "natural jurisprudence" as part of larger system that included "natural religion" and "private morality." These more or less matched the tripartite division of duties (to God, self, and others) as later written about by Thomas Gisborne, whom Percival viewed as an

important authority on practical ethics. Professional ethics logically fell under the duty to others, and hence under the division of "jurisprudence." Haakonssen, *Medicine and Morals*: 102–107. Ryan would later argue that medical ethics should be subsumed under the larger category of medical jurisprudence.

42. Haakonssen, *Medicine and Morals*: 141–145.

43. Haakonssen, *Medicine and Morals*: 129. Modern critics explain the incompleteness and lack of philosophical foundations in Percival's *Medical Ethics* as due to the fact that the origins of the work lay in the response to a specific local problem. Haakonssen claims that the deficiencies are less in the work itself and more in the eye of today's reader: "Percival claimed in his preface that his work was up-to-date, comprehensive, original, and that it had the stamp of approval of the most eminent physicians and philosophers. The relevant public seems to have accepted this, and it undoubtedly did so because the work was in fact a judicious mixture of the traditional and the innovative" (129).

44. J.V. Pickstone and S.V.F. Butler, "The Politics of Medicine in Manchester, 1788–1792: Hospital Reform and Public Health Services in the Early Industrial City," *Medical History* 28: 227–249, 1984.

45. Thomas Percival, *Medical Ethics; or, a Code of Institutes and Precepts adapted to the Professional Conduct of Physicians and Surgeons…* (Manchester: Russell, 1803; modern reprint by the Classics of Medicine Library, Birmingham, AL: 1985): 1. "Urbanity and rectitude" might sound off-putting to the modern reader; Haakonssen argues that in Percival's day this simply meant the "whole duty of the upright Christian character" (Haakonssen, *Medicine and Morals*: 127).

46. Percival, *Medical Ethics*: 6.

47. Percival, *Medical Ethics*: 7.

48. Robert B. Baker and Laurence B. McCullough, "What is the History of Medical Ethics?" In: *Cambridge World History of Medical Ethics*, ed. Robert B. Baker and Laurence B. McCullough, Cambridge, UK: Cambridge University Press, 2009:3–15. Baker and McCullough emphasize the historical problem this creates for arguing (as Ryan was in fact the first to do) that "medical ethics" represents a long historical tradition embracing both the Hippocratic Oath and today's bioethics. Percival himself wrote as if previous works on "medical ethics" were in no way authoritative for his project.

49. Haakonssen, *Medicine and Morals*: 136.

50. At the end of the eighteenth century, the existing "gentleman's" code of honor, especially in matters such as dueling, was coming under criticism from practical moralists like Gisborne. There was a call for a new code of gentlemanly conduct more suited to the role of reason and religion. Percival felt this debate important enough to insert a discussion of dueling into his Appendix; Percival, *Medical Ethics*: 214–28. "[Gregory] had been attracted to the ideal combination of liberal education and independent fortune in the make-up of the professional gentleman. Percival's vision not only was much more middle-class, it also had a different institutional focus in the hospital. The pursuit of medical science and its application to practical medical care were gradually united within the one institutional framework, and the profession's mastery of this new medicine could indeed make it into a gentlemanly elite, unsurpassed by any previous generation of scholarly or philosophical physician" (Haakonssen, *Medicine and Morals*: 139).

51. Robert B. Baker, "The Discourses of Practitioners in Nineteenth- and Twentieth-Century Britain and America," in *Cambridge World History of Medical Ethics*, ed. Robert B. Baker and Laurence B. McCullough, Cambridge, UK: Cambridge University Press, 2009:446–464.

52. Haakonssen, *Medicine and Morals*: 154.

53. Haakonssen, *Medicine and Morals*: 154–158. Haakonssen quotes Percival: "Every man who enters into a fraternity engages, by a tacit compact, not only to submit to the laws, but to promote the honour and interest of the association, so far as they are consistent with morality, and the general good of mankind" (Haakonssen, *Medicine and Morals*: 123, quoting Percival, *Medical Ethics*: 45–46). Therefore the implied social contract, according to Haakonssen, was

aimed in two directions – physicians owed duties both to the profession and to other physicians, and to society generally for the honest and faithful carrying out of their duties. Later critics who claim that Percival's ethics has no sense of duties owed to patients and is simply an etiquette of how doctors should treat each other omit the latter part of this contract, in her view. We are grateful to Robert Baker and Laurence McCullough for offering a somewhat different interpretation of Percival's ethics, noting that as a Unitarian, Percival would have held more socially egalitarian views than Haakonssen attributes to him.

54. Haakonssen, *Medicine and Morals*: 149.

55. We are grateful to Robert Baker and Laurence McCullough for a novel reading of this passage in Percival's ethics, that does not seem to have been considered by other recent writers on the history of the ethical debate on truthful disclosure. They suggest that Percival intended all along that the patient should be told of the dire prognosis. He merely intended that the family members, rather than the physician, be the ones to make this communication, arguing that such words coming from the physician would have much more dire emotional consequences. We suggest in turn that this alternative reading presumes that as a matter of fact, it was customary for families to tell seriously ill patients the truth about their prognoses at the end of the eighteenth century, if the physician left this office to their discretion. We know of no empirical data to indicate one way or the other.

56. Haakonssen, *Medicine and Morals*: 167–173. It is important to recall that Percival was not defending deception as general medical practice, but the much more specific issue of whether one should tell the patient a possibly shocking truth when the patient is in the midst of a medical crisis and is presumed to be in an especially vulnerable state. Percival, *Medical Ethics*: 166–167.

57. Haakonssen, *Medicine and Morals*: 132.

58. Haakonssen, *Medicine and Morals*: 134. We are grateful to Robert Baker and Laurence McCullough for the reminder that Percival, as a Unitarian, was very careful not to confuse his own ethical code with any sort of sectarian Christian document. Rev. Thomas B. Percival might well have been a rebellious son who broke with his father over religion, and it would be a mistake to attribute the younger Percival's religious views to his father.

59. Specifically, he claimed that three measures could manage the potential rivalry among the orders: (1) a strong faculty committee; (2) improved dispute arbitration; and (3) new, detailed consultation guidelines. Ryan reprinted these relevant passages. Haakonssen, *Medicine and Morals*: 141–145.

60. Percival, *Medical Ethics*: 9; quoted in Haakonssen, *Medicine and Morals*: 152.

61. Percival, *Medical Ethics*: 125–126.

62. Haakonssen, *Medicine and Morals*: 148–154.

63. England General Register Office, Certified Copy of an Entry of Death, District of St. Giles in the Fields of St. George Bloomsbury, copy certified 6 January 2004.

64. "Death of Dr. Ryan," *Provincial Medical and Surgical Journal*, 1 (1840–1841): 207. A handwritten "0" or "6" could easily be confused one for the other, perhaps accounting for this discrepancy.

65. Confirmed by electronic mail communication, Tina Craig, Royal College of Surgeons, to HB, 1/13/04. We will see later, however, that there seem to have been two Michael Ryans, each MRCS, practicing in Kilkenny in 1824, and the University of Edinburgh lists a number of Michael Ryans as having matriculated during the period 1816–1822; so some confusion cannot be entirely excluded.

66. The Wellcome Library catalogue entry for Ryan's medical school thesis agrees with the MRCS date of 1819 and adds the useful information about the minimum age requirement.

67. Adrian Desmond, *The Politics of Evolution: Morphology, Medicine, and Reform in Radical London*, Chicago, IL: University of Chicago Press, 1989: 427.

68. Electronic mail communications, Paul Gorry to HB, 8/6/04.

69. "Insolvent Debtor's Court, November 2." *The Times of London*, 4 November 1836: 6.

70. "Essays, Papers, &c. Published by Dr. Ryan," endpaper advertisement included in Ryan, *Prostitution in London with a Comparative View of That of Paris and New York, as Illustrative of the Capitals and Large Towns of All Countries; and Proving Moral Depravation to Be the*

Most Fertile Source of Crime, and of Personal and Social Misery; with an Account of the Nature and Treatment of Various Diseases, Caused by the Abuses of the Reproductive Function. London: H. Bailliere, 1839.

71. See below for Ryan's writings, including *A new practical formulary of hospitals … Translated from the new French ed. of Milne Edwards and P. Vavasseur,* London: John Churchill, 1835. Besides the translations of Latin passages within *Manual of Medical Jurisprudence* (below), Ryan also translated William Harvey's celebrated work on the circulation of the blood, in a series of articles printed in the *LMSJ,* 1832–1833.

72. Electronic mail communication, Robert Baker to HB, 8/2/04. Baker and McCullough have written that Ryan had an "Irish-Catholic education"; Baker RB, McCullough LB. "What Is the History of Medical Ethics?" In: *Cambridge World History of Medical Ethics,* ed. Robert B. Baker and Laurence B. McCullough, Cambridge, UK: Cambridge University Press, 2009:3–15. Admittedly, Ryan might have had a Catholic-influenced education even though he and his family were not of that faith; in fact we know no details of his education prior to medical school. We are grateful to Harold Vanderpool for pointing out several phrases in Ryan's ethical writings that suggest a close knowledge of Catholic teachings.

73. Rev. Richard Boyle Bernard, born Sept. 4, 1787, was second son of the first Earl of Bandon. He served as rector of Glankeen, 1817–1822; rector of Shankill, Leighlin, 1826–1833. He was Dean at Leighlin from 1822 until he died on March 1, 1850, leaving a number of munificent charitable bequests to church institutions. Personal communication, Dr. R. Refaussé, Librarian, Church of Ireland, to HB, 2/15/04.

74. Against these arguments, Robert Baker has urged that Ryan quoted with apparent approval the Roman Catholic teaching on abortion and on the destruction of fetal life to save the live of the mother; electronic mail communication, Baker to HB, 8/2/04. Ryan was, in that context, surveying the extant opinions of various authorities in Europe; *A Manual on Midwifery* (1828): ii, 269–270. We think it much more likely that a non-Catholic physician should have quoted from Catholic opinions on these subjects, than that a Catholic person would have dedicated his M.D. thesis to a non-Catholic clergyman.

75. Michael Ryan, *A Manual of Midwifery and Diseases of Women and Children…,* 4th ed. London: (self-published), 1841: 7.

76. Ryan, *Prostitution in London* (1839): 322.

77. Eoin O'Brien, ed., *The Charitable Infirmary, Jervis Street, 1718–1987: A Farewell Tribute,* Dublin: Anniversary Press, 1987; J.D.H. Widdess, *The Royal College of Surgeons in Ireland and Its Medical School, 1784–1984,* Dublin: Royal College of Surgeons in Ireland, 1984.

78. In a number of his books, beginning in 1824, Ryan gave as one of his qualifications on the title page, "Member of the Association of Fellows and Licentiates of the Royal College of Physicians in Dublin" (or in some cases, the "King's and Queen's College of Physicians in Ireland"). The current keeper of records, however, states that no evidence of Ryan's membership now exists; electronic mail communication, Robert Mills to HB, 10/9/03.

79. Porter, *Greatest Benefit*: 306–320.

80. The archives at Edinburgh list two Michael Ryans, one from "Barrisoleigh, Tippy." matriculating in 1818–1819, one from "Co. Tipperary" matriculating in 1819–1820. Electronic mail communication, Irene Ferguson to HB, 11/27/03.

81. Ryan discussed his Dissertation in his *A Manual on Midwifery…* (London: Longman and Co., 1828):127. At this point there is some confusion in the Edinburgh archives. The Ryan who graduated in 1821 is described further: "Asst Surgeon 51 Foot 1825; Surgeon 1841. Died at Ireland Island, Bermuda, 1853." This would not fit our Michael Ryan who, so far as we know, never had any military connections and died in 1840. It appears that the Archives mixed up the correct graduation date and thesis title of our Michael Ryan with the later biography of a different Michael Ryan. The second Ryan is described in the catalog of the Wellcome Library as "Michael Ryan of Cork, d. 1853." The Library holds a copy if his M.D. thesis from Edinburgh, titled "De Variis Ictericis Morbis," 1819. The Edinburgh archives includes a Michael Ryan of Cork who matriculated in 1817–1818. It seems quite possible

that such a transposition of records occurred: electronic mail communication, Irene Ferguson to HB, 1/13/04.

Further confirmation as to the correct identity of our Ryan is provided by the minutes of the Royal College of Physicians of London, to whom Ryan applied in 1830 for the status of Licentiate (the highest recognition possible to someone who had not graduated from Oxford or Cambridge): "Dr. Michael Ryan produced a Diploma from the University of Edinburgh by which he was created Dr. of Medicine on 1st of Augt. 1821." Comitiis Minoribus Ordinariis, 7 May 1830. This Michael Ryan is almost certainly our Ryan, as he applied for this status around the date when our Ryan was known to have taken up residence in London and was seeking appointments as a lecturer (for which the RCP Licentiate status would be very useful). By contrast, the Michael Ryan of Cork, who apparently served as a military surgeon, would have no reason to seek an RCP qualification.

82. "Death of Dr. Ryan," *Provincial Medical and Surgical Journal*, 1 (1840–1841): 207. The professor of midwifery at Edinburgh between 1800 and 1839 was James Hamilton (1767–1839), son of the preceding professor, Alexander Hamilton. The younger Hamilton was best known for his *Practical Observations* (1839). He was said to have been very kindly toward his patients and notably egalitarian in how he treated patients of different social classes, but rude and argumentative toward his colleagues, against two of whom he filed lawsuits at various times. Alexander Grant, *The Story of the University of Edinburgh during its First Three Hundred Years* (London: Longmans, Green & Co., 1884): ii:417–419. Ryan possibly emulated his favorite professor in more than his choice of specialty.

83. Ryan, *Tentamentum Medico-Physicum Inaugurale, de Genere Humano ejusque Varietatibus…*, Edinburgh: P. Neill, 1821. The titles of the sections within the dissertation are: Generis humani historia; de generatione; de partu humano; de infantia; de forma generis humani externa; de calore; de frigore; de situ; de colore; de statura.

84. The *List of the Members of the Royal College of Surgeons in London* is available in the Wellcome Library, London for the years 1822, 1825, and 1833. The 1822 edition lists one Michael Ryan, with the address, "Burris O'Leigh, co. Tipperary." The 1825 edition lists *two* persons, each named Michael Ryan and each practicing in Kilkenny (see note 85). The 1833 edition lists our Michael Ryan in Hatton Garden, London, and the second Michael Ryan is no longer listed.

85. Electronic mail communication, Declan Mcauley to HB, 1/21/04. This Ryan listed his credentials in the directory as both an M.D. from Edinburgh and MRCS status, consistent with our Michael Ryan. As previously noted, there would appear to have been two Michael Ryans practicing in Kilkenny as of 1825 if the *List* of the Royal College of Surgeons is to be believed. Ryan later wrote in a case report of a 21-year-old woman whom he treated on 8 July 1823 for angina pectoris (with the highly toxic-sounding remedy of prussic acid): "Leeches and blisters were applied to the left side, and many internal remedies administered, by my late much respected namesake, but with only temporary advantage." "Remarks on the Use of Hydrocyanic Acid in Angina Pectoris and other Diseases of the Heart, and in Bronchitis, Phthisis, and Dyspepsia," *London Medical and Physical Journal* 51 (1824): 369. This suggests the presence of another physician named Michael Ryan in the same city. We are grateful to Paul Gorry for unearthing the obituary of this second Michael Ryan, M.D., who appears to have died November 26, 1822; death notice in the *Leinster Journal*, Kilkenny, 30 Nov 1822. (We assume that the Royal College of Surgeons had not yet received notice of this Dr. Ryan's death in time to omit him from the 1825 edition of their List.) The other Michael Ryan also had a son named Michael Ryan who in 1837 was living in 47 Gower Place, Euston Square, London, according to a deed dated 30 March 1837, in which the son agreed to sell his share of the property he had inherited on the death of his father; Registry of Deeds, Dublin, 1837 6 295. This other Michael Ryan living in London was apparently not a physician and seems to have had no connection with our Michael Ryan.

86. From a list of "essays, papers, &c. published by Dr. Ryan," included in the printer's endpapers to Michael Ryan, *A Manual of Midwifery and Diseases of Women and Children…*, 4th ed. London: (self-published), 1841. These reportedly appeared in journals including the

London Medical and Physical Journal and the *Transactions of Dublin College of Physicians*.

87. The title page bore the imprint, "Kilkenny: Printed and published by J. Reynolds, for Hodges, M'Arthur, and J. Cummings, Dublin..., 1824."

88. Ryan, *Treatise on the Most Celebrated Mineral Waters* (1824): 18–19. These chemical analyses were totally qualitative; Ryan apparently lacked either the know-how or the equipment to conduct quantitative chemical analyses.

89. Ryan, *Treatise on the Most Celebrated Mineral Waters* (1824): 20.

90. Ryan, *Treatise on the Most Celebrated Mineral Waters* (1824): 21.

91. Ryan, *Treatise on the Most Celebrated Mineral Waters* (1824): ix–xii.

92. Ryan, *Treatise on the Most Celebrated Mineral Waters* (1824): viii.

93. Roderic a Castro (Rodrigo de Castro, 1546–1627), *Medicus-politicus: sive de officiis medico-politicis tractatus, quatuor distinctus libris: in quibus non solum bonorum medicorum mores ac virtutes exprimuntur, malorum vero fraudes et imposturae deteguntur ...* (The Politic Physician: or a Treatise on Medico-Political Duties ...), Hamburg, 1614; quoted in Winfried Schleiner, *Medical Ethics in the Renaissance* (Washington, DC: Georgetown University Press, 1995): 71. The quotation, however, was in common use, and Ryan may have taken it from another source. We are indebted to Debra Nails for tracing various forms of the quotation back as far as Plato (*Republic* 595b–c) in Greek, and in Latin to Roger Bacon, *Opus Magnum* I..vii ("Amicus est Socrates magister meus sed magis est amica veritas"). Cervantes included in *Don Quixote* II. Chap. 51, "Amicus Plato, sed magis amica veritas." Isaac Newton was said to have headed remarks in his commonplace book, "Amicus Plato amicus Aristoteles magis amica veritas" (reportedly in turn quoting Walter Charleton, b. 1619).

94. "Death of Dr. Ryan," *Provincial Medical and Surgical Journal*, 1 (1840–1841): 207.

95. Ryan published "Delirium Tremens, Treated as a Nervous Disease," in *The Lancet* 2 (1827–1828): 791–793. The article describes a case that he treated, beginning 23 July 1827, and contains information on the status of the patient 1 year later. Ryan is described in the masthead as "one of the Physicians to the Central Infirmary, Greville Street, Hatton Garden, Lecturer on Midwifery." While the article does not specifically say so, every indication suggests that he had treated this patient in London as part of his infirmary practice. Had he begun treating the patient before leaving Ireland, it is not clear how he would know about the patient's later clinical course. This case report would therefore suggest July, 1827 as the latest possible date for his arrival in London. However, when Ryan was later hauled into Debtor's Court (1836), it was reported that he had left Ireland in 1828, and that at the time he had an income of £156 annually from property there; "Insolvent Debtor's Court, November 2." *The Times of London*, 4 November 1836: 6.

96. The advertisement of these lectures appears in the endpapers of Ryan, *A Manual on Midwifery; or a Summary of the Science and Art of Obstetric Medicine, including the Anatomy, Physiology, Pathology, and Therapeutics, Peculiar to Females; Treatment of Parturition, Puerperal, and Infantile Diseases; and an Exposition of Obstetrical-Legal Medicine*. London: Longmans, 1828 (following page 353). In those years, a number of lecturers unaffiliated with medical schools offered lectures on various topics in their private rooms. All lecturers, whether or not affiliated with medical schools, received reimbursement by selling individual tickets to their lecture courses.

97. Ryan, *Manual on Midwifery* (1828): i.

98. Ryan, "An Essay on the Natural, Chemical, and Medical History of Water, in its simple and Combined States; including an Account of the Chemical composition, and Medical Effects, of the principal Mineral Waters in the United Kingdom of Great Britain and Ireland, and also the Continent of Europe," *London Medical and Physical Journal* 54 (1825): 442–461. Ryan had authored this paper while still in practice in Kilkenny.

99. Ryan, *Remarks on the Supply of Water to the Metropolis* (1828): xii.

100. Ryan, in fairness, did add two additional pages at the end of his earlier work, "Observations on Bathing" (45–46).

101. This formula was set for the most part by the requirements for certification by the Society of Apothecaries and the Royal College of Surgeons. Since students bought tickets individually

for each course of lectures or term of clinical experience in the wards or dispensaries, there was no incentive for the student to attend any lectures that were not part of the prescribed curriculum leading to the LSA and MRCP certificates.

102. "Introduction," Robert B. Baker, Arthur L. Caplan, Linda L. Emanuel, and Stephen R. Latham, eds., *The American Medical Ethics Revolution: How the AMA's Code of Ethics Has Transformed Physicians' Relationships to Patients, Professionals, and Society*, Baltimore, MD: Johns Hopkins University Press, 1999: xix. We have, however, been unable to find any documentation to confirm Baker and colleagues' statement that Ryan claimed to be a *professor* of medical ethics, or that he was a professor affiliated with University College, London.

103. Ryan claimed to have originated at least the term "obstetrician" to replace the awkward usages, "male midwife" or the French "accoucheur"; *Manual on Midwifery*: v–vi.

104. Ryan, *Manual on Midwifery* (1828): ii.

105. Ryan, *Manual on Midwifery* (1828): ii.

106. Ryan, *Manual on Midwifery* (1828): ii.

107. Ryan, *Manual on Midwifery* (1828): endpaper advertisement (following 353).

108. Ryan, *Manual on Midwifery* (1828): iii–iv.

109. Ryan, *Manual on Midwifery* (1828): iv–v.

110. [Anonymous review of Ryan, *A Manual on Midwifery*,] *London Medical Gazette* 2 (1828): 714–715, quote p. 715.

111. [Anonymous review of Ryan, *A Manual on Midwifery*,] *London Medical Gazette* 2 (1828): 714–715, quote p. 715.

112. The same reviewer noted as much when he came to review the 3rd edition of this volume that appeared in 1831. He did, however, repeat and even expand his compliment, that in the book could be found "more information on the subject of midwifery … than in any other work of three times its size to which we can refer." [Anonymous review of Ryan, *A Manual of Midwifery*, 3rd ed.] *London Medical Gazette* 9 (1831–1832): 52.

113. *Robson's London Commercial Directory* … London: Robson, Blades & Co., 1828; *Robson's Classification of Trades and Street Guide* … London: Robson, Blades & Co., 1829; *Robson's London Commercial Directory* … London: Robson, Blades & Co., 1830. His lecture notices appeared in "Practice of Physic and Midwifery," *Lancet* 1 (1829–1830): 18. Admission to one course of lectures in either subject cost the student 3 guineas (3 pounds, 3 shillings).

114. For example, John Snow, coming to London as an apprentice-trained unknown from Yorkshire in 1836, was able to use the Westminster Medical Society as an effective springboard to a career in medical research; Peter Vinten-Johansen, Howard Brody, Nigel Paneth, Stephen Rachman, and Michael Rip, *Cholera, Chloroform, and the Science of Medicine: A Life of John Snow*, New York: Oxford University Press, 2003: 69–72, 84–86.

115. Ryan, *Manual of Medical Jurisprudence*, 1st ed. London: Renshaw and Rush, 1831 [title page].

116. Kaspar Hauser (d. 1831) was found in 1828, apparently aged about 16 or 18 but talking like a 3–4 year old child, and gave evidence of having been imprisoned for most of his youth. His case attracted widespread attention in Europe and there were rumors that he was actually the son and rightful heir of the Prince of Baden. Stanhope, a friend of Baden's family, initially became very interested in Hauser, befriended him, and helped to provide for his support. After a while Stanhope apparently became bored and convinced himself that Hauser was a fake. Stanhope's interventions had the result of cutting Hauser off from those who were best able to care for him and placing him among others who treated the youth badly. A good overview is to be found at http://www.mysteriouspeople.com/Hauser1.htm (accessed August 25, 2004).

117. The other dedications are to Rev. Boyle Bernard in his M.D. dissertation, already noted; and to Dr. John Elliotson, professor of medicine at University College, London, in the *Physician's Vademecum*, 1836.

118. "Dr. Ryan also made some observations on the catechu and cardamoms," "Medico-Botanical Society," *Lancet* 1 (1835–36): 341 (society meeting of Nov. 24, 1835). The *Lancet* for December 10, 1836 (*Lancet* 1 (1836–37): 414) describes a meeting (no date given) at which Ryan made some extended comments on croton oil and buchu leaves.

119. "Medical Society of London," *London Medical Gazette* 3 (1828–1829): 646–647. "Dr. Ryan did not think the case very peculiar. It was not, he thought, an hour-glass contraction. He concurred with Mrs. Ashwell and Waller. Dr. Milligan some time ago gained great reputation because he turned when no one else could. This he affected by opium." The case being discussed was one of difficult labor; "turning" referred to version of the fetus, to change the fetal position to affect an easier delivery.

120. Ryan, "To the Editor of the *London Medical Gazette*," *London Medical Gazette* 3 (1828–1829): 684. We have not searched through all the reports of medical society meetings in the *Lancet* and *London Medical Gazette* to see whether Ryan is listed as a speaker on any other occasion. The journals of that time frequently gave very full reports of these meetings, hiring medical students to serve as short-hand reporters.

121. On radicalism and reform generally during this period see Adrian Desmond, *The Politics of Evolution: Morphology, Medicine, and Reform in Radical London* (Chicago, IL: University of Chicago Press, 1989).

122. Desmond, *Politics of Evolution:* 15, 170–171.

123. Ryan explained in the Preface to the first edition, *Manual of Medical Jurisprudence* that the text was originally published in the "lacunae" of *London Medical and Surgical Journal* as an aid for students (x).

124. James F. Clarke, *Autobiographical Recollections of the Medical Profession*, London: J. and A. Churchill, 1874: 134–135.

125. *Lancet* 2 (1831–1832): 92, 21 April 1832. At this time Wakley favored a contagious explanation of the transmission of cholera, as did the Board of Health, while Ryan favored a miasmatic, non-contagionist hypothesis. As the epidemic progressed the majority of medical opinion swung into the non-contagion camp, giving Ryan numerous opportunities to crow over Wakley's downfall; see for instance *LMSJ* 1 (NS) (1832): 411–413.

126. *LMSJ* 1 (NS) (1832): 17–18. Ryan here referred to the Anatomy Act, and blamed Wakley and the *Lancet* for retarding its passage in a desirable form.

127. *LMSJ* 1 (NS) (1832): 289–291.

128. "Letter from Dr. Ryan," *Lancet* 1 (1831–1832): 222–226.

129. *LMSJ* 7 (1831): 433, 512, 522–526.

130. "Drs. Gordon Smith, Ryan, and A. Thomson," *Lancet* 1 (1830–1831): 72–73, quote p. 72. Smith died in 1833, while incarcerated in the Fleet Prison. The author of his obituary in *Gentleman's Magazine* described Smith as having become "involved in pecuniary difficulties, combined somewhat with irregular habits," leading to his imprisonment, "where he gradually sunk and expired"; H.W.D., "Obituary: John Gordon Smith, M.D., F.R.S.L.," *Gentleman's Magazine*, September 1833, 278–279. In fairness to Ryan, perhaps Smith's "irregular habits" manifested themselves several years before his death and caused him to act unpredictably. The obituary author credited Ryan and Smith jointly with having prevailed upon Apothecaries Hall to include the subject of medical jurisprudence in their required examination, thus compelling all medical schools to provide courses of lectures in that "neglected but truly important subject."

131. "Drs. Ryan and Gordon Smith," *Lancet* 1 (1830–1831): 106–108, quote p. 108. It is quite surprising to the modern reader that a supposedly legitimate medical journal of that day would devote this many columns to a spat over who stopped speaking to whom first.

132. Ryan, *Prostitution in London* (1839): 433.

133. Ryan, *Prostitution in London* (1839): 433.

134. Ryan, *Prostitution in London* (1839): 434.

135. Ryan, *Prostitution in London* (1839): 434. Indications of the sort of plot to malign his work that Ryan imagined–and perhaps evidence that he was not far wrong in so thinking–are provided by an exchange in *The Lancet* in 1839. An anonymous (as always) *Lancet* reviewer had criticized a volume, *A Manual of Diseases of the Eye; or, Treatise on Ophthalmology*, by S. Little, revised and enlarged by Hugh Houston, as "got up for the express purpose of bespattering with praise Michael Ryan, Esq., in the execution of which laudable task the original work has been most disgracefully disfigured"; *Lancet* 1 (1838–1839): 494. Ryan

could never resist such an opening, and wrote back a lengthy letter defending Houston's book. He claimed, "The article of which I complain was surreptitiously concocted, by persons who are most ungrateful to me for many acts of kindness, and now, without just cause, my personal enemies; and whose conduct, if exposed, would do them little credit in the estimation of the profession"; *Lancet* 1 (1838–1839): 706. Wakley took delight in appending an editorial comment: "Dr. Ryan, we are well aware, may have a sufficient ground for believing that persons who were at one time connected with him in the publication of a medical journal are now his bitter and malignant enemies; but he may be assured, with reference to the notice of Mr. Houston's work, ... that it was not the production of his secret foes, but of the sub-editor of this Journal"; *Lancet* 1 (1838–1839): 706.

136. In anticipating negative reviews of *Prostitution in London*, Ryan proved prescient. The anonymous reviewer for the *British and Foreign Medical Review* insisted that he was no prude, and that Parent-Duchatelet's *On Prostitution in Paris* had indeed been very favorably reviewed by that journal. Ryan's volume, however, was of a very different nature: "The greater portion of the volume is filled with details of the most disgusting kind ... without any necessary connexion to the professed subject of the work, unredeemed by the slightest tincture of utility, and only calculated ... to pander to the vilest and most depraved tastes"; [Anonymous review of Ryan, *Prostitution in London*], *British and Foreign Medical Review* 7 (1839): 540.

137. None of the discussion of the case in *The Lancet* or the *LMSJ* mentions Dr. Ramadge's full name; but it is given as "Francis H. Ramadge" in an editorial, "Dr. Ramadge and St. John Long," *London Medical Gazette* 8 (1831): 117–120. In *A Manual on Midwifery* (1828) Ryan in numerous places speaks glowingly of a colleague at the Central Infirmary in Greville Street, Dr. Ramadge, who maintained an excellent collection of pathological specimens. This Dr. Ramadge appeared in several of the annual lists of London medical lecturers in *The Lancet*. Presumably the Dr. Ramadge of the present libel case was a different individual.

138. Letter from Dr. Ramadge addressed to "[John] Long the Quack" as printed in *Lancet* 2 (1830–1831); No. 398 (16 April 1831): 90–93; No. 400 (30 April 1831): 154–156.

139. "London Medical Society: Expulsion of Dr. Ramadge," *Lancet* 2 (1830–31): 251–252.

140. The original version, "Tweedie v. Ramadge," appeared in *Lancet* 2 (1830–31), no. 419 (10 Sept. 1831). The version with Ryan's editorial comment appended would have appeared in *LMSJ* 7 (1831), which we were unable to examine in the original. That version in turn was reprinted later in *Lancet* 2 (1831–32): 408–9, which we use as our source.

141. A summary account of both trials (Ramadge v. Wakley; Ramadge v. Ryan et al.) may be found in *Lancet* 2 (1831–32): 408–409. Ryan published a brief account of the trials, with little commentary; *LMSJ* 1 (NS) (1832): 705–6. See also *The Times of London*, 27 June 1832, 7 November 1832, 16 November 1832, and 21 November 1832.

142. "Subscription for Dr. Ryan," *Lancet* 1 (1832–33): 351–352. The list of donors included staff physicians from many of the major London hospitals, fellow medical editors and authors, and a number of professors from Dublin; Earl Stanhope, president of the Medico-Botanical Society, contributed £5, one of the larger donations. (Two other donors were names frequently seen in meeting reports of the Medico-Botanical Society: Drs. Sigmond and Hancock.) The total raised at that time was approximately £160; it is not known how much of the judgment Ryan himself eventually had to pay, and how much this contributed to his later problems with debt.

143. Ryan's account of the quarrel, typically, depicted him as having wished to remain silent, and attacking Renshaw only after the latter had circulated false advertisements. Ryan accused Renshaw of interfering in editorial decisions, and trying to convince Ryan to publish more articles on various nostrums so as to boost sales of advertisements. He claimed that as he was the sole proprietor of the *LMSJ* when he first entered into partnership with Renshaw, the dissolution of that partnership left him again as proprietor and retaining all rights to the *Journal. LMSJ* 6 (NS) (1834): 153–155.

144. Clarke, *Autobiographical Recollections* (1874): 135–136.

145. Desmond, *Politics of Evolution:* 170.

146. "Insolvent Debtor's Court, November 2." *The Times of London*, 4 November 1836: 6. The account mentioned in passing that a "Mr. Wakley" was also a creditor of Ryan, but did not further identify this person. It seems unlikely that Ryan would have become financially indebted to his bitter editorial rival. Since the dispute at the time of this hearing was over a few books, it seems somewhat questionable that this should have been sufficient grounds to send Ryan to prison, though there may have been an earlier legal proceeding of which we could find no record.

147. Zachary Cope, "The Private Medical Schools of London, 1746–1914." In *The Evolution of Medical Education in Britain*, ed. F.N.L. Poynter, Baltimore, MD: Williams and Wilkins, 1966: 89–109.

148. For at least 1 year after Ryan's departure, some lectures were still offered at the Central Infirmary and Dispensary; *Lancet* 1 (1829–1830): 12. So the dissolution of the school was not, apparently, the immediate reason for Ryan's setting himself up to lecture on his own.

149. Elizabeth Epps, ed. *Diary of the Late John Epps*. London: Kent & Co., 1875: 181.

150. Desmond, *Politics of Evolution:* 166–170.

151. Ryan viciously attacked phrenology as a materialist movement that saw the human mind as nothing but a secretion of the brain, that died with the body, and thus implicitly denied the immortality of the soul, see especially *Manual of Medical Jurisprudence*, 2nd ed. London: Sherwood, Gilbert, and Piper, 1836: 14–20, reprinted in this volume. Epps wrote a book on phrenology and was sympathetic to the movement. His writing on phrenology was "chiefly intended to prove that phrenology was intimately connected with religion"; Elizabeth Epps, ed., *Diary of the Late John Epps*, London: Kent & Co., 1875: 168, quoting the editor of the *Monthly Gazette of Practical Medicine*. (Epps was the author of *Internal Evidences of Christianity Induced from Phrenology*, Edinburgh: J. Anderson, 1827.) Dermott wrote a book in which he argued that the mind and the soul were two distinct concepts so that the former could be material while the latter was immortal and immaterial; *A Discussion on the Organic Materiality of the Mind, the Immateriality of the Soul, and the Non-Identity of the Two* (London: Callow & Wilson, 1830). The three must have had some interesting conversations.

152. Cope, "The Private Medical Schools of London, 1746–1914," 105–106.

153. This comment highlights the stereotypes with which the Irish were viewed in that day, and the difficulties Ryan very likely experienced in being accepted in London medical circles. Any behavior of Ryan's that seems excessively self-promoting must be interpreted within that context as he might well have felt that he had to work especially hard to overcome these prejudices. We have discovered no evidence that Ryan was of the Roman Catholic faith, and have seen some evidence that he was Anglican (Church of Ireland); if so, he would at least not have had to deal with religious prejudice on top of ethnic prejudice. It may be of interest that despite the various *ad hominem* attacks launched against Ryan by his rivals of *The Lancet*, we have discovered no blatant ethnic slurs.

154. Indirect support for the assertion that medical practice was not Ryan's strong suit comes from the anonymous obituary in the *Provincial Medical and Surgical Journal*: "With all these avocations, it was not to be expected that Dr. Ryan could have had leisure to toil much in the practice of his profession; the zeal and delight which he took in intellectual pursuits must have left him little time and have given him but little relish for what he appropriately designated 'the trade of his profession;' and often may be seen among his writings allusions to the degraded position of the majority of its members, who, in place of prosecuting medicine as a science, dealt in it as a trade." "Death of Dr. Ryan," *Provincial Medical and Surgical Journal* 1 (1840–1841): 207.

155. The gold-headed cane was, at the time, a distinctive accessory for physicians.

156. Clarke, *Autobiographical Recollections* (1874): 133–134.

157. We are grateful to Jason Glenn for pointing out that "Milesian" refers to the mythical account of the origins of the Irish people from the sons of Mil Espáine.

158. Clarke, *Autobiographical Recollections* (1874): 136.

159. The word "vade-mecum" (literally, "come with me," Latin) would probably be rendered today as "companion."

160. Endpaper advertisements, *Prostitution in London*, 1839.

161. The notice announcing this volume appears in endpaper advertisements for Ryan's books and lectures, appended at the back of the second (1836) British edition of *Manual of Medical Jurisprudence*; see these on-line at http://books.google.com/books?id=vToEAAAAQAAJ& printsec=frontcover&dq=michael+ryan+manual+of+medical+jurisprudence#PPA556,M1. The copy of this volume that we possess lacks these endpaper advertisements.

162. "Metropolitan Hospital." Archives in London and the M25 Area. http://www.aim25.ac.uk/ cgi-bin/search2?coll_id=5160&inst_id=51 (accessed January 23, 2005).

163. "Voluntary Hospitals," http://www.nhshistory.net/voluntary_hospitals.htm (accessed January 23, 2005).

164. "Metropolitan Hospital." The hospital was forced to move several times between 1836 and 1885 as land was taken up by expanding railways. In the 1880s it changed its name to "Metropolitan Hospital," and finally closed in 1977.

165. Minute books of the General Committee, Metropolitan Free Hospital, 10 March through 7 April 1836 and 11 May 1836; electronic mail communication, Samantha Farhall (St. Bartholomew's Hospital archivist) to HB, 1 February 2005. The archives of the Metropolitan Free Hospital are now lodged with those of St. Bartholomew's Hospital.

166. For example, "Senior Physician to the Metropolitan Free Hospital," title page of *Prostitution in London* (1839).

167. Ryan, *Manual of Midwifery* (4th ed. 1841): 35.

168. "Death of Dr. Ryan," *Provincial Medical and Surgical Journal* 1 (1840–1841): 206–207.

169. Minutes, General Committee, Metropolitan Free Hospital, 2 December 1840; electronic mail communication, Katie Ormerod (St. Bartholomew's Hospital archivist) to HB, 4 February 2005.

170. Clarke, *Autobiographical Recollections* (1874): 136.

171. Clarke admitted, "The entire work is written from memory, for I have had no documents or notes to refer to." Clarke, *Autobiographical Recollections* (1874): vi.

172. England General Register Office, Certified Copy of an Entry of Death, District of St. Giles in the Fields of St. George Bloomsbury, copy certified 6 January 2004. We cannot tell whether the Mr. Earles, surgeon, who signed the death certificate was the same as the supposedly unqualified "Mr. Earles" whom Ryan apparently sent to substitute for himself at the Metropolitan Free Hospital. Presumably someone listed on a death certificate as a "surgeon" had at least an MRCS qualification.

173. Minutes of the Metropolitan Free Hospital MHA 1/1 1836–1850, entry for 16 December 1840, Archives, St. Bartholomew's Hospital. We are grateful to Samantha Farhall and Katie Ormerod of the St. Bartholomew's Archive for their assistance in locating and copying this material. Despite the rather obvious hint contained in the report from the medical staff, the Committee of the Hospital simply had no spare cash to address the needs of Ryan's widow and children. They proceeded at that meeting to officially express their regret and condolences, and to place an advertisement for a replacement physician.

174. "Death of Dr. Ryan," *Provincial Medical and Surgical Journal* 1 (1840–1841):206–207. "Dec. 11. Aged 41, Dr. Michael Ryan, editor of the Medical and Surgical Journal. He was an amiable and clever man, and has left a young family wholly unprovided for." *Gentleman's Magazine* 15 (1841): 105. Another, very brief obituary notice stated that Ryan left four infant children; *London Medical Gazette*, January 1, 1841: 560.

175. "Subscription for the Widow and Children of the late Dr. Ryan," *Lancet* 1 (1840–1841): 903. The list of contributors is very limited; previous to this notice £137 had already been collected. In this short list were included Sir J. Clarke (certainly not James F. Clarke), Mr. Copeland, Dr. Crucifix (probably a pseudonym), Plymouth Medical Society, Dr. Ramsbotham, and Mr. Churchill (probably the medical publisher, John Churchill). The treasurer of the fund was Dr. Klein Grant, who lived at 18 Charlotte Street and was thus a near neighbor of the Ryans's; and lectured on Therapeutics at the nearby North London School (20 Charlotte Street); "North London School" (catalogue of courses), *Lancet* 1 (1838–39): 13.

176. See the Bibliography for the list of articles published in *LMSJ* which were later collected into the *Manual of Medical Jurisprudence*, 1st ed. (1831).

177. The negative review appeared as "*A Manual of Medical Jurisprudence ... by Michael Ryan...*" *Lancet* 1 (1831–1832): 137–143. Wakley printed one letter he received from Ryan in response to this review, and added at the end of his own reply to that letter that he had received a follow-up letter from Ryan, "occupying no less than seven closely written pages of letter paper," and accompanied by numerous positive notices of his book clipped from other periodicals. Wakley stated that, with apparently newly-found concern for the patience of the reader, he would not publish this second letter. [Wakley, Thomas.] "Letter from Dr. Ryan," *Lancet* 1 (1831–1832): 222–226.[0]

178. Michael Ryan, *A Manual of Medical Jurisprudence and State Medicine, Compiled from the Latest Legal and Medical Works, of Beck, Paris, Christison, Fodere, Ofila, etc.* ... 2nd ed. London: Sherwood, Gilbert, and Piper, 1836: v.

179. Chester R. Burns, "Reciprocity in the Development of Anglo-American Medical Ethics," in *The Codification of Medical Morality*, ed. Robert Baker, Boston, MD: Kluwer, 1995: 135–143; quote p. 137. To our reading, Burns may have over-emphasized the degree to which Ryan could be said to have an overarching theory of the relationship between medical ethics and law.

180. Ryan, *Manual of Medical Jurisprudence* (2nd ed): 3. Ryan's reliance on the Bible as his main source of ethical wisdom, coupled with the political prominence of Evangelical Anglicanism during the era when he was writing, make it reasonable to wonder whether Ryan was himself an Evangelical or influenced by those beliefs. We are grateful to Harold Vanderpool for identifying several passages in Ryan's works that would argue against an Evangelical background. These passages include Ryan's frequent reference to the Bible as the "sacred book," and his insistence that denying the immortality of the soul inevitably leads to immorality.

181. Ryan, *Manual of Medical Jurisprudence* (2nd ed): 6. Ryan elected not to address problems with reconciling his assertion of the great esteem in which the populace holds medicine and physicians, with his later complaints of poor behavior among physicians, or of the tendency of the public to ignore the advice of wise physicians and instead to seek out quacks.

182. Ryan, *Manual of Medical Jurisprudence* (2nd ed): 8.

183. Ryan, *Manual of Medical Jurisprudence* (2nd ed): 8.

184. Robert B. Baker and Laurence B. McCullough, "Introduction." In: *Cambridge World History of Medical Ethics*, eds. Robert B. Baker and Laurence B. McCullough, Cambridge, UK: Cambridge University Press, 2009:3–15. Burns also alluded to Ryan's priority in organizing his discussion of ethics within a historical framework; Chester R. Burns, "Reciprocity in the Development of Anglo-American Medical Ethics," in *The Codification of Medical Morality*, ed. Robert Baker, Boston, MD: Kluwer, 1995: 135–143. The way Ryan organized the Introduction to this volume–listing the works on the topic published in each historical period–suggests that a historical approach to any given topic came naturally to him and was not unique to his approach to medical ethics, per se.

185. An additional bit of evidence in support of this assertion is found in Ryan's letter to the editor of *The Lancet*, dated May 21, 1828, introducing his "Introductory Lecture to the Theory and Practice of Midwifery": "Another reason that induces me to send this communication is, that it contains a concise, though perfect, history of the rise, progress, improvement, importance, and utility of this branch of medicine [midwifery or obstetrics], and also the universal sanction it has received from mankind in the past ages, and present most enlightened times."*Lancet* 2 (1827–28): 394. With only a slight change in wording, this could have served as a preface to the ethics portion of the *Manual of Medical Jurisprudence*.

186. Ryan, *Manual of Medical Jurisprudence* (2nd ed): 10.

187. Ryan, *Manual of Medical Jurisprudence* (2nd ed): 11–12.

188. Ryan, *Manual of Medical Jurisprudence* (2nd ed): 13. Ryan may be referring to Friedrich Hoffmann (1660–1742), who among other topics wrote works on medical ethics (*Medicus Politicus*, 1738) and hygiene.

189. Ryan, *Manual of Medical Jurisprudence* (2nd ed): 14.

190. Ryan, *Manual of Medical Jurisprudence* (2nd ed): 15–16.

191. Ryan, *Manual of Medical Jurisprudence* (2nd ed): 19.

192. The full title is: *Medicus-politicus: sive de officiis medico-politicis tractatus, quatuor distinctus libris: in quibus non solum bonorum medicorum mores ac virtutes exprimuntur, malorum vero*

fraudes et imposturae deteguntur ... (The Politic Physician: or a Treatise on Medico-Political Duties...) Winfried Schleiner has noted that Roderici á Castro Lusitani (1564–1627), to give one variant spelling of his name, was one of a group of "Lusitani," Jewish physicians who had nominally converted to Christianity to escape persecution in the anti-Semitic Europe of their day. This group wrote on morality from a secular-humanistic perspective. Winfried Schleiner, *Medical Ethics in the Renaissance* (Washington, DC: Georgetown University Press, 1995): 49–86. It is ironic that Ryan, given his emphasis on how all morality arises from Christianity, would have adopted Castro as a source for such a large portion of his text. According to Schleiner, later scholarship has tended to downplay or deny the fact that this group of physicians were Jews; hence Ryan might simply have been unaware of the fact.

193. Winfried Schleiner, *Medical Ethics in the Renaissance* (Washington, DC: Georgetown University Press, 1995): 50, 86.

194. Ryan, *Manual of Medical Jurisprudence* (2nd ed): 35.

195. Ryan, *Manual of Medical Jurisprudence* (2nd ed): 36.

196. While Ryan, as an avid medical reformer, argued strongly against the dogmatism and unmerited privileges of the medical corporations, he has nothing here but praise for the "admirable" Statutes of Morality of the Royal College of Physicians; *Manual of Medical Jurisprudence* (2nd ed): 67.

197. Ryan, *Manual of Medical Jurisprudence* (2nd ed): 47. Apparently mindful of the charges of plagiarism launched by the *Lancet* reviewer of his first edition, as well as of his own legal difficulties in the Ramadge case, Ryan took pains to claim that the works of Gregory and Percival were in the public domain and that he had every legal right to reprint them (46–47).

198. Ryan also had severe criticism for the anonymous editor of the most recent edition of Percival's *Medical Ethics,* whom he accused of adding "inappropriate comments"; *Manual of Medical Jurisprudence* (2nd ed): 47. The edition in question was *Medical Ethics; or, a Code of Institutes and Precepts, Adapted to the Professional Conduct of Physicians and Surgeons...by the late Thomas Percival ...,* London: W. Jackson, 1827. A review in *The Lancet* noted that the editor had omitted Percival's supplementary notes and illustrations, replacing them with extensive, sarcastic comments on contemporary medical quackery; [review of *Medical Ethics...*, by Thomas Percival], *Lancet* 12 (series 2) (1826–1827): 696.

199. We are grateful to Robert Baker and Laurence McCullough for stressing this point. This also seems further to reinforce Chester Burns's observation that Ryan made important contributions to medical ethics by situating it within the field of medical jurisprudence (note 186).

200. Ryan, *Manual of Medical Jurisprudence* (2nd ed): 72.

201. Thomas Percival, *Medical Ethics,* Manchester, S. Russell, 1803: 14–15.

202. Robert B. Baker and Laurence B. McCullough, "Introduction." In: *Cambridge World History of Medical Ethics*, eds. Robert B. Baker and Laurence B. McCullough (Cambridge, UK: Cambridge University Press, 2009:3–15, quoting Henry K. Beecher, *Research and the Individual: Human Studies*, Boston, Little, Brown, 1970:218.

203. Ryan, *Manual of Medical Jurisprudence* (2nd ed): 70–71.

204. U.S. National Commission for the Protection of Human Subjects of Biomedical and Behavioral Research. *The Belmont Report: Ethical Principles and Guidelines for the Protections of Human Subjects.* Washington, DC: U.S. National Commission..., 1978. The distinction is concisely explained by Robert Levine, "Clarifying the concepts of research ethics," *Hastings Center Report* 9 (3) (1979): 21–26. We must recall that Gregory, Percival, and Ryan all wrote in an era when the concept of informed consent was virtually unknown within medicine; so all imagined that experimenting on patients would be done without their explicit consent.

205. In an era in which we have focused so much on informed consent as the central component of research ethics, we might ask whether Ryan suggests anything akin to patient consent as a requirement for the ethical conduct of research. One might argue that any use of a form of contract theory implicitly invokes the concept of consent, since parties to a contract can modify the contract by mutual, explicit agreement. Martin Pernick, reviewing the history of informed consent in nineteenth and twentieth centuries medicine, argued that physicians in Ryan's day advocated educating and informing the public about medical matters, and in specific contexts

indicated considerable respect for the patient's decision on accepting or rejecting specific treatments. But the underlying logic focused much less on any concept of respect for patients' rights or autonomy, and more on the prevalent concept that respecting patients' choices could be therapeutic; Pernick, "The Patient's Role if Medical Decisionmaking: A Social History of Informed Cosent in Medical Therapy," in: President's Commission for the Study of Ethical Problems in Medicine and Biomedical and Behavioral Research, *Making Health Care Decisions: The Ethical and Legal Implications of Informed Consent in the Patient-Practitioner Relationship; Volume Three: Appendices : Studies on the Foundations of Informed Consent,* Washington, DC: U.S. Government Printing Office, 1982: 1–35. Consider, for example: "The *feelings* and *emotions* of the patients, under critical circumstances, require to be known and to be attended to, no less than the symptoms of their diseases. Thus, extreme *timidity*, with respect to venæsection, contraindicates its use, in certain cases and constitutions. Even the *prejudices* of the sick are not to be contemned, or opposed with harshness. For though silenced by authority, they will operate secretly and forcibly on the mind, creating fear, anxiety, and watchfulness." Thomas Percival, *Medical Ethics,* Manchester, S. Russell, 1803: 14–15; quoted in Ryan, *Manual of Medical Jurisprudence* (2nd ed.): 75.

206. Ryan's unexplained assertion that animal experimentation had replaced human experimentation became a bone of contention between him and *The Lancet.* The negative reviewer for that journal expressed frank incredulity that anyone could suggest that animal experiments would suffice to establish the curative powers of any treatment intended for human beings. Ryan on his side invoked the Golden Rule once again and asked whether the reviewer would volunteer himself to be the subject of the experiments. The exchange is nicely summarized by Robert B. Baker, "The Discourses of Practitioners in Nineteenth- and Twentieth-Century Britain and United States" (Chapter 36, Section VIII), in *Cambridge World History of Medical Ethics,* eds. Robert B. Baker and Laurence B. McCullough, Cambridge, UK: Cambridge University Press, 2009:446–464. Baker also reminds us that Gregory had proposed a Golden Rule test for medical experimentation: "I would ask that a man if he have done [dangerous Experiments] to his Child, if he would not do it to his own Child, why should he indanger the lives of other people"; *John Gregory's Writings on Medical Ethics and Philosophy of Medicine,* ed. Laurence B. McCullough, Dordrecht: Kluwer, 1998: 249. Ryan must certainly have appreciated this appeal to his favorite moral rule, and one wonders why he did not highlight it in his own discussion.

207. Ryan, *Manual of Medical Jurisprudence* (2nd ed): 80.

208. On Rush's medical ethics see Haakonssen. *Medicine and Morals*: 187–226. See also Chester R. Burns, "Reciprocity in the Development of Anglo–American Medical Ethics," in *The Codification of Medical Morality,* ed. Robert Baker, Boston, MD: Kluwer, 1995: 135–143.

209. Ryan, *Manual of Medical Jurisprudence* (2nd ed): 105. The volume sent to Ryan appears to have been John D. Godman, *Addresses Delivered on Various Public Occasions,* Philadelphia, PA: Carey, Lea, and Carey 1829. Godman (1794–1830) lectured at a number of U.S. medical schools and was considered a first-rate anatomist and naturalist.

210. Ryan, *Manual of Medical Jurisprudence* (2nd ed): 105.

211. Ryan, *Manual of Medical Jurisprudence* (2nd ed): 125.

212. Truly comprehensive reform would not occur till 1858, when the traditional distinctions among physicians, surgeons, and apothecaries were abolished in favor of uniform educational standards.

213. Ryan, *Manual of Medical Jurisprudence* (2nd ed): 130–131.

214. Ryan, *Manual of Medical Jurisprudence* (2nd ed): 131.

215. Ryan, *Manual of Medical Jurisprudence* (2nd ed): 133.

216. As we have already seen, Ryan was probably not motivated by any compunction about duplicate publication. He was never loath to borrow liberally from already published works when he came to write a new book. Even had he been finicky about not duplicating his earlier writing, he could at least have called readers' attention to these earlier passages, if he truly believed that "medical ethics" described their contents.

217. Ryan, *Manual on Midwifery* (1828): 238.

218. Ryan, *Manual of Midwifery* (1841): 226.

219. Ryan, *Manual of Midwifery* (1841): 255.

220. Ryan mentions that the view of the Sorbonne theologians was still the official stance of the Roman Catholic Church; *Manual of Midwifery* (1841): 255. This passage apparently led Robert Baker and Chester Burns to conclude that Ryan himself was Catholic. We offer reasons above (note 74) to discount this view.

221. Ryan, *The Philosophy of Marriage in its Social, Moral, and Physical, Relations; with an Account of the Diseases of the Genito-urinary Organs, Which Impair or Destroy the Reproductive Function, and Induce a Variety of Complaints; with the Physiology of Generation in the Vegetable and Animal Kingdoms; Being Part of a Course of Obstetric Lectures Delivered at the North London School of Medicine, Charlotte Street, Bloomsbury, Bedford Square.* [unnumbered edition "from the last London edition."]. Philadelphia, PA: Barrington & Haswell, 1848:3.

222. Ryan, *Philosophy of Marriage*, 31. Ryan seems to have used the term "nature" as a shorthand for God. In a later chapter discussing reproduction within the animal kingdom, he wrote, "I have described, in my Lectures on Midwifery and the Diseases of Women and Children, published in the London Medical and Surgical Journal, 1836, the function of reproduction in all the classes in zoology, with a view of showing the analogy of all, and the consummate wisdom of the Creator in regenerating organized beings"; *Philosophy of Marriage*: 172.

223. Ryan, *Philosophy of Marriage*: 53–57, 116–17, 126–28, 246.

224. Ryan, *Philosophy of Marriage*: 128.

225. The posters Ryan alludes to appear to have advocated a device similar to a contraceptive sponge. They were distributed in 1823, rather than 1822, and appear to have been the work of Francis Place, whose volume, *Illustrations and Proofs of the Principle of Population: Being the First Work on Population in the English Language Recommending Birth Control*, had appeared in 1822 (a reprint appeared in 1994, published in London by Routledge/Thoemmes). We note that Place and his compatriots were "so-called" Malthusians because even though they invoked Malthus's theory of population, Malthus himself, an Anglican clergyman, disapproved of birth control for the same reasons argued by Ryan; John C. Caldwell, "Malthus and the Less Developed World: The Pivotal Role of India," *Population and Development Review* 1998; 24: 675–96.

226. Ryan, *Philosophy of Marriage*: 19.

227. Malthus argued that the Poor Laws ought to be abolished, leaving virtually all of the poor to swim or sink on their own; he believed that guaranteeing relief for the poor actually worked against their interests by perpetuating and expanding poverty; T. Robert Malthus, *An Essay on the Principle of Population as It Affects the Future Improvement of Society with Remarks on the Speculations of Mr. Godwin, M. Condorcet, and Other Writers.* 1798. (http://www.faculty.rsu.edu/~felwell/Theorists/Malthus/Essay.htm#35) Ryan by contrast sought to reform the Poor Laws but not to abolish them completely.

228. Ryan, *Philosophy of Marriage*: 63, 64, 65.

229. It must be noted that Ryan occasionally speaks approvingly of early marriages. Condemning the dissoluteness of city life, whilst praising the morality of those living in the countryside, he writes, "It is proved beyond doubt that the population increases much more in the country, and in villages, than in large cities. The citizens pass their youth in dissipation, and marry late for the sake of interest. But in the country, illegitimate unions cannot occur without exposure, as everyone knows the conduct of his neighbour; a man marries early in life, he is stranger to luxury and effeminacy, and his offspring is generally healthful, vigourous, and numerous." *Philosophy of Marriage*: 128.

230. James Gregory (1753–1821), son of John, wrote *Conspectus Medicinae Theoreticae ad Usum Academicum (Views on the Theory of Medicine…)*, which appeared in numerous editions from 1780 through the 1830s.

231. Ryan, *Philosophy of Marriage*, 153. The same passage may be found in *Manual of Midwifery* (1841): 87.

232. However, he did acknowledge that even healthy parents could produce defective infants. He believed that moral and physical traits had hereditary and environmental components.

For example, he quoted extensively from Moore's *Marriage Customs of All Nations* in *Philosophy of Marriage*; Moore attributed the different characteristics of nations to climatic variances.

233. Ryan, *Manual of Midwifery*: 34–35; he cites "Wrolick's Essay on the Diversity of the Pelvis, Bull. De Sciences Med. Fev., 1827. Velpeau, 1835." Alfred Armand L.M. Velpeau (1795–1867) was the author of *Traité complet de l'art des accouchemens*, published in Paris in 1835. It is unclear whether Ryan is quoting or just referencing Wrolick and/or Velpeau. For more information on Velpeau, see http://www.general-anaesthesia.com/images/velpeau.html.

234. Ryan appears to use the terms "race" and "nation" interchangeably.

235. Ryan, *Philosophy of Marriage*: 83.

236. Ryan, *Philosophy of Marriage*: 10. Ryan responded to the anticipated censure from these prudes with typical self-promotion, boasting that earlier editions of his works had garnered approval from "enlightened individuals, including statesmen, judges, divines, lawyers, and all classes of society"; *Philosophy of Marriage*: 16.

237. Ryan, *Prostitution in London* (1839): 168.

238. A.-J.-B. Parent-Duchatelet, *De la Prostitution dans la Ville de Paris : Considérée Sous le Rapport de l'Hygiène Publique, de la Morale et de l'Administration; Ouvrage Appuyé de Documents Statistiques Puisés dans les Archives de la Préfecture de Police; et Précédé d'une Notice sur la Vie et les Ouvrages de l'Auteur par Fr. Leuret*, Paris: J.-B. Baillière, 1837. Ryan may initially have intended merely to translate Parent-Duchatelet's work, but then decided rather to expand it considerably by adding observations on prostitution in London and New York. For more on Parent-Duchatelet, see Ann Elizabeth Fowler La Berg, *Mission and Method: The Early Nineteenth-Century French Public Health Movement* (Cambridge History of Medicine Series), Cambridge: Cambridge University Press, 2002.

239. Ryan, *Prostitution in London:* 42.

240. Ryan's critique of the gendered, unjust effect of the poor laws co-existed comfortably with his conviction that that men and women should behave in a manner dictated by their reproductive physiology; *Philosophy of Marriage*: 108, 236. He echoed the predominant view of his era, that "...the present mode of female education is highly injurious to health, predisposes to spinal curvature, and consequently, to deformity of the hip and other bones, thereby often rendering parturition highly dangerous and fatal"; *Philosophy of Marriage*: 236. Yet he expressed no doubts that women could act as obstetricians and midwives, provided they received adequate training; *Philosophy of Marriage*: 209, 230.

241. Ryan, *Philosophy of Marriage*: 27.

242. Ryan might appear at first glance to be inconsistent in arguing for specific legislative reforms, since he advised the physician, as part of his survey of "ethics of the present period," to avoid all involvement in politics; Ryan, *Manual of Medical Jurisprudence*, 2nd ed.. (1836): 69–70. It is likely that Ryan distinguished between the physician's personally becoming involved in politics (or worse yet, engaging in political harangues with his patients), as opposed to offering his expert advice to legislators via the medical press or other means.

243. Ryan, *Philosophy of Marriage*: 28–29. Ryan quoted an unnamed "public journal," describing the case "of a mother and her three daughters who are toiling for only 8 s. a-week, is, we are sure, not a solitary one, and that many unfortunate girls are literally compelled, at the close of their ill-requited honest daily work to walk the streets for money to procure the necessities of life, is a notorious and a very deplorable fact" (28). Ryan further noted that unwed mothers would not be able to secure employment as domestics – another line of work available to women – because potential employers considered them a likely immoral influence upon their families and home.

244. Ryan, *Prostitution in London*: vi.

245. Ryan, *Philosophy of Marriage*: 27. Ryan also wanted to marshal religion and medicine in the efforts to curb prostitution because he believed they could be relied on to inspire fear in men, restraining them from engaging in venereal excesses and abuses that resulted in disease; *Prostitution in London*, 392–393.

246. Ryan, *Prostitution in London*: 85–86.

247. Ryan, *Prostitution in London*, 39.

248. Ryan, *Philosophy of Marriage*: 27.
249. Ryan, *Prostitution in London*: 82.
250. Ryan, *Prostitution in London*: 87. In Belgium, it appears, proposals had recently been made to encourage the use of contraceptives among prostitutes, in hopes of reducing the incidence of unwed births even if the evil of prostitution itself appeared to be immune to elimination. As one would expect from his comments on the Malthusian contraceptive proposals, Ryan utterly opposed any such measures on the grounds that contraception was contrary to Divine edict.
251. Ryan, *Prostitution in London*: 9.
252. Ryan, *Prostitution in London*: 433.
253. Ryan, *Philosophy of Marriage*: 46–47. His attacks on "empirics" or quacks were generated both by compassion for the patient, and also by his desire to stand up for what he considered proper medical science and appropriate training and qualifications; *Prostitution in London*, 3.
254. See generally Robert B. Baker, Arthur L. Caplan, Linda L. Emanuel, and Stephen R. Latham (eds.) *The American Medical Ethics Revolution*. Baltimore, MD: Johns Hopkins University Press, 1999.
255. Jukes Styrap, *A Code of Medical Ethics*, reprinted in *The Codification of Medical Morality*, ed. Robert Baker, Boston, MA: Kluwer, 1995: 149–171. See also Peter Bartrip, "An Introduction to Jukes Styrap's *A Code of Medical Ethics* (1878)," in Baker, *Codification*: 145–148. Bartrip notes that despite Styrap's efforts, the British Medical Association refused to adopt his Code as an official statement, but by 1896 the *British Medical Journal* had pronounced it "the usually accepted authority on ethics in the BMA" (145).
256. Chester R. Burns, "Reciprocity in the Development of Anglo-American Medical Ethics," in *The Codification of Medical Morality*, ed. Robert Baker, Boston, MA: Kluwer, 1995: 135–143.
257. We base this conclusion on two brief works that appeared at mid-century. W. Fraser published a 14–page booklet, *Queries in Medical Ethics, Read before the Medico-Chirurgical Society of Aberdeen, 5th April, 1849* (apparently a reprint from the *London Medical Gazette*, 2 (1849): 181–187, 227–232). This took the form of a list of 27 ethical queries with answers ranging from a single paragraph to a page or so. Fraser generally followed Percival's aphoristic style, but referred only once to Percival, and never to Gregory or to Ryan. Perhaps more pertinent was an anonymous *Review of Works on Medical Ethics*, attributed to John Brown (1810–1882), Edinburgh: Murray and Gibb, 1850, a 12-page pamphlet reprinted from the *Monthly Journal of Medical Sciences*. Brown offered a reading list of about a dozen items for the student of medical ethics, and included Gisborne, Gregory, Percival, and the new American Medical Association code, but omitted any mention of Ryan's work.
258. Robert M. Veatch, *Disrupted Dialogue: Medical Ethics and the Collapse of Physician-Humanist Communication (1770–1980)*. New York: Oxford University Press, 2005.
259. Veatch, *Disrupted Dialogue*: 72–74.
260. Veatch admits that all through the years of "disrupted dialogue," religious ethicists, most notably Roman Catholics, were engaged in writing moral guidance for the physicians who happened to be of their religious sect. So Veatch could, alternatively, have classified Ryan in this manner. Such would still be inaccurate because clearly Ryan saw himself as writing an ethics for all physicians, not just for those of one denomination. We have already seen reasons to doubt that Ryan himself was Catholic.
261. The medical school calendar began at the beginning of October; there was traditionally a long winter session (October–April) followed by a short summer session (May–July). Medicine (physic) and midwifery as major topics would typically be offered during winter, while medical jurisprudence was viewed as a subsidiary topic to be covered during summer. The date shown is that of the beginning of the academic year. The 1828 information is taken from endpaper advertisement, Ryan, *A Manual on Midwifery* (1828) (following page 353). In 1829, *The Lancet* began to print a complete listing of all London medical school courses and independent lectures in its initial number for each volume (issued around October 1, as *The Lancet* tied its publication schedule to the schools' calendar), and the remaining information is from those listings.
262. The *Lancet* medical school listings (25 September 1830) show Ryan both as affiliated with the New School in Brewer Street and as a private lecturer, but no specific address is given for

the latter. He is shown as lecturing on Medicine and Midwifery at the school, and Medicine, Midwifery, and Medical Jurisprudence privately. *Lancet* 1 (1830–1831): 16, 18.

263. The *Lancet* listing for October, 1831 lists Ryan as lecturing at two different sites – 9 Gerrard Street (the Westminster Dispensary school) and 61 Hatton Garden (his private residence); *Lancet* 1 (1831–13): 14.

264. From catalogues of the Wellcome Institute Library, London; National Library of Medicine, Bethesda; title pages of available copies; and booksellers' lists. The information on later editions appears incomplete, as editions are alluded to in some texts that are not accounted for in the available catalogues. Besides the titles listed here, the Yale University medical library contains a volume, "The anatomical exercitations of … on the motion of the heart and circulation of the blood [by William Harvey], translated from the original Latin, by M. Ryan, M.D. To which is prefixed, a biographical sketch of the illustrious author," 1832–1833. However, this appears to be a series of articles by Ryan published in the *LMSJ* and then privately bound, rather than a true book publication.

265. The 1839 or 3rd London edition of this work ("Considerably Enlarged and Improved") has the variant title, *The Universal Pharmacopœia, or, a Practical Formulary of Hospitals both British and Foreign, Including All Medicines in Use, Translated from the Last Edition of Mm. Mile Edwards, and P. Vavasseur.*

266. We find this volume mentioned both in the catalogue of the Wellcome Library, London, and in an end-paper advertisement for books published by John Churchill in Delabere Blaine, *Outlines of the Veterinary Art…*, London: Longman, Orme & Co., and J. Churchill, 1841. The 1841 advertisement does not mention the number of the current edition. The Wellcome Library copy is a second edition, dated 1838. We have been unable to find notice of the first edition.

267. *The London Catalogue of Books Published in Great Britain, with their Sizes, Prices, and Publishers' Names, 1816 to 1851* (London: Thomas Hodgson, 1851) (http://books.google. com/books?id = 6a8YAAAAIAAJ&pg=PT938&dq = % 22Obstetric + aphorisms% 22 + ryan, accessed December 14, 2007) lists as a work of Michael Ryan "Obstetric Aphorisms," a 32mo (very small dimension) volume published by Brown. In the endpapers of one copy of *Prostitution in London*, in a list of Ryan's works, mention is made (with no further details) of a volume, "Obstetric Aphorisms on the management of natural and difficult Parturitions, Puerperal Diseases, and the Physical management of Infants" (http://books.google.com/ books?id = zHsEAAAAQAAJ&pg = PP 9 &dq = % 22obstetric + aphorisms% 22 + ryan#PPP9,M1, accessed December 14, 2007). As the Wellcome Library and other libraries with extensive historical collections appear to lack a volume with this title, we suggest that this is a variant title of *The Obstetrician's Vademecum, or Aphorisms on Natural and Difficult Parturition…* by Denman as edited and updated by Ryan.

Part II
Selected Writings by Ryan

Explanatory Note on Selections from Ryan's Works

We have attempted to reprint the following selections from Ryan's writings in a form as closely as possible identical to the original publications. The pagination in the original is noted by pairs of numerals within square brackets, so that, for example, "[76/77]" indicates the point within the text of the page break between pages 76 and 77. The original footnotes are retained and are identified as *"Author's note."* In the original, these are not numbered but are indicated by typographic symbols such as asterisks. By contrast, notes that we have added for this edition are designated by *"Editors' note."* In a few instances, a footnote is neither Ryan's nor ours; rather, Ryan copied them verbatim from the original source that he is quoting. These notes occur in Chapters IV (where Ryan extensively quotes Percival's *Medical Ethics*) and V (where Ryan quotes extensively from a lecture by Dr. John Godman). Such notes are labeled with the original source from which Ryan copied them.

Many of the Editors' notes offer translations of Latin passages in the original text. Ryan assumed that these passages would be understood by the medical students and physicians who made up his intended audience. We have tried to follow the policy of providing a translation in those instances where Ryan's text does not, by itself, proceed to render the meaning of the Latin passage in English. We are immensely grateful to Mary Hope Griffin for her assistance in wrestling with the debased form of language that nineteenth century British medical practitioners took to be Latin.

When Ryan is merely copying material from other sources, we have taken the liberty of omitting some of the Latin passages for which translations are immediately provided in the text, and quotations from other authors that do not add significantly to the meaning of the passages.

I. A Manual of Medical Jurisprudence

A

MANUAL

OF

MEDICAL JURISPRUDENCE,

AND

STATE MEDICINE,

COMPILED FROM THE LATEST LEGAL AND MEDICAL WORKS,

OF BECK, PARIS, CHRISTISON, FODERE, ORFILA, ETC.

CONTAINING

PART I

MEDICAL ETHICS OF ANCIENT AND MODERN TIMES.

PART II

LAWS RELATING TO THE MEDICAL PROFESSION, FROM PARIS AND FONELANQUE, SCULLY, WILCOCK, AND CHITTY.

PART III

MEDICAL JURISPRUDENCE AND STATE MEDICINE FROM THE MOST CELEBRATED ANCIENT AND MODERN MEDICO-LEGAL WRITERS.

PART IV

LAWS RELATING TO THE PRESERVATION OF PUBLIC HEALTH—MEDICAL POLICE—STATE MEDICINE—PUBLIC HYGIENE.

INTENDED FOR THE USE OF

LEGISLATORS, BARRISTERS, MAGISTRATES, CORONERS, PRIVATE GENTLEMEN, JURORS, AND MEDICAL PRACTITIONERS.

By MICHAEL RYAN, M.D.,

Member of the Royal College of Physicians in London; Member of the Association of Fellows and Licentiates of the King's and Queen's College of Physicians in

Dublin; Member of the Royal Colleges of Surgeons in London and Edinburgh; Consulting Physician to the East London Midwifery Institution; Lecturer on Medical Jurisprudence in the Medical School, Gerrard Street, Soho Square, &c. &c.

Second Edition considerably Enlarged and Improved.

LONDON:
SHERWOOD, GILBERT, AND PIPER,

PATERNOSTER ROW.

1836.

[i/ii]

LONDON:

JOHN HADDON AND CO. DOCTORS' COMMONS.

[ii/iii]

CONTENTS

[iv/v]

PREFACE

TO THE

FIRST EDITION

The object of the author of these pages, is to give a concise and comprehensive view of the received Principles of Medical Jurisprudence, and to collect the scattered and isolated facts from the standard works of legal and medical writers. He is inclined to hope that this volume contains all that is valuable in the systematic works upon the subject; and numerous other topics of vast importance to medical practitioners, which have no place in similar productions. In proof of this assertion, he refers the reader to the ethical and legal parts of this work, and to the articles on medical evidence, adulterations of alimentary matters, and public medicine. He hopes that the promulgation of Medical Ethics, or the institutes of professional conduct, will contribute in no small degree to maintain and support the honour, dignity, character, and utility of the profession, by impressing the minds of medical students with a sense of the noble and virtuous principles which have always characterised their predecessors, and which ought ever to distinguish the scientific cultivators of medicine.

Another novel and important feature in this production is, the exposition of the laws relating to the different orders of the faculty in these kingdoms, which cannot fail to prove instructive and useful to medical practitioners, by informing them of the immense power, vast influence, and high privileges, conferred upon them by the legislature, in exempting them from the performance of many civil duties, and in deeming their evidence conclusive in an immense number of civil and criminal proceedings, which affect the lives, the liberty, the honour, reputation, and property, of every class of society. For many facts mentioned in this part of the subject, the author is indebted to the works of Dr. Paris, and Mr. Fon- [v–vi] blanque, of Mr. Willcocks, of Mr. Scully, the celebrated Irish barrister; of Mr. Phelan and Mr. Chitty.

The medical part has been compiled from the standard systems on Forensic Medicine, both domestic and foreign, and illustrated by the opinions and experience of the author. The works of Drs. Male, Gordon Smith, Duncan, Paris, Beck, Christison, Foderè, Mahon, Chaussier, Orfila, Briand, Sedillot, &c. have been laid under large contribution, and the extracts duly acknowledged.

The chapter on medical evidence has been condensed from the valuable lectures delivered in the London University by Professor Amos, as published in the *London Medical Gazette*. Under this head are considered the powers of coroners, and *the propriety and expediency of appointing medical men to the office of coroner*. The addition of the forms of medical certificates, and the method of examining recruits, were considered useful for the guidance of young practitioners.

It has been the author's anxious wish to compress the fullest information in the smallest space, and in the most familiar language, in order to simplify the subject, and render it intelligible to every class of medical practitioners, to barristers, solicitors, coroners, magistrates, private gentlemen, and general readers. Every medico-legal fact which can become the subject of judicial inquiry, is accurately detailed, so that this Manual may be fairly looked upon as a text-book for the practitioners of the law and medicine. It is intended to save both classes much trouble and research. Whether the author has executed his task in a satisfactory manner, must be decided by his contemporaries; but he consoles himself with the reflection, that his design and intention were good, had he sufficient ability to execute it as he wished.

In exposing the absurd distinctions, the defective state of the laws relating to the profession, and the gross abuses of its constituted authorities, a love of freedom and of equality, with an ardent desire to promote the interests of his favourite science, and of humanity, have impelled him to declare the truth, however unpalatable it may be in certain quarters, or to the different orders of the faculty. His motto has been, *"amicus Socrates, amicus Plato, sed magis amica veritas."* [vi/vii] He has not been the advocate of any party, of any order, of any corporation, but the advocate of the whole profession. He has examined the whole of the charters and statutes relating to the practice of medicine in these countries, and has found them defective, contradictory, oppressive; and, since the legislative Unions between England, Scotland, and Ireland, highly unjust and impolitic. In proof of this statement, the reader has only to recollect the fact, that a Scotch physician, surgeon, or apothecary, cannot practise legally in England; an English member of the faculty cannot practise in Ireland; and an Irish member can neither practise in England nor in Scotland. Such is the state of the law, which was just and right when the three countries were distinct nations, but is unjust and preposterous since they have formed one united empire. The author has, therefore, considered that there never was a more auspicious period than the present, for exposing the absurd state of the laws relating to the faculty in the United Kingdom, when the propriety of constitutional, ecclesiastical, and legal reforms, is the subject of national discussion; and when medical reform is loudly demanded by the whole profession, with the exception of the few interested monopolists. It need scarcely be stated, that the principles and practice of medicine and therapeutical agents, are identically the same in every part of these countries.

Medical jurisprudence is at length included in the course of study required by the Apothecaries' Company of London; but it is strangely excluded in the course required by the Royal College of Surgeons. Had not the University of London appointed a professor of Forensic Medicine, chiefly at the recommendation of the philanthropic Dr. Birkbeck, this branch of medical science would as yet have no place in the course of medical education required in England.

Nevertheless, the most casual observer must be convinced, by daily examples, that the ignorance of ethics and medical jurisprudence, impedes or arrests the career of the medical practitioner, and frequently destroys professional reputation altogether. If proof were demanded of the validity of this position, I need only refer to the newspaper reports of legal [vii/viii] proceedings, in which we daily observe ample attestations of the fact. We peruse the most absurd and unscientific medical evidence, more especially in the reports of coroners' inquests, which could never have appeared, had the witness possessed a proper knowledge of forensic medicine, or had the coroner been a medical practitioner. Such displays of ignorance excite the pity and contempt of the scientific portion of the faculty, and the ridicule of the legal profession, and of the public. Mr. Amos has well illustrated this defect, in his admirable cautions to medical witnesses, in which he has cited many cases to prove, that medical evidence has been censured by the bench, ridiculed by the bar, derided by the auditory, and has entailed obloquy and disgrace upon the unfortunate witness.

Medical practitioners should be aware, that all the rising barristers of our criminal courts attend lectures on legal medicine; and often does forensic fame arise from the ability with which an advocate examines a medical witness. A knowledge of medico-legal science is almost as indispensable to the one as to the other; and the coroner who is ignorant of it, is evidently incompetent to discharge his duty to the public, or to secure impartial justice to the accuser or accused. This has been incontrovertibly proved in the article on medical evidence.

It has been erroneously stated by some writers, that the science of medical jurisprudence, is nothing more than the application of the elementary branches of medicine, to the elucidation of judicial investigations; and, consequently, that a scientific medical practitioner must necessarily be a good jurist. This is not correct, inasmuch as the most scientific physicians and surgeons have too often proved to be the worst jurists; because they could not derive the requisite information on medico-legal science from the common systems of medicine or surgery, as it is only to be obtained from works exclusively devoted to the subject. In no lectures or works, except those upon this science, is a student informed of the laws relating to his profession; of his rights, privileges, and immunities; of the cases, civil and criminal, on which be is liable to be called to give evidence: of the received opinions [viii/ix] upon these cases; of the danger of wounds, and contusions, and of injuries prejudicial to health or destructive of life; of the analysis or mode of detecting the numerous poisons; of the manner of giving evidence, or of his ethical duties in public and in private practice. It is therefore manifest, that medical jurisprudence is a distinct science, and one of the greatest importance and utility to the members of the medical and legal professions. If it could be taught by the professors of the elementary branches of medical education, there would be no need of a separate professorship; which, therefore, exists in all the medical schools of these and foreign countries.

In conclusion, the author has to remind the critical reader, that in attempting to compress the extensive information comprehended in the narrow limits of these pages, brevity of expression, and too much conciseness, may have rendered the style occasionally obscure or inelegant. This perhaps may be pardoned in a work chiefly intended for medical students, and which was originally published in one of the periodicals (*The London Medical and Surgical Journal*), to fill up lacunae, and arranged as the suddenness of the occasion demanded; which in general afforded little time for attending to the beauties of style, to euphonious sentences, or to the other qualities of literary composition. If the work supply a want, or contribute to the maintenance of the character and utility of medicine, or in any way benefit the interests of mankind, or the administration of justice, the object of the author will be fully attained. [ix/x]

PREFACE

TO THE

SECOND EDITION

The favourable reception which this work has received in this and other countries has compelled the author to endeavour to improve a new edition. He has accordingly enlarged many chapters introduced much new matter, and carefully revised every part of his work. He has referred to the last editions of all standard works on the subject of which he treats, and appended all that is valuable since his former edition. He has considerably enlarged the first part, on Medical Ethics, by introducing the codes of continental Europe and America, and also those of Dr. John Gregory and Dr. Percival. Under this head, he has commented on Medical Education, Degrees, Diplomas, Medical Appointments – Success – Reputation – Eminence – Moral and Physical Medicine – Clinical Medicine – Rules for Prescribing Medicines – Action of Medicines on the Economy – Posology, or Fixation of Doses of Medicines – Pharmacology.

Part II. Laws relating to the Medical Profession is brought down to 1836. Few additions are made in this section, as the author is of opinion that the whole of the laws relating to Medical Polity in this country, will be remodelled during the present or next Session of Parliament.

Part III. Medical Jurisprudence is considerably enlarged, and in the opinion of the author, improved. Several additional Medico-legal questions have been made in this section.

Part IV. Laws for the Preservation of Health-Medical Police-State Medicine-Public Hygiene. This section [x/xi] occupies 60 pages of new matter. It comprises a vast number of subjects of great interest to public health and happiness.

Chapter I. Laws for the preservation of Public Health – Quarantine, Boards of Health—Contagious Diseases – Disinfection – Purity of Air, Water, and Situation – Guardians of the Public Health – College of Physicians, as inspectors of Apothecaries' Shops, officers of Public Charities – Commissioners of Paving, Sewers, Cleansing the Streets, &c – Framers of the Bills of Mortality.

Chapter II. Inhumation – Burial of the dead in Cities – Searchers and Inspectors—Danger of Exhumation and Opening of Graves – Premature Interment – Uncertainty of the Signs of Death – Account of Individuals buried alive—Vivisections – Proper period of Inhumation – Custom in France and England.

Chapter III. Signs of Real Death; aspect of the Face; Absence of Heat, and Lividity of the Skin; Absence of Circulation and Respiration; Cadaverous Rigidity of Stiffening; Physiological Causes of it; Physical Proofs of Death; Surgical Proofs; Incisions, Decapitation, &c.

Chapter IV. Putrefaction of Animal Matters; Putrefaction in the open Air, Water, and different Earths; modified by Age, Constitution, Sex, State of Thinness or

Obesity; Mutilation or Integrity of the Body; Genus and Duration of Disease; Phenomena before Death; Period Of Inhumation; Appearance and Cause Of Insects; Humidity and Dryness of Earth; Chemical Composition of Earth; Depth of the Grave; Naked or Clothed state of the Body; Atmospheric Influences; Conversion of the Body into Adipocere; Dr. Fletcher's Physiological Views of Putrefaction, 1836.

Chapter V. Nuisances Legally and Medically considered; Trades and Manufactures; Filth in the Streets; Noises by Day or Night; Transitory Nuisances; Physical Effects of; Arrangement of by Paris and Fonblanque, including the Principal Trades and Manufactures; Purity of Air, Water, and Situation.

In addition to the new matter, there is prefixed a history of the rise, progress, and present state of Public Medicine, and also a copious Index, which may, perhaps, be considered a [xi/xii] Medico-Legal Dictionary. It is the fullest hitherto published in our language. Such are the features of this work. The author extracted from all available sources, both domestic and foreign; added the result of his own experience; and has presented to his readers a variety of information, not so far, as his researches enable him to state, to be found in any other elementary work in one volume, of equal size, on Medical Jurisprudence and State Medicine. He has not prefixed or added a bibliography, but leaves his readers to form their own conclusions on the number of works to which he has referred, and which he has duly acknowledged.

Great Queen Street
St. James's Park, Westminster
February, 1836

[xii/xiii]

Introduction

Medical Police, Political Medicine, State Medicine, Public Hygiene, Police of Health, and Medical Jurisprudence, comprise the acts of a legislature or Government, and magistracy, for the conservation of public health, and also the enactment of laws for the regulation of the practice of the medical profession, and the duty of medical practitioners in aiding the legislature in forming just laws, and public tribunals in the administration of justice.

Medical Jurisprudence, or Legal Medicine, is a science by which medicine and its branches, are rendered subservient to the construction, elucidation, and administration of the laws for the preservation of public health. This term is considered by some writers as best calculated to express, in the most comprehensive manner, the application of the medical sciences to the purposes of law.

It has been divided into Forensic, Legal, Judiciary, and Judicial medicine, comprising the opinions and evidence and into Medical Police or State Medicine, comprising all medical opinions and precepts which inform the legislature and magistracy in constructing the laws, and in enforcing them for the preservation of the public health. Both these divisions are included by the Germans in the term *State Medicine*. I have employed both in the construction of this work, as I have enumerated the laws relating to the profession, and to the preservation of public health, together with all medicao-legal inquiries. I have preferred this arrangement to any other proposed by medico-legal writers in this country, as it is the most comprehensive. M. Foderè, of Strasburg, who has executed the most elaborate modern work on State Medicine, has divided the science [xiii/xiv] into Medical Jurisprudence and Public Hygiene,[1] and published three octavo volumes on each subject. The term Medical Jurisprudence is incorrect, as it may imply a knowledge of the laws relating to medical subjects, or a knowledge of the medical sciences as applied to legislation, forensic inquiries, and the Police of Health. It may also unite the legal and medical sciences, and in this sense it is employed by some modern writers (Paris and Fonblanque). Dr. Beck used the term in his excellent work in the restricted sense, and defined Medical Jurisprudence, Legal Medicine or Forensic Medicine, a science which applies the principles and practice of the different branches of medicine to the elucidation of doubtful questions in courts of justice

H. Brody et al. (eds.), *Michael Ryan's Writings on Medical Ethics*,
Philosophy and Medicine, Vol 105,
© Springer Science + Business Media B.V. 2009

(Elements of Medical Jurisprudence, 1823). He designed to offer another treatise on Medical Police, which has not as yet been published.

Dr. Gordon Smith, our first systematic author on this science, followed the example of Dr. Beck, and published his work entitled, "The Principles of Forensic Medicine, Systematically arranged and applied to British Practice;" it was also his intention to write a separate production on Medical Police, which he did not accomplish. I have included Medical Ethics and Jurisprudence, and State or Public Medicine, in one work, and thus differ from M. Foderè, Dr. Beck, Dr. Gordon Smith, and Dr. Paris. This was the arrangement of some ancient medico-legal authors: Medicus-Politicus, sive de officiis Medico-Politicis tractatus, quatuor distinctus libris: in quibus non solum bonorum medicorum mores ac virtutes exprimuntur, malorum vero fraudes et Imposturæ deteguntur: verum etiam pleraque alia circa novum hoc argumentum utilia atque jucunda exactissime proponuntur: opus admodum utile medicis, ægrotis, ægrotarum assistentibus, et cunctis aliis literarum, atque adeo politicæ disciplinæ cultoribus. Roderici à Castro, Lusitani Philos. ac Med. Doct. per Europam Notissimi. Hamburgi, 1614.[2]

I have also included the codes of Ethics of Dr. John Gregory and Dr. Percival, of the Royal College of Physicians in London, and of the Americans.

The history of Hygiene is coeval with the origin of society. The first wants of man were aliment, and he was instructed in what was salutiferous and noxious to his condition. He soon discovered the influences of exertion, repose, sleep, waking, &c., upon his constitution. The knowledge of the laws, [xiv/xv] morals, and police of the people relating to the preservation of health, constitutes public hygiene or medicine. It necessarily existed in the remotest ages. The ancients were of opinion, that there was an intimate and mutual dependence between the moral and physical states, and of the necessity of enacting laws for the promotion of temperance and wisdom, as well as for the punishment of excesses and of crimes.

All eastern nations had their systems of medical police; we trace it in their legislation, their manners and customs, and in the rules of their public police.

The hygiene of Moses embraced three principal objects: the prohibition of certain aliments, ablutions for certain impurities, and the sequestration of certain diseases reputed contagious, as lepra, &c. The Levitical law, ch. xiii. xiv., commanded the priests to visit the houses infected with the plague of leprosy, or with other contagious disease; to examine the inhabitants, to establish quarantine, to purify the houses, to shut them up, and pull them down in certain cases. We find in Deuteronomy, ch. xxii., that in questions of doubtful virginity, the elders were to be consulted, and enjoined to deliver judgment according to the physical evidence of the case.

It is easy to conceive, the object of legal purifications in warm climates, where the tendency of animal foods to putrefaction, and the copious perspiration of the people, were the causes of insalubrity, which frequent ablutions destroyed. The chief hygienic means of the ancients, in addition to those already mentioned, were gymnastics, baths, and repose after the latter.

The principal rules relative to public police among the ancients, referred to the salubrity of habitations, the locality of towns, cities, aind streets.

The Father of Medicine wrote a Treatise on Air, Water, and Situation, and this work has influenced all civilized nations for nearly 3,000 years.[3] It led to the draining of marshes, and improvement of situations. Most persons know that the Roman emperors Julius and Augustus Cæsar, ordered the Pontine marshes to be dried; to which Horace alluded:

> **** Sterilis diu palus, aptaque remis
> Vicinas urbes alit, et grave sentit aratrum.[4]

Hygiene was, as already observed, cultivated long before the time of Hippocrates; and the rules regarding it, were deduced from the universal laws of the animal economy, and referred [xv/xvi] to age, constitution, intelligence, to wants, or pleasures.

Moses enforced laws as to the use of all aliments, fruits, grains, herbage, bread, milk, fish, flesh, wine, &c. According to Herodotus, the Egyptians invented beer, which was subsequently mentioned by Moses,[5] and they were the most healthful individuals. The Egyptian law ascribed to Menes, ordained, according to Plutarch, that no pregnant woman should suffer afflictive punishment. The Jews made a distinction between mortal and dangerous wounds; and had laws relating to virginity, marriage, adultery, embalming and interring the dead, and other subjects of public medicine. Joseph found the Egyptian priests physicians, and he ordered his father's body to be embalmed (Gen. 1. 6. A.C. 1747).

The ancient Greeks highly estimated the power of morals in preserving health, and perfecting the species; and we find this opinion prevalent in all ancient nations. Their primary object was to give to the country a robust people and vigorous defenders. It was this principle that gave origin to customs, which, in our time, appear to be inhuman and barbarous. Thus the destruction of delicate infants by the Spartans – a horrible custom of the Chinese at present with respect to their female offspring.

The laws of Lycurgus on the physical education of infants and of girls to the time of marriage, with a view of transmitting a good constitution to their offspring, and many more hygienic ordinances, afford ample proof of the state of public medicine among the ancient Greeks. The physical legislation of Pythagoras and Plato was also intended for the conservation of the public health.

It is remarkable, however, that there are scarcely any traces of the union between law and medicine in the works the ancient Greek physicians. There is some portion of hygiene and state medicine scattered through the voluminous writings attributed to Hippocrates; and some crude speculations on the nature and growth of the infant, the period of pregnancy; and the abolisliment of some rude obstetricy. The opinions of the father of physic on the perpetuation of the species, the animation of the fœtus, and on many other physiological questions had great influence on all ancient legislators. Those of Aristotle have had a like tendency; and even to this time. Both these authors held that the fœtis inanimate in the womb for a certain time, during which it was no crime to cause abortion. The canon law was founded [xvi/xvii] on this opinion, and has influenced all legislators to the present period.[6] The above authorities also maintained, that pregnancy might be protracted, beyond the ordinary term of 9 months; which is now the received opinion of physiologists.

Medical men, both among the Greeks and Romans, were consulted by the magistrates most frequently on questions of *medical police*, as on the salubrity of cities, and on the means of assuaging the virulence of epidemic diseases: and to subjects of that nature the public functions of the *Archiater,* or state physician, were confined in the lower ages of the empire. That they were summoned to give evidence in a court of justice no where appears; and had they been called, their testimony, in ages when human anatomy was proscribed, must, in a majority of difficult cases, have been of comparatively little value.

The laws of ancient Rome were chiefly enacted on the same grounds as those of Greece; but they embraced more important medico-legal questions. In the reign of Numa Pompilius, before the Christian era, 600 years, and 140 before the time of Hippocrates, it was enacted, "de inferendo mortuo," (lex regia)[7] that the bodies of all women who were in the last months of pregnancy, should be opened immediately after death, so that infants might be saved if possible. History also informs us, that the elder Scipio Africanus, Marcus Manlius, and the first Cæsar,[8] were brought into the world by incision of the abdomen, or gastro-hysterotomy; and hence the origin of the term, Cæsarian operation. It was also enacted in the Twelve Tables, A. C. 452, that the infant in the mother's womb was to be considered as living, and all civil rights secured to it. The legitimacy of an infant was limited to 300 days after the death or absence of his putative father, or within ten months of the supposed time of impregnation; but the period was afterwards extended to eleven months. Most of the Roman laws were framed in accordance with the opinions of the ancient philosophers and physicians, and many legal decisions were given; *propter auctoritatem doctissimi Hippocratis.*[9]

Several laws were constructed during the reign of Severus, Antonine, Adrian, and Aurelius, on the authority of Hippocrates and Aristotle; the crime of procuring abortion was limited to those cases in which the fœtus had exceeded 40 days, as before that time it was not considered a living [xvii/xviii] being;[10] and the Emperor Adrian extended the term of pregnancy from ten months – the period of legitimacy according to the Decemvirs – to eleven, in accordance with the physiological opinions of the times.[11] It is, also recorded by Suetonius, that the bloody remains of Julius Cæsar, when exposed to public view, were examined by a physician named Antistius, who declared that out of twenty-three wounds which had been inflicted, one only was mortal, which had penetrated the thorax between the. first and second ribs. Gerike has collected some curious instances of the inspection of murdered bodies by medical witnesses, from the writings of Suetonius, Tacitus, and Plutarch. The body of Germanicus was also medically inspected; and, by indications conformable to the superstitions of the age, it was decided that it had been poisoned.[12] The bodies of Agricola and others were also examined by physicians, according to Tacitus.

It does not appear that there was any positive law which required the inspection of wounded bodies by medical practitioners, or that legislators required medical opinions before making laws. It is, however recorded, that before the time of Philumenos, A.D. 80 – who first studied obstetricy at Rome – the midwives were ordered by the prætors to examine pregnant women in judicial inquiries. Augustus had previously favoured the profession of medicine, A.D. 10, and exempted its members from public burthens and taxes. About 100, valetudinaria and veterinatia were established in the Roman camps, for wounded soldiers and horses.[13]

Galen was born in 131, and was physician to the gladiators at Pergamos, A.D. 159. He alluded to some questions. of legal medicine; he remarked the difference which subsists between the lungs of an adult and the fœtus; admitted the legitimacy of seven months children, and offered remarks on the manner of detecting simulated diseases. He related the case of a Greek lady who escaped punishment for adultery on the opinion of a physician. She had borne an infant very unlike the supposed father; and the physician accounted for this, by referring its resemblance to that of a picture which hung in the chamber of the mother.

The promulgation of the Justinian Pandects, A.D. 529, [xviii/xix] which laid the foundation of the legal codes of most of the modern states of Europe, may be considered the origin of the connexion between the sciences of law and medicine, or of forensic medicine. Under the titles, *De Statu Hominum*; *De Pœnis et Manumissis*; *De Sicariis et Veneficis*, of the *Lex Cornelia*; *De Inspiciendo Ventre custodiendoque Partu*, of the *Lex Aquilia*; *De Hermaphroditis*; *De Impotentia*; *De Muliere quæ peperit undecimo mense* – some questions in medical jurisprudence are discussed; but the decisions of the judges were directed to be guided on such matters, not on the oral testimony of living Physicians, examined on the particular point then under consideration, but *Propter auctoritatem doctissimi Hippocratis*.

If the Justinian code was an improvement in many respects on more ancient usages, in *one* point of medical jurisprudence, it must be allowed to be more erroneous than that which preceded it, namely, on the time of utero-gestation.

Aulus Gallus informs us, that a decree of Adrian acknowledged the legitimacy of a child born eleven months after the death of its reputed father; and this period has been adopted by Justinian (Novella 39 c. 2, Nov. 89).

From this period to the year 555, the Roman empire was overrun by the Goths and Vandals. The most savage customs now prevailed in the west of Europe, the principles of legal medicine were totally neglected, and the trial by the ordeals of fire, water, and duel, were introduced as appeals to the Deity, in cases which, to rude minds, appeared above human investigation.

The trials by ordeal in the dark ages of modern Europe, when the decision of the most important questions was abandoned to chance or to fraud, when carrying in the hand a piece of red hot iron, or plunging:the arm in boiling water,[14] was deemed a test of innocence, and a painful or fraudulent experiment, supplanting a righteous award, might consign to punishment the most innocent, or save from it the most criminal of men, have ever been deemed a shocking singularity in the institutions of our barbarous ancestors. We are ready to admit the justice of this charge generally; and yet we fancy that, upon some occasions, we are enabled to discern through the dim mist of credulity and ignorance, a ray of policy that may have been derived from the dawning of a [xix/xx] rude philosophy. Trials by ordeal, as we are informed by Mr. Mill, hold a high rank in the institutes of the Hindus. It appears that there are no less than nine different modes of trial, but that by water in which an idol has been washed, and the one by rice, are those which we shall select, as well calculated to illustrate the observations which we shall venture to offer.

The first of these trials consists in obliging the accused person to drink three draughts of the water in which the images of the sun and other deities have been washed; and, if within fourteen days he has any indisposition, his crime is considered as proved.

In the other species of ordeal alluded to, the persons suspected of theft are each made to chew a quantity of dried rice, and to throw it upon some leaves or bark of a tree; they from whose mouth it comes dry, or stained with blood, are deemed guilty, while, those who are capable of returning it in a pulpy form, are at once pronounced innocent.

When we reflect upon the superstitious state of these people, and at the same time, consider the influence which the mind, under such circumstances, is capable of producing upon the functions of the body, it is impossible not to admit, that the ordeals above described are capable of assisting the ends of justice, and of leading to the detection of guilt. The accused, conscious of his own innocence, will fear no ill effects from the magical potation, but will cheerfully acquiesce in the ordeal; whereas the guilty person, from the mere uneasiness and dread of his own mind, will, if narrowly watched, most probably discover some symptoms of bodily indisposition, before the expiration of the period of his probation. In the case of the ordeal by rice, a result, in correspondence with the justice of the case, may be fairly anticipated on the soundest principle of physiology. There is, perhaps, no secretion that is more immediately influenced by the passions than that of saliva. The sight of a delicious repast to a hungry man is not more effectual in exciting the salivary secretion, than is the operation of fear and anxiety in repressing and suspending it. If the reader be a medical practitioner, we refer him for an illustration to the feelings which he experienced during his examination before the medical colleges; and if he be a barrister, he may remember with what a parched lip he gave utterance to his first address to the jury. Is it then unreasonable to believe, that a person under the influence of conscious guilt, will be unable, from the dryness of his mouth, to surrender the rice in that soft state, which an innocent individual, with an undiminished supply of saliva, will so easily accomplish? – *Paris and Fonblanque.* [xx/xxi]

It was supposed, even in this country, that the agitation of a murderer, on seeing his victim, was sufficient to prove his guilt; and, in some cases, he was obliged to place his hand on the dead body. It is also a popular opinion, that the countenance of a murdered corpse will change when gazed on by the murderer, or any of his near relatives. The following extraordinary cases are noticed by Dr. Beck, from Hargrave's State Trials, vol. x. Appen. p. 29 – reign of Charles II:

An ancient and grave person, *minister to the parish were [sic] the fact was committed*, sworn to give evidence, according to custom, deposed, "that the body being taken up out of the grave, thirty days after the party's death, and lying on the grass, and the four defendants being present, were required each of them to touch the dead body. Okeman's wife fell upon her knees, and prayed God to show tokens of her innocency. The appellant did touch the dead body, whereupon the brow of the dead, which before was of a livid and carrion colour (in terminis, *the verbal expression of the witness*), began to have a dew, or gentle sweat arise on it, which increased bv degrees, till the sweat ran down in drops on the face; the brow turned to a lively and fresh colour; and the deceased opened one of her eyes, and shut it again; and this opening of the eye was done three several times. She likewise thrust out the ring or marriage finger three times, and pulled it in again, and the finger dropped blood from it on the grass." Sir Nicholas Hyde, chief-justice, seeming to doubt the evidence, asked the witness, Who saw this besides you? – Witness. I cannot swear what others saw. But, my Lord (said he), I do believe the whole company saw it, and if it had been thought a doubt, proof would have been made of it, and many would have attested with me. Then the witness, observing some admiration in the auditors, spake further, "My Lord, I am minister of the parish, and have long known all the parties, but never had occasion of displeasure against any of them, nor had to do with them, or they with me; but as I was minister, the thing was wonderful to me. But I have no interest in the matter, but as called upon to testify the truth; and this I have done." [This witness was a very reverend person, as I guessed, of about seventy years of age. His testimony was delivered gravely and temperately, to the great admiration of the auditory.] Whereupon, applying himself to the chief-justice, he said, "My Lord, my brother here present, is minister of the next parish adjacent, and I am sure saw all done that I have affirmed." Therefore, that person was also sworn to give evidence, and did depose in every point – 'the sweating of the brow, the change of the colour, thrice opening the [xxi/xxii] eye, the thrice motion of the finger, and drawing it in again:' only the first witness added, that he himself dipped his finger in the blood which came from the dead body, to examine it, and beswore he believed it was blood. I conferred afterwards with Sir Edward Powell, barrister-at-law, and others, who all concurred in the observation. And for myself, if I were upon oath, can depose that these depositions (especially the first witness) are truly reported in substance" (Ibid. p. 29).

"In the trial of Standsfield, for the murder of his father, a similar charge was brought. It is stated, that when the son was assisting in lifting the body of his father into the coffin, it bled afresh, and defiled all his hand. The opposite lawyers observe, that "this is but a superstitious observation, without any ground either in law or reason. Carpsovius says, he has seen a body bleed in the presence of one not guilty, and not bleed when the guilty were present." They assign, with great probability, as the cause of this bleeding, the fact, that the surgeons had made an incision about the neck, and that the motion of the body, in removing it, caused the fresh hemorrhage from that part. Hargrave, vol. 4, p. 283 –

I will refer those who are curious on this subject, to Metzger (p. 328); and Valentini Novellæ, App. 3. *De stillicidio sanguinis in hominis violenter occisi, cadavere conspicui, an sit sufficiens præsentis homicidæ indicium?*"

The many causes of sudden or suspicious death, could be very imperfectly discovered, even by medical practitioners, in the dark ages, when prejudices and superstition prevented the study of anatomy. It was no sooner than the year 1315, that Mondini gave the first public demonstrations on the human subject in the University of Bologna; and it was not until 1334, that human dissection was permitted in France, in the celebrated medical school of Montpellier. About this period, Frederick II. of Naples, authorized one annual dissection in his capital. This legalization of anatomy, was the revival of medical science in Europe.

The first cases in which medical witnesses were summoned to assist the judge, were after the establishment of the inquisition by Pope Innocent III. A.D. 1204, when torture was inflicted to extort confession. This mode of arriving at truth, was too favourable to screen the indolence or incapacity of the judges, not to be adopted by other tribunals. Accordingly the rack, and other modes of inflicting the extreme of suffering compatible with life, passed into the codes of almost all European nations, from which they are not yet altogether expunged; and medical men were sum- [xxii/xxiii] moned by the tribunals to superintend these horrible scenes, to mark how far fiend-like ingenuity could protract human sufferings, without allowing the victim the last refuge of human misery. Accordingly, we read in the works of the early medical jurists, observations on the mode of applying the torture in criminal trials.

It is humiliating to acknowledge, that even now a species of torture is enforced by British laws in the army and navy, and in civil cases, by means of the tread-mill, too often it is to be feared without proper discrimination of age, sex, or state of health.

But medical jurisprudence, as a science, cannot date farther back than the middle of the sixteenth century. The Emperors Rudolf, Sigismund, Albert, and Maximilian I., had successively attempted reforms of the criminal codes of Germany, by the introduction of a uniform system of legislation; but this idea was first reduced into practice by George, Bishop of Bamberg, who published, in 1507, a German penal code, drawn up and systematized by John, Baron of Schwartzemburg, which was speedily adopted in the petty States of Bayreuth and Anspach, and afterwards in Brandenburg. The other States of Germany, from narrow views of self-interest, and the jealousy of neighbouring innovations, for some time resisted the introduction of this code; and even Charles V., in the plenitude of his power, made two unsuccessful attempts to persuade the diet of Nuremberg to adopt the project of a similar code; but his perseverance finally prevailed, and the diet of Ratisbon, in 1532, proclaimed, as the law of the empire, the criminal code since well knowm by the title of *Constitutio Criminalis Carolina*. This celebrated body of jurisprudence, first published in 1553, was founded on that of Bamberg, and both were remarkable as requiring the evidence of medical men in all cases where their testimony could enlighten the judge, or assist the magistrate, as in cases of personal injuries, murders, pretended pregnancy, abortion, infanticide, hanging, drowning, poisoning, &c. This is the true era of the dawn of legal medicine; and we must regard Baron Schwartzemberg as the father of medical jurisprudence, and Germany as the country which gave it birth.

To the same nation we must attribute the glory of having produced the man who first dared to throw the shield of medical science over the victims of a barbarous and dark fanaticism.

The belief in the powers and influence of witches and sorcerers was in full force in the sixteenth century.

It is computed, that in Lorraine, nine hundred persons, of [xxiii/xxiv] both sexes, were burnt alive in fifteen years for the imputed crime of sorcery; and that, in the Electorate of Treves alone, within a few years, six thousand five hundred individuals had perished in the flames for the same imaginary crime. In various parts of Germany and France, instances of supposed demoniacal possession were perpetually occurring; and at Freidberg public prayers were ordered to assuage this dire calamity. Weiher, physician to William, Duke of Cleves, had the boldness openly to impugn these superstitious notions, in a curious work entitled, *De Præstigiis Dæmonum et Incantationibus*, printed at Basil, in 1568. He undertook to prove, that magicians and demoniacs ought to be considered as unfortunate persons, subject to hysteria and hypochondriasis, and that they should rather excite pity for their infirmities, than be obnoxious to punishment; and he ridiculed the ordinary modes of persecution to which these unhappy beings were subjected.

This attack on a popular superstition, and on a powerful and lucrative engine of clerical influence, aroused alike indignation and vengeance against the daring innovator; and Weiher was himself arrested as a magician, dragged before a tribunal, and owed his liberty and life to the fluence and earnest solicitations of his ducal patron. Weiher was too enlightened for the age in which he flourished; for we find his opinions attacked in no measured terms by two of his contemporaries, Erastus and Scribonius.[15]

The belief in diabolical illusions and witchcraft was entertained in this country so late as the seventeenth century. Sir Thomas Browne, the author of the Religio Medici, who was called on by Lord Chief Baron, Sir Matthew Hale, to give evidence in the cases of two unfortunate persons indicted for having bewitched two children, and caused them to have fits, who were tried and executed at Bury St. Edmonds; deposed, "He was clearly of opinion that the fits were natural, but heightened by the devil, co-operating with the malice of the witches, at whose instance he did the villanies," and he added, "that in Denmark there had been lately a great discovery of witches who used the very same way of inflicting persons by conveying pins into them." This relation of Sir Thomas Browne, says the historian of the case, made that good and great man, Sir Matthew Hale, doubtful; but he would not so much as sum up the evidence, but left it to the jury with prayers that the great God of heaven would direct their [xxiv/xxv] hearts in that weighty matter. The jury accordingly returned a verdict of guilty; and the executions were amongst the latest instances of the kind that disgrace the English annals.

Sir Thomas Browne was born in 1605, was a celebrated physician in London in 1646, and was knighted by Charles II.

The publication of the Caroline Code very naturally attracted the attention of the medical profession, and many of its members commenced the study of forensic medicine. The kings of France soon felt the necessity of a similar code. In 1556, Henry II. promulgated a law, by the virtue of which death was inflicted on any woman who concealed her pregnancy, and destroyed her offspring. In 1606, Henry IV. presented letters patent to his first physician, authorizing him to appoint two surgeons in every city and important town, whose exclusive duty it should be to examine all wounded or murdered persons, and make reports thereon. This law was amended by Louis XIV in 1667, who declared that no report should be received, unless sanctioned by one of these surgeons. These were named medical councillors to the king, they were often appointed by court intrigue, and became so corrupt that they were suppressed in 1790. There are still, however, district medical reporters in France, as will appear hereafter (Foderè op. cit.). In 1692, physicians were associated with surgeons.

Various detached treatises on different branches of legal medicine appear to have been produced towards the close of the sixteenth century. Ambrose Paré wrote on monstrous births, simulated diseases, and on the art of drawing up medico-legal reports, 1575. In 1598 Severin Pineace (or Pinæus) published at Paris his treatise *De notis Integritatis et Corruptionis Virginum*; a book still quoted on that subject.

The earliest systematic work on legal medicine is unquestionably the book *De Relationibus Medicorum* of Fortunato Fidele, published in 1598 in Sicily. As might be expected in his age and country, the opinions of this physician are greatly warped by his servile deference to the dicta of the canon law, and his submission to its clerical expounders. It consists of four books, of which the following may be received as an outline of the contents, viz. I. On Public Food; the Salubrity of the Air; Pestilence. II. Wounds; Pretended Diseases; Torture; Injuries of the Muscles; Medical Errors. III. Virginity; Impotence; Hereditary Diseases; Pregnancy; Moles; the Vitality of the Fœtus; On Birth; Monsters. IV. Life and Death; Mortality of Wounds; Suffocation; Death by Lightning and Poisoning. [xxv/xxvi]

The rapid progress of anatomy in the commencement of the seventeenth century, by the labours of Sylvius, Vesalius, Fallopius, and Eustachius, had a surprising influence on the progress of medico-legal investigations, which became speedily apparent in the publication of the great and invaluable *Quæstiones Medico Legales* of Paulo Zacchia; which appeared in successive volumes from 1621 to 1635.

Zacchia was one of the most eminent physicians of Italy. His celebrity brought a vast body of facts on medical jurisprudence under his cognizance; and his work, which has gone through many editions, is still regarded as the great magazine of medico-legal lore, illustrated by immense erudition, and by a subtile and refined discrimination. Though Zacchia be tinctured with the superstition of his age, in what relates to demoniacal possession, and some other subjects, yet his judicious remarks on the gross injustice and impropriety of classing as demoniacs those females in whom suppressed catamenia had produced mania, or men in whom frenzy had originated in a melancholic temperament, place his physiological reasoning and his medical indications in a favourable point of view; and, perhaps, we ought to place his recommendation in such cases, not to neglect the prayers of the church, while suitable remedies are prescribed, as much to his anxiety to avoid offence, as to his faith in the efficacy of the former.

With the imperfections that were unavoidable in an age in which physiology was in its infancy, chemistry imperfect, and ere our illustrious countryman Harvey had demonstrated the circulation of the blood, the work of Zacchia will ever be considered as one of the landmarks of medical jurisprudence, and remain a stupendous monument of his learning and sagacity. It was written at different times; abounds in repetitions, and occasionally in contradictions; circumstances which render it better suited for the consultation of the learned than the information of the student.

The following is an outline of the contents of the first volume. *First Book*: Age; Legitimacy; Pregnancy; Superfœtation and Moles; Death during Delivery; Resemblance of Children to their Parents. *Second Book*: Dementia; Poisoning. *Third Book*: Impotence; Feigned Diseases; The Plague and Contagion. *Fourth Book*: Miracles; Rape. *Fifth Book*: Fasting; Wounds; Mutilation; Salubrity of Air, &c. The second volume is filled with casuistical questions. This work is highly esteemed as one of great authority. Zacutus Lusitanus, in alluding to its value, exclaims: emi, vidi, legi, obstupui.[16] It is justly considered the most celebrated of all the Italian works on the subject. [xxvi/xxvii]

The next important step in the history of our science is due to Harvey, who investigated the difference between the lungs of the fœtus and of an infant that had breathed, and pointed out how this fact might be available in alleged cases of infanticide. It should, however, be admitted, that long before, the difference in colour, consistence, and weight between the fœtal and adult lungs of animals, had been remarked by Galen, but who had not applied it to any practical use.

Almost at the same time appeared two valuable treatises by Melchior Sebiz, at Strasburg, viz. *De Notis Virginitatis* (1630) and *Examen Vulnerum* (1638). In the first he maintained that the existence of the hymen was the indisputable mark of virginity; an inference which was warmly denied by Orazio Augenio, and defended by the celebrated Pietro Gassendi (*Opera* VI.), &c. In the second he drew an important distinction between wounds necessarily fatal, and those which become so incidentally.

We must give the next important step to the Danish physician, Thomas Bartholin, who carefully investigated the period of human utero-gestation; and, having confirmed the opinions of Galen and Harvey respecting the difference between the lungs before and after the first respiration, proposed the hydrostatic test for solving the question, and pointed out the best method of rendering it available in medico-legal investigations (*De Pulmonum substantia et Motu.* Hafniæ 1663). The rationale of this process was more fully explained by Swammerdam in a tract, published at Leyden, in 1677 – and its first practical application was made in 1682, by Jan Schreyer, after it had been investigated by Thurston and Carl Rayger.

This process is the celebrated *Docimasia Pulmonum*; which, till within a comparatively recent period, was considered as an irrefragable test of the question whether or not the child had breathed; but which has given rise to a keen controversy which I shall have afterwards an opportunity of examining. See *Infanticide*. About this period the work of Ambrose Paré, published 1575, which was regarded for nearly a century the only standard authority in France, was superseded by the more comprehensive treatises of Gendri, of Augers, in 1650, of Blegni, of Lyons, in 1684, and of Devereux, of Paris, in 1693. Louis is, however, with great justice, considered as the first who promulgated a just idea of the science to his countrymen. He flourished about the middle of the last century, and will be noticed in the history of that period.

While these important steps were in progress, Germany, the country where medical jurisprudence took its rise, had [xxvii/xxviii] instituted public prelections on this important branch of medical and forsenic education. About the middle of the seventeenth century, Michaëlis gave the first lectures on this subject at Leipsic, which were followed by those of Bohn, Professor of Anatomy and Surgery in that city, who, before the end of that century, had published his valuable works *De Vulnerum Renunciatione*, and *Dissertationes Medicinæ Forensis* (1689), and these were speedily followed by a tract *De officio Medici duplici, Clinico et Forensi* (1704). In the same period appeared the interesting investigations of Gottfried Welsch, and of P. Amann, on the fatality of wounds.[17] The celebrated work of Licetus, *De Monstris,* appeared in 1669 at Amsterdam; and it would be unpardonable to omit the vast accessions to medico-legal knowledge unfolded in that noble pathological collection, the *Sepulchretum* of Bonnet. (Lugd. 1700, 3 tom. folio). The mode of conducting medico-legal investigations obtained some attention in France during the seventeenth century, but its institutions were extremely imperfect compared to those of Germany even to a much later period.

The following were among the German authors of the eighteenth century: John Bohn, De Renunciatione Vulnerum, 1689, 4to. Amsterdam. Valentini, Pandectæ Medico-Legales, 4to. Francof. 1702. Fred. Boerner, Prof. Med. Wirtemburg, 1723.

Several Dissertations. Kannegeiser, Inst. Med. Leg. Michael Alberti, Prof. Med. Hall. – Systema Jurisprudentiæ Medicæ Schneeberg, 4to. 1725. tom. vi. Zittman, Medicina Forensis, 4to. Francofurti. Richter, Decisiones Medico-Forenses. Teichmeyer, Institutiones Med. Leg. 4to. Jenæ 1740. Stark, De Medicinæ Utilitate in Jurisprudentia, 4to. Helmont, 1730. Hebenstreit, Anthropologia Forensis, 8vo. Lipsiæ, 1753. Ludwig. Institutiones Medicinæ Forensis. Fazellius, Elementa Medicinæ Forensis.

The first German work of authority was Bohn's (1689), in which he attempted to discriminate what wounds were necessarily fatal. The next was Valentini's *Pandects*, 1702. Michele Bernardo Valentini published his *Pandectæ Medico-legales* in 1702 (Francofurte), and his *Novellæ* in 1711, which were incorporated, in 1722, into his *Corpus Juris Medico-legale*. This great work contains an excellent review of all that had been done before his time, and is considered by some as equal in value to the work of Zacchia. Valentini was fully aware of the importance of his subject, and strenuously insists, [xxviii/xxix] in the preface, on the necessity of cultivating this branch of medical knowledge. Several Professorships for teaching juridical medicine were about this period founded in the German Universities; and very useful though sometimes too diffuse treatises on the subject issued from the German press. The number of successive authors becomes now so great, that I cannot, in this sketch, attempt to mention all, far less to characterise them.

Zittman, Boerner, Kannegeiser, and Teichmeyer, each published systems of various, yet conspicuous merit. The *Institutiones Medicincæ Legalis* of the latter, long formed the manual of the student and the text-book of the professor. The clear and forcible reasoning of Stoerck (*De Medicinæ Utilitate in Jurisprudentia*, Helmont 1730) vindicated the high importance of this branch of knowledge. This author very ably advocated the importance of medical knowledge in legal investigations. But the *Systema* of Alberti, the professor of medical jurisprudence in the University of Halle, in six quarto volumes, was the most laborious and complete work of that century. The writings of this learned person are obscured by his devoted attachment to the mysticism of the Stahlian school; yet the industry with which he has collected an immense body of facts, renders his work a precious mine of medico-legal information. His example appears to have stimulated other physicians to publish collections of cases; and curious additions were made to our knowledge by Loesve, Reichter, Budæus, Fritch and Wolff, Hermann, Clauder, Herzog, and Parmeon, which are particularly valuable to those who may be called to exercise the profession in Germany.

The next author of celebrity was Plenck, who published his "Elementa Medicinæ et Chirurgiæ Forensis, ["] 1781. Then followed the lectures of the illustrious Haller, on Juridical Medicine, which were based on the Institutes of Teichmeyer, but greatly amplified and improved. These celebrated lectures were afterwards published in three volumes, octavo, entitled, "Vorlesungen über die gerichtliche Arzneiwissenchaft, ["] 1782. Daniel published his Bibliotheca of State Medicine at Halle, 1784, entitled "Bibliothek der Staatsarzneykunde." Several other important German works appeared towards the close of the eighteenth century, some illustrative of particular questions of legal medicine, and some elementary productions which deserve notice. Among these are Sikora's "Conspectus Medicinæ Legalis,"

Pragæ, 1792; Loder's first lines, "Anfangsgründe der Medicinischen Anthropologie und der Staatsarzneykunde," 1793; Metzer's [xxix/xxx] system, "System der Gerichtlichen Arzneiwissenschaft," Koningsb., 1793; Latin, by Keup, 1794; Muller's Delineations, "Entwurf der Gerichtlichen Arzneiwissenchaft," Francf.

Those who may be desirous of obtaining the titles of the German works or essays, omitted in this sketch, most of which refer to wounds, poisoning, and infanticide, will find the necessary information in the Bibliotheque Medicale of Plouquet, in the works of Struvius and Goelicke, and in the "Collectio Opusculorum Selectorum ad Medicinam forensam Spectantium, Curante. Dort," F. C. T. Schlegel, Leipsiæ, 1785–1800. Dr. Beck gives a list of the titles inserted by Schlegel; but these, and a vast number of others, will be found in the last works to which I have referred, and these are in the possession of most medical practitioners throughout the world – The Cyclopædia of Practical Medicine, London, 1834; and Forbes' Medical Bibliography, 1835.

The eighteenth century commenced with happier prospects for this branch of science, and the press teemed with important works on legal medicine. As early as 1,700 we find an admirable treatise on the diseases of artificers by Bernardino Ramazzini, of Padua. His attention had been called to this subject by observing how frequently a fatal asphyxia occurred to persons employed in clearing sewers and privies; and this accomplished man extended his inquiries to the more remarkable diseases of artizans of various denominations. Ramazzini's work, *De Morbis Artificum Diatriba* is still a standard work, and has been commented on by several writers even so late as 1822. It has also been translated into many languages, and is a standard work in all civilized countries.[18]

During the last century little had been done, in Italy, towards the advancement of medical jurisprudence or police. Beccaria's work appeared in 1749, entitled "Scriptura Medico-legalis," and soon after Bonni's "Instituzioni Teorico Pratiche di Chirurgia. ["]

The science had made little progress in France until after the middle of the eighteenth century. In 1788 Professor Louis commenced the publication of his Memoirs on the Certainty of the Signs of Death, on drowning, on the means of distinguishing suicide from homicide in cases of hanging. [xxx/xxxi] His consultations on the cases of Calas, Montbaillet, Syrven, Cassagneux, and Baronet, which are preserved in the Causes Cèlébres, are models of medico-legal reasoning, and exalt him as a jurist. He also examined, most minutely, the signs of real and pretended pregnancy, and was the first who publicly taught jurisprudence in France.

The Courts of France had decided for the legitimacy of several persons born twelve, and even thirteen months, after the death of their putative father. The famous case of Villeblanche, born 320 days after such an event, called forth the Mémoire of the celebrated Louis, President of the Royal College of Surgeons in Paris. In this dissertation, Louis attacked the legitimacy of those pretended cases of retarded birth with powerful arguments, in which he was seconded by Astruc and Bouvart; but was vehemently opposed by Le Bas and Antoine Petit. This gave rise to very able publications on both sides, in which Bertin Pouteau and Vogel bore a part; but the victory remained with Louis and his adherents.

Winslow discussed the moral, political, and religious relations of the Cæsarean Operation; Lorry investigated the subject of survivorship with success. Salin examined the question of poisoning, and undertook to prove that one Lamotte, who had been buried for 67 days, had died from the effects of corrosive sublimate.[19] This inquiry led to important results. About this period Lafosse endeavoured to distinguish the phenomena caused by death, from the effects of injuries inflicted on the living body. He also described the positive signs of pregnancy and parturition.

In the year 1789 Professor Chaussier read his excellent Memoir before the Academy of Dijon, on the great irnportance of juridical medicine, and in it he treated of several important forensic questions. In it he enforced the necessity of the careful personal examination, by the medical man, in all cases of violence, as blows and wounds; and pointed out the precautions he should adopt in such visits. He gave admirable models of reports, and shewed the attention which was necessary to arrive at truth: and to enforce his precepts still more, next year he delivered a full course of lectures on legal medicine to numerous pupils. Towards the close of the last century the preceding authors, in conjunction with Professor Mahon, compiled the elaborate articles on Medical Jurisprudence in the Encyclopoediæ Methodique. Such were the materials, says Foderè, which enabled me to publish my first systematic work in 1796. The subject now claimed the [xxxi/xxxii] attention of the existing government, and became generally patronised.

In 1796, Foderè published the first edition of his celebrated work, in three octavo volumes, entitled, "Les lois eclairèes par les Sciences Physiques, ou Traite de Medicine Legale, et d'Hygiene publique." This learned professor was the author of many other works of high reputation.

In the present century, France has taken the lead in this science. The great work of Foderè was soon followed by the establishment of a professorship of forensic medicine in Paris, another at Strasburg, and a third at Montpellier. The first was conferred on Mahon, who obtained great celebrity as a teacher; but the posthumous publication of his lectures has not confirmed his reputation. The arrangement of the whole is defective, and the matter is very slovenly put together, being in a great measure made up of excerpts from French authors, and translations from German writers. It is no doubt, the duty of a public teacher to collect his facts from all sources, and it is his duty rather to aim at giving a fair view of the present state of his subject than to affect originality; but those who take the trouble of examining the small volume of Mahon, will be surprised how little pains he has taken to clothe the facts in his own words.

The chair of Montpellier was conferred on René, whose advanced age rendered him little able to perform its duties.

The first Professor at Strasburg was Noel, who is said to have left interesting manuscript lectures, which, I believe, have not been published.

Several valuable works appeared in the early part of the present century. Vigné, of Rouen, published his valuable production in 1805. In 1807, the posthumous work of Professor Mahon was edited by Fautrel. *Medicine Legale et Police Medicale de P. A. O. Mahon,* Prof. de Med. Leg. &c. avec quelques notes de Fautrel. About this date, Belloc, a surgeon at Agen, published a small but sensible treatise, *Cours de Medicine Legale, Theorique et Pratique,* de J. J. Belloc, Chirurgien à Agen.

In 1808, Marc translated from the German, the Manuel of Rose on Medico-legal Dissection, and enriched it with many original observations; to which he also subjoined two instructive dissertations – one on *the docimasia pulmonum*, and the other on *death by drowning. Manuel d'Autopsie cadaverique Medico-Legale, &c.* In 1812, Ballard translated from the German, of Metzger's *Principles of Legal Medicine,* already noticed, and his notes are replete with information.

In 1813, Professor Foderè enlarged his work, which may [xxxii/xxxiii] be considered a new one, in six octavo volumes; four on legal medicine, and two on medical police. The renowned author divides his work into three parts; viz. the *First,* comprehending subjects of a mixed nature, or those, which admit of application to civil as well as criminal cases, "*Medecine Legale mixte.*" The *second* exclusively relating to criminal jurisprudence, "*Medecine Legale Criminelle;*" and the *third,* to medical police, "*Medecine Legale Sanitaire.*"

The work opens with a learned introduction, in which the importance of the science is fairly examined, and its history pursued with much detail, from its origin, to the period at which the author wrote. The qualifications of the forensic physician are also considered, and the different circumstances opposed to the success of his labours, enumerated and appreciated. Then follow in succession the subjects of the first division, viz. the different ages of human life, puberty, minority, majority, with the anomalies to which the natural growth and development of the body are liable. Personal identity and resemblance. The relative and absolute duration of life. The grounds of prohibition in testatorship, such as habitual, periodical, and temporary insanity; suicide; deaf and dumb state; somnambulism; intoxication. The qualifications of testators and witnesses. Marriage and divorce. Pregnancy, true and false. Parturition, and the signs denoting the death of the fœtus in utero. Paternity and filiation. Premature and retarded births. Monsters. Hermaphrodites. Survivorship. Signs of real and apparent death. Treatment of the different varieties of Asphyxia. Certificates of exemption, and diseases which exempt. Feigned, dissimulated, and imputed maladies.

The *second* division commences with the third volume, and includes, in their respective order, chapters on the examination of bodies found dead. The distinction of assassination from suicide. Wounds. Poisoning. Rape. Abortion. Concealment and substitution of the offspring; and Infanticide.

The *third* division, with which the fifth volume commences, successively treats of the preservation of the human species, and of the means of remedying its physical degeneracy. Contagions, hereditary, and epidemic diseases, and the precautions to be adopted against them. The medical police of cities, with regard to aliment, arts, manufactures, and attention to the sick. Military and naval hygiene; and, lastly, the medical police of hospitals and prisons.

This is universally allowed to be the most valuable systematic treatise in the French language; it evinces great research, learning, judgment, and ability; but it is not calcu- [xxxiii/xxxiv] lated to instruct the British jurist. "It is unnecessarily prolix and minute, and is adapted only to the judicial courts of the continent." In 1817, Bertrand published his valuable manual. It is not calculated for British practice in courts of justice, though it is replete with scientific information.

The next work of great value which has appeared in France since 1800, is the system of Toxicology of M. Orfila. *Toxicologie Générale considérée, sous les Rapports de la Physiologie, de la Pathologie, et de la Medicine Legale*; and another—*Leçons faisant Partie du Cours de Medicine Legale, de M. Orfila*, 1821. This celebrated Professor has published many other valuable works, which are laid under contribution in the following pages. His Traité des Exhumations, &c. 1833, is largely quoted in the article Putrefaction, in this work. The author is now the most celebrated medical jurist in existence. In 1821, Professor Capuron published his admirable work relating to all obstetric inquiries. – *La Medecine Legale, relative à l'art des Accouchemens, par J. Capuron*, Doct. en Med., Professor des Accouchemens, &c. This production contains every thing valuable in the branch of medicine, of which it treats.

The various essays in the four voluminous French dictionaries of Medicine, and in the many periodicals of France, abound with medico-legal information. The treatises on infanticide, by Lecieux, on the proper manner of examining bodies for legal inquiries, by Renard; on the perforation of the stomach, by Laisné; on ecchymosis and sugillation, by Rieux, published in one volume in Paris, in 1819, ought to be in the possession of every one engaged in the practice of medicine. The works of Briand, 1828 – *Manuel complet de Medecine Legale, &c.,* and of Sedillot, 1830, contain all the valuable conclusions of French jurists, on legal or forensic medicine. The treatise of M. Devergie, just published, January, 1836, ought to be in every medical library.

In Germany, this century has produced the excellent compends of Schmid and Müller (Landshut 1804), Metzger (Kœnigsberg, 1805), Masiers (at Rostock, 1810), and of Wilberg (Berlin, 1812.) Rose published a very valuable tract on Medico-legal dissection. The useful Toxicologia of Plenck appeared in one volume, at Vienna, in 1801. An useful compend on Pharmaco-chemical Medical Police, which forms the best treatise we have on the preparations and adulterations of food – on the proper kinds of culinary vessels – on the venom of snakes, insects, &c. and the police of apotheca- [xxxiv/xxxv] ries' shops, was published by Professor Remer of Kœnigsberg in 1811, and was translated into French by Lagrange and Vogel, in 1816.

During this century, Italy has contributed some admirable treatises. That of Tortosa is the most scientific and elaborate of his country in the present age. It was written under the sanction of Caldani Franck, of Pavia, and Plouquet of Turin. It only embraces forensic medicine, and is entitled – *Istituzioni di Medicina Forense, di Guiseppe Tortosa,* Professore Medico della Commissione Dipartimentale di Sanita del Bacchiglione, 1809. It is divided into three parts, viz. (1) Comprehending all the principal objects of *Ecclesiastical* jurisdiction. (2) Subjects relating to the *Civil* courts. (3) Those which relate to the *Criminal* courts. The subdivisions of each part are arranged in the following order. Part I. Conjugal Impotence – Conjugal Rites – Monstrous Births – Hermaphrodites – Magic – of Persons possessed of Spirits – Miracles – Ecclesiastical Fasting. Part II. Age – Pregnancy – Birth – Superfœtation – Cæsarean Operation – Simulated and Dissimulated Diseases. Part III. Of Deflowering – Sodomy – Torture – Legal Examination of Wounds, and Dead Bodies – Poisoning – Infanticide – Homicide by wounding – Fœticide – Accidental Death.

Another valuable work is by Barzelotti, entitled *Medicina Legale*, in the form of questions, which embraces the most important points of medical jurisprudence. The best edition of it is by Rossi of Bologna, which was published in 1823.

During the present century, America has contributed to advance the Science of Medical Jurisprudence. In 1823, Dr. Theodoric Romeyn Beck published his *Elements of Medical Jurisprudence*, of which two editions have since appeared in this country. This work is a monument of research, industry, and talent. The author confines himself to Medical Jurisprudence or Forensic Medicine, and excludes State Medicine, or Medical Police. The work consists of 19 chapters: I. Feigned diseases; II. Disqualifying diseases; III. Impotence and Sterility; IV. Doubtful sex; V. Rape; VI. Pregnancy; VII. Delivery; VIII. Infanticide; IX. Legitimacy; X. Presumption of Survivorship; XI. Age and Identity; XII. Mental Alienation; XIII. Persons found dead; XIV. Wounds on the Living Body; XV. Poisons; XVI. and XVII. Mineral Poisons; XVIII. Vegetable Poisons; XIX. Animal Poisons. There is also an Appendix. All the subjects treated by Dr. Beck, are considered most minutely and generally illustrated by numerous cases which [xxxv/xxxvi] occurred in courts of justice. Dr. Duncan, jun. and Dr. Male, two of our best medical jurists, have declared the work as one of the best hitherto published in this or in any other country. I fully agree with them in this opinion, and have accordingly laid it under great contribution in arranging the following pages. So far as it extends, there is no work in the English language equal to it for laborious research and scientific infor- mation. It is, however, confined to forensic medicine, and necessarily omits a vast number of questions on State Medicine, a knowledge of which is indispensably necessary to the medical practitioner; and which the justly celebrated author intended to elucidate in another work on Medical Police. The present production is an admirable system of forensic medicine, which, as Dr. Duncan observed, "embraces almost every valuable fact relating to it. Each of its diversified depart- ments has been investigated so minutely, that few cases can occur in practice, on which it will be necessary to seek elsewhere for farther information." This criticism is fair, candid, and just; but only refers to one department of State Medicine, and not to the whole. As a book of reference and authority, and as a report of medico- legal decisions, the work is unequalled in our language. It ought to have a place in every medical library. It is not, however, a perfect production. It contains a vast number of judicial opinions and law cases, of little or no interest to the great mass of medical practitioners in this country. It omits many subjects of interest to the profession, such as Medical Ethics, the laws relating to the education and Practice of the Medical Profession, and the laws relating to Public Health. These, it is true, are subjects of Medical Police or Public Medicine, but they are as deeply important to all engaged in the practice of Medicine, as pure forensic questions. This being my firm conviction, I have accordingly introduced them in this work.

Such is the history of medical jurisprudence and state medicine, from the earliest age to the present period, except as regards this kingdom, and this I shall now introduce.

It is a matter of no small surprise, that this science has scarcely been patronised in this country. Britain, which has so distinguished itself in almost every science; which has made legislation its study and its boast, has, hitherto, neglected a branch of medicine, so highly cultivated, in all other civilized nations. In proof of this statement I must observe, that the science of medical jurisprudence formed no part of medical education in this country until the present century; nor was there a single work published on the subject before the last quarter of the past century. Many medico-legal [xxxvi/xxxvii] questions were discussed in our periodicals, but no work appeared in the United Kingdom, expressly treating of the subject, previous to the small and imperfect production of the first Professor Dease, of Dublin, the author of the work on Injuries of the Head, which was published in 1783, entitled, "Remarks on Medical Jurisprudence," dedicated to Lord Clonmel, Chief Justice of Ireland. "This was the first attempt to write on medical jurisprudence in our native language."[20]

The next in succession was also an imperfect production, by Dr. Samuel Farr, in 1788, entitled, "Elements of *Medical Jurisprudence*," and was an abstract of the work of Faselius, already noticed. In or about this period the first Professor Duncan, of Edinburgh, delivered a private course of lectures on the subject, and was chiefly instrumental in directing the attention of the profession to its importance. In 1808 Dr. Robertson published a *Treatise on Medical Police,* in two volumes. In 1815, Dr. Bartley, of Bristol, published a most imperfect work, entitled, a *Treatise on Forensic Medicine.* The first respectable and original production that this country contributed to medical jurisprudence, was "*An Epitome of Juridical or Forensic Medicine, for the Use of Medical Men, Coroners, and Barristers,*" by Dr. Male, of Birmingham, who is justly considered the father of the science in Great Britain and Ireland. This work was published in 1816. It embraces the following subjects: Poisons, wounds and contusions, infanticide, pregnancy, abortion and concealed birth, pretended delivery, rape, hanging and strangulation, drowning, dangerous inebriety, insanity, pretended diseases, imputed diseases, apparent death, impotence, hermaphrodites. These subjects are described in 1,999 octavo pages.

It is true, that there are many valuable discussions on various medico-legal questions in the writings of Mead, Monro, the Hunters, Denman, Percival, and others; but these eminent individuals have no claim to rank among our authors on medical jurisprudence. Dr. John Gregory's Duties and Qualifications of a Physician, and Dr. Percival's Medical Ethics, contain much valuable information on the moral duties of the profession, and which naturally comprise the rules that ought to guide them in public and private practice, and consequently in state and forensic inquiries. There are also admirable essays on certain medico-legal questions, such as Dr. Hunter's "On the Uncertainty of the Signs of [xxxvii/xxxviii] Murder in the Case of Bastard Children," 1783. Dr. John Johnstone's essay on *Medical Jurisprudence in Cases of Madness*, 1800; Dr. Haslam's admirable essay on Medical Jurisprudence, as relates to Insanity according to the Law of England, 1817; re-published with other Tracts by Dr. Cooper, Philadelphia, 1819; and Dr. Hutchinson's elaborate "*Dissertation on Infanticide*," 1819.

The next, and by far the most comprehensive work that had appeared in this country, to the date last-mentioned, was Dr. Gordon Smith's, entitled, Principles of Forensic Medicine, systematically arranged, and applied to British Practice, 1821. This work rapidly passed through three editions. Its contents were arranged as follows: (1) Questions relating to the extinction of life. (2) Questions arising from personal injuries not involving a fatal issue. (3) Disqualifications for the discharge of offices, or the exercise of social functions. (4) Miscellaneous questions. This production was most favourably received by the profession. The author commenced a course of lectures, and a few followed his example. In 1823, the conjoint treatise of Dr. Paris and J. S. M. Fonblanque, Esq., Barrister at Law, was published in three volumes, octavo, entitled, "*Medical Jurisprudence*."

In 1825, Dr. Gordon Smith published his "Analysis of Medical Evidence," in 1827 Dr. Lyall his little volume "on the Duration of Pregnancy;" in 1829 Dr. Gordon Smith his "Hints for the Examination of Medical Witnesses," and the same year, Mr. Forsyth his "Synopsis of Modern Medical Jurisprudence, Anatomically, Physiologically, and Forensically illustrated, for the Faculty of Medicine, Magistrates, Lawyers, Coroners, and Jurymen."

Towards the end of 1831, the first edition of this work appeared, entitled, "A Manual of Medical Jurisprudence, compiled from the best Medical and Legal works, comprising an account of: I. The Ethics of the Medical Profession; II The Charters and Statutes relating to the Faculty; and, III, All Medico-legal Questions, with the latest decisions: being an analysis of a course of lectures annually delivered in London, and intended as a compendium for the use of Barristers, Solicitors, Magistrates, Coroners, and Medical Practitioners."[21] This work contained an account of medical ethics, the laws relating to the medical profession, and all forensic subjects noticed by preceding writers. It was very favourably received by the public as well as the medical press, and republished by Professor Griffith in America, as will appear by the notices appended to the present edition.

The last work, before the present edition of this, published [xxxviii/xxxix] in England, is by Mr. Chitty, the celebrated and voluminous legal author, entitled, "A Practical Treatise on Medical Jurisprudence, with so much of the Anatomy, Physiology, Pathology, and the Practice of Medicine and Surgery, as are essential to be known by Members of Parliament, Lawyers, Coroners, Magistrates, Officers in the Army and Navy; and Private Gentlemen, and all the Laws relating to Medical Practitioners, with explanatory Plates, Part 1. ["]

I feel it due to the Profession to which I have the honour to belong, to state, that the present part of his work is almost entirely medical, anatomical, physiological and pathological, with some legal references; all admirably arranged and executed, and most instructive to non-medical readers. It is a valuable treatise, and one well calculated to interest the legislature, the legal profession, the magistracy, and private individuals, and cannot fail to advance the interests of medical jurisprudence and state medicine.

In the preceding sketch of the history of MEDICAL JURISPRUDENCE AND STATE MEDICINE, I have necessarily confined myself to our monographic and systematic writers on the subject; and I deemed it a digression to allude to all essayists, and to those who have elucidated certain parts of it. Among these are the late Professor Duncan, of Edinburgh, the first public lecturer on medical jurisprudence in this country, whose valuable essays and reviews in *the Edinburgh Medical and Surgical Journal*, have contributed largely to advance the science in this country. His successor, Professor Christison, has given us a work on Toxicology, that will transmit his name to the latest posterity, as one of the most laborious, erudite, and experienced medical jurists that this or any country has produced. This work appeared in 1829, entitled, *"A Treatise on Poisons, in relation to Medical Jurisprudence, Physiology, and practice of Physic,"* the third edition of which is lately published.

The medico-legal writers in the *Cyclopædia of Practical Medicine*, and in our numerous medical periodicals, have also elucidated many questions of forensic medicine, and are entitled to much praise for their contributions. I shall briefly allude to the writers in the *Cyclopædia of Medicine* in alphabetical order: Dr. Apjohn, of Dublin, on Spontaneous Combustion and Toxicology; Dr. Arrowsmith, of Coventry, On Infanticide; Dr. Beatty, of Dublin, on Impotence, Persons found Dead, Rape, Doubtful Sex, Survivorship, and Death after Wounds; and Dr. Montgomery, of Dublin, on Personal Identity, on the Signs of Pregnancy and Delivery, and on Legitimacy. [xxxix/xl]

There are numerous other recent writers in the French Medical Dictionaries, Dictionnaire de Medecine et de Chirurgie Pratiques, and the Dictionnaire de Medecine, in the American Cyclopædia of the Medical Sciences, and in the celebrated German Dictionary of Medicine, Encyclopädisches Wörterbuch der Medicinischen Wissenchaft, now in course of publication. The Traité des Exhumations Juridiques, and the Leçons de Chimie appliquée à la Medicine Pratique et à la Medicine Legale, by M. Orfila, 1836, and Nouvelles Rescherches, sur les Secours à donner aux Noyés et Asphyxiés, by M. Marc, 1836, are valuable additions to medico-legal literature.

Notes

1. *Author's note*: Traité de Medecine Legale, et d'hygeine Publique ou de Police de Santé adapté aux Codes de l'Empire Francais, &c. Par F. E. Foderè Docteur in Med. Professeur de Med. Legal. et de Police Medicale al a Faculté de Medicine de Strasbourg. T. vi. 1813.
2. *Editors' note*: A full translation of the title is: "The State Physician: Treatise concerning the office of State Physician in four separate books in which not only are good, medical mores and virtues described, but even deceptions of evil and deceit are revealed: truly also a new argument is most precisely proposed by another method both useful and agreeable. The full work being fit for a physician, student of illness, attendant upon the ill, and by all others of letters, and also by supporters of state education."
3. *Author's note*: Hippocrates flourished and wrote before Christ, 460 years.
4. *Editors' note*: "[O]r the lake, for a long time infertile and fit for oars, now maintains its neighboring cities and feels the heavy plow." Horace, "The Art of Poetry: To the Pisos," in *The Works of Horace*, trans. C. Smart. New York: Harper & Brothers, 1863.
5. *Author's note*: Leviticus ch. x. Numbers ch. vi.
6. *Author's note*: The present law of this country awards a different punishment against the perpetrators of abortion before and after quickening. See INFANTICIDE.
7. *Editors' note*: "concerning reducing death (royal law)."
8. *Author's note*: Pliny, Hist. Nat. L. vii.
9. *Editors' note*: "according to the authority of the most learned Hippocrates."

10. *Author's note*: See INFANTICIDE.

11. *Author's note*: Foderè. Introduction, p. xiv.

12. *Author's note*: Foderè, op.cit.

13. *Author's note*: Hospitals were first established in Jerusalem by Eudocia, wife of Theodosius, A.D. 440; but medical practitioners did not attend their inmates.

14. *Author's note*: Priests were among the earlier chemists, and it is asserted, that they frequently instructed the accused, either from a conviction of his innocence, or from less disinterested motives, in some of those means of resisting the action of fire, by which modern jugglers are still enabled to amuse and astonish the vulgar.

15. *Author's note*: Erasti Disputatio de Lamiis. *Basil*, 1572. 4to. Scribonius de Sagarum Natura et Potestate, *Helmstadt*, 1584. 4to.

16. *Editors' note*: "Buy! Look! Read! Be amazed!"

17. *Author's note*: Welschii Rationale Vulnerum Letbalium judicium, 1660. Ammanni Irenicum Numæ Pompilii c. Hippocrate, 1698. – Praxes Vulerum Lethalium, 1701.

18. *Author's note*: This production had led the late Mr. Thackrah to publish his instructive and valuable work on the same subject, which passed through two impressions in a year. (The Effects of Arts, Trades, and Professions, and of Civic States and Habits of Living, on Health and Longevity: with suggestions for the removal of many of the Agents which produce Disease and shorten the duration of Life. Second Edition, 1832.)

19. *Author's note*: Recueil Periodiqtie de la Sociéte de Med. T. vii.

20. *Author's note*: Dr. Gordon Smith's Analysis of Medical Evidence, 1825. Appendix, p. 181.

21. *Editors' note*: The inconsistent punctuation in the title is taken directly from the original.

Chapter I
Origin of Medicine – Veneration for Its Cultivators

All medicine is derived from God, and without his will it cannot exist or be practised. Hence the healing art, if disunited from religion, would be impious or nothing. Illness requires us to implore the Deity for assistance and relief, and humbles human pride. The seeds of the art, the wonderful cures, and the powers of remedies, are in the hand of God. He has beneficently supplied various remedies, and pronounces with our tongues, the fate, life and death, of man. Whence, we see the dignity of medicine, and what reverence is due to the Divine Author of it. Sacred history confirms this sentiment, "Every cure is from God." "The Most High created medicines out of the earth." Every thing we enjoy are the gifts of God: none but the impious ever doubted this truth; none but fools dared to deny it.

It is recorded that Jesus, the son of Sirach, was one of the first who attributed the origin of medicine to the Deity. And we also read in Scripture, "Honour the physician for the need thou hast of him, for the Lord hath created him."

"All medicine is from God, and without him it cannot exist or prosper; our art, disunited from religion, is either impious or nothing."[1] Such is the first precept in the moral statutes of the Italian Universities, and it is that of Roderic a Castro, in his Medico-Politica, and of the profession in all countries.

The fate of the sick and the success of medicines are in [3/4] the hand of the Deity: "in him we live, move, and have our being;" and our curative means, and our knowledge of the nature and treatment of diseases, are subservient to his divine will and pleasure. He works with his own, and not with our hands; his power begins where ours ceases.[2] Without him man can do nothing good for himself or his species. His first duty is, therefore, due to the Creator and Conservator of the universe. In the first and unerring history of the world, we find that human aid was given during parturition by the midwives of Egypt, which proves the early practice of obstetric medicine. Some even contend that Adam was the first obstetrician. We likewise discover, that Solomon wrote very largely on the nature of animals and plants, from "the hyssop that grows out of the wall to the cedar of Lebanon;" and we are informed by profane authors of his time, that his writings were copied by the Greeks, Arabians, and finally by the Romans. Again we find, that Joseph commanded the physicians to embalm the body of his father (Gen. I. A.C. 1769). Some have pretended that Moses was initiated into the medicine of the Egyptians, and that to them he was indebted for a knowledge of the precepts he laid down in his writings.

The antiquity of medicine is proved by reference to the sacred volume, in which we find many of its precepts inculcated, the natural history of animals and plants recorded, diseases and their remedies described, and hygienic rules laid down for the preservation of health and prevention of human infirmities.[3]

The instinct of man and animals directs them to select what is salutary both in health and disease, and to avoid that which is pernicious. It is recorded, that the first of our species was, soon after his creation, rendered liable to diseases; and as he was also gifted with the knowledge of the nature of all the things that surrounded him, he was necessarily apprized of their noxious and sanative properties. Medicine was, therefore, almost coeval with man; and must have occupied his [4/5] attention in the first days of his existence. The vicissitudes of season, the varieties of climate, the influence of the circumambient atmosphere, the action of surrounding bodies, and the construction of the human fabric, must have rendered diseases nearly coetaneous with mankind. The presence of bodily infirmity produces pain, and impels man to seek immediate alleviation, and to employ means for that purpose, either by instinct, experiment, or spontaneous exertion. The many accidents and injuries to which he was exposed in the early and rude ages, must have frequently obliged him, to suppress hæmorrhage, to remove the deformity of dislocation, and to adjust the painful fracture. Thus necessity conceived the art of medicine, reason nourished it, long use promoted it, and experience at length completed it, and made it absolute. The foundations of this art among mankind, were first laid by chance, instinct, and unforeseen events; these were improved by the success and recollection of former experiments; the results of observations, experiments, and remedies were carefully recorded, and a comparison was instituted between events already observed, with those of daily occurrence. The sick were exposed in the streets and highways; and inquiries made of passengers if they knew any remedy. The names of diseases, their remedies and events, were handed down from father to son among the ancestral traditions; and were at length inscribed on the walls and paintings of the public temples: a system of medical instruction was formed from these sacred records, which all were obliged to follow in the cure of diseases; a law alone that was the cause of the extermination of thousands, until the healing art was practised by certain individuals, as a peculiar calling. Such was the manner of cultivating medicine in the early ages, even after the deluge, by the Egyptians, Babylonians, Grecians, Germans, Indians, Assyrians, Chaldeans, and Portuguese; and afterwards in the islands of Cnidos, Cos, and Epidaurus.[4]

Ancient medicine, like all the early transactions of mankind, is a mixture of monsters, giants, demi-gods, and fables; and, therefore, I shall not notice the legends and dark archives of antiquity relative to the first epoch of the healing art, except by adverting to Ovid's allusion to Apollo, as the [5/6] inventor of physic, in the following lines, which now grace the armorial bearings of our society of apothecaries:

Inventum medicina meum est, opiferque per orbem Dicor; et herbarum subjecta potentia nobis.[5]

All the ancient heathens agree in one point, that some Deity was its inventor, which is also confirmed by Hippocrates (*Liber de Vet. Med.*), and by Cicero, in *Tusc. Lib.* iii. "Deorum immortalium inventioni consecrata est ars medica."

The concurrent testimony of historians of all ages proves it to be the noblest and most useful of human pursuits; and hence the esteem and veneration universally entertained for its cultivators by mankind. The dignity of medicine arises from the nobleness of its subject and its end; its subject is the human body, which excels all other material bodies; its end is health, which is the greatest temporal blessing of man. The importance and dignity of medicine were felt and acknowledged in all ages and countries, both civilized and savage, because the severity of pain and dread of death, were almost coeval with the existence of the human race.

The heathens also highly esteemed the healing art. Democritus said it was the sister of philosophy; the latter removed affections of the mind, the former diseases of the body. Hippocrates said the ancient cultivators of medicine ascribed its origin to a divine source, in which he concurred, as also did Galen. This was likewise the belief of the primitive Christians: St. Austin observes, "Medicina non invenitur unde ad homines manare potuerit, nisi a Deo" (*De. Civ. Dei*).[6] But the healing art was rendered pre-eminent by the divine Redeemer having practised it, while he avoided all other human pursuits.

The excellence and pre-eminence of the healing art were admitted by the Roman orators and moralists. Cicero said, "In nulla re homines propius ad Deos appropinquant, quam salutem hominibus dando."[7] Seneca observes, "quædam pluris esse quam emuntur; emis a medico rem inestimabilem vitam ac valetudinem bonam. ["]8

"A physician of genius," says Monfalcon, "is the most magnificent present that nature can make the world." Me- [6/7] dicine was encouraged, cultivated, and practised by kings, princes, and pontiffs; the highest, wisest, and best of men. In proof of this position it may be stated, that Solomon, Saphoris, and Gyges, kings of Persia; Habidus, and Mithridates, kings of Pontus; Mesue, king of Damascus; Avicenna, prince of Cordova; Isaac, the adopted son of the king of Persia; Nicholas V., and John XXII., Roman pontiffs, illustrated medicine by their writings. Homer records the great esteem entertained for Machaon and Podalirius, the sons of Esculapius, in the Grecian army; Virgil, that for Japis, physician to Æneas; and Silius Italicus, that for Synalaus, the physician to Hannibal. All civilized nations conferred the highest privileges and honours on the practitioners of medicine. They were exempt from the performance of all civic duties; they were supported by the state in many countries; and ample, nay, prodigal rewards were bestowed upon them throughout the civilized world. The history of medicine affords abundant proof of this assertion.

All wise and prudent emperors and kings, duly estimated the utility and excellence of the healing art, and were extremely desirous of having learned and experienced physicians, to preserve their own and their subjects' health; and hence the honours, immunities, and privileges bestowed in every civilized nation upon the faculty. It is unnecessary to enumerate the vast number of temples dedicated to the early founders of physic, or refer to the rank so signally conferred on Hippocrates by the Athenians, or to the honours bestowed by all nations in succeeding ages on the profession. I may briefly mention, that the court physicians and surgeons of our own country have had titles and emoluments amply conferred upon them; and such is the custom of all nations in Europe.

Medical practitioners are those men to whom we confide our health, which is above all earthly concerns. To them are intrusted the existence of those that are most dear to us, and in their hands are placed the lives of our nearest connexions, and of the friends to whom we are most attached. All gradations of society are alike dependent on them, and must, sooner or later, require their assistance; for, from the earliest period of life to the last moment of existence, their [7/8] skill may be exerted to preserve health, to arrest the progress of disease, or to smooth the approach of death. It is to physicians that all classes commit their health, which is above all treasure, the father confides his child, the husband his wife, the monarch as well as the peasant; they preserve the feeble infant from all the dangers to which it is exposed on coming into the world; their cares protect infancy, childhood, adolescence, manhood, and old age. At every period of life, man calls for the assistance of a physician, and he rarely implores in vain. The confidence placed in him is unbounded: health and life are committed to his care, and on him depend the comfort or misery of families. Hence it is, that no class of men enjoys the respect of every rank of society to such an unlimited degree, as the practitioners of the healing art; and hence it is that they have ever acquired the esteem and veneration of mankind.

That this confidence and esteem should be fairly merited, the father of medicine, and all his eminent successors, to the present period, required an oath of their disciples, the principal obligations of which were, the cultivation of every virtue that adorns the human character. A code of professional duties, or ethics, was arranged, which all were obliged to obey, and which still governs those who have been properly educated. But there never was a period in medical history, in which ethics was so neglected and violated as in this "age of intellect," nor the dignity of the science so degraded and disregarded. It is therefore necessary to inform the rising members of the profession, of those virtuous and noble principles which regulated the professional conduct of their predecessors, and procured that unbounded confidence and universal esteem, which was bestowed on them by society in every age and country. I shall, therefore, describe the ethics of the founder and father of medicine, and those of his successors to the present time, with the hope of exciting my readers to imitate their example, and practise those precepts which have always characterized the erudite and scientific portion of the profession.

Notes

1. *Author's note*: "Omnis medicina a Deo est. Cœlitus delapsa non sine Dei consilio vivit agitque. Hine ars nostra sine religione, vel impia vel nihil." – La Politica del Medico nell' esercizio dell' arte sua. Dal celebre Alessandro Knipps Macoppe, Professore di Medicina nell' J. R. Università di Padova. Milano, 1826.

2. *Author's note*: "Deus propriis non nostris agit manibus. Incipit artis Dei potentia, ubi tua desinit." *Op. cit.*
3. *Author's note*: Gen. xl. Exod. xxi. Lev. xii. Ecelias. xlvii.
4. *Editors' note*: This paragraph very closely follows Ryan's "Introductory Lecture to the Theory and Practice of Midwifery," *Lancet* 2 (1827–28): 394–400, quoting p. 395.
5. *Editors' note*: "The art of medicine is my invention, and the healing power of herbs, which are declared useful throughout the world" (loosely quoting Ovid, *Metamorphosis*, 1:523).
6. *Editors' note*: "Medicine would not have been invented unless humans were able to derive it from God." The reference is to St. Augustine, not Austin.
7. *Editors' note*: "Humans approach the gods in no things more than when giving health to people."
8. *Editors' note*: "Some things are worth more than they seem; you buy from a doctor inestimable things—life and good health."

Chapter II
Medical Ethics of Hippocrates

The duties and qualifications of medical practitioners were never more fully exemplified than by the conduct of Hippocrates, or more eloquently described than by his pen.[1] He admitted no one to his instructions without the solemnity of an oath, the chief obligations of which were, "the most religious attention to the advantages and cure of the sick, the strictest chastity, and most inviolable secrecy about private or domestic matters, which might be seen or heard during attendance, and which ought not to be divulged."[2]

The father of physic strongly inculcated the necessity of the cultivation of piety and virtue; and held that his disciples should excel in religion and morals. He also maintained that they should acquire the most perfect knowledge of every form of disease, and of the best mode of treatment. He considered calumny and illiberality disgraceful, and the disclosure of the errors of a contemporary highly culpable. He was of opinion, that the morals of a medical practitioner should be excellent and unexceptionable, conjoined with gravity and humanity. He ought to be correct in every custom of life; and demean himself honourably and politely towards every rank in society, and thus will he promote the glory of his profession. To these precepts nature is the best guide. He is to retain in his recollection all remedies, their mode of preparation and application, and the use of all mechanical means which are employed for the cure of diseases. This is the beginning, middle, and end of medicine. Let him be cautious in his prognosis, and predict only those events sanctioned by observation and experience. In his approach to the sick, let his [9/10] countenance be mild and humane, not rough, proud, or inhuman; and let him evince a sincere desire to afford relief; and employ all remedies with diligence and caution. He should be ready to answer all questions, establish constancy in perturbations of mind, allay tumult by reason, and be ever ready to afford relief in all emergencies. He is never to exhibit an improper or dangerous remedy, even to a common malefactor; but try those medicines approved by the majority of the profession. When a patient desires popular remedies, he is to be cautioned against them, but left to his own discretion. But if attendance is commenced without remuneration, the sick must not be abandoned. Any discussion relative to pecuniary matters is injurious, especially in acute diseases. When disease is rapid, there is no time to arrange concerning reward; it has no influence on a good practitioner, who is only

H. Brody et al. (eds.), *Michael Ryan's Writings on Medical Ethics,*
Philosophy and Medicine, Vol 105,
© Springer Science + Business Media B.V. 2009

anxious to preserve life, and enjoy the more noble gratification, the universal esteem of mankind. It is much better to accuse those cured of disease of ingratitude, than deny them aid when in danger.

Some, on account of friendship or acquaintance, expect attendance gratuitously, but these deserve neglect. In all cases the best remedies ought to be employed; and the reward should be in accordance with the custom of the faculty, and with the wealth or means of the sick. In some cases aid is to be afforded gratuitously, which will ensure more renown than if remuneration had been awarded. When an opportunity offers in the case of a stranger or necessitous individual, afford succour immediately. Some men are so avaricious as to extract riches from the most indigent; but these men seldom prosper, while the humane and good practitioner is considered an honour to his art. Consultations cannot be refused, even to the most necessitous. "This I affirm by an oath, that one medical man should never invidiously calumniate another, or rob him of his merit, or diminish the confidence of his patient." Such were the leading features of the code of ethics, inculcated by the immortal founder of physic;[3] the chief obligations of which have been [10/11] recommended by all his successors down to our own time.

It must be conceded that the first duty of a medical practitioner in common with his species, is to God, for his sovereign majesty, supreme excellence, and infinite goodness. The history of medicine, of all other sciences, comprises the most intimate acquaintance with the works of nature, and elevates the mind to the most sublime conceptions of the supreme Being, and expands the heart with the most pleasing ideas of Providence. It nearly extends through the whole range of the creation; and no other profession requires so extensive a knowledge of the works of Providence. The objects which engage the attention of the medical practitioner are the influence of the sun, moon, and heavenly bodies, the laws of their uninterrupted revolutions and various movements; the various productions of the earth, including the vegetable and mineral kingdoms, the innumerable variety of living creatures that fill the air, the earth and the waters; and the microcosm of the human body, with its wonderful organs, functions, and immortal principle. Now when we consider the creation, and conservation of all these objects, and their subserviency to human happiness, we can never reflect without the profoundest veneration upon the attributes of that Being from whom all these things have proceeded. We cannot help acknowledging what an exertion of benevolence creation was, of a benevolence how minute in its care, how vast in its comprehension!

The most eminent of the faculty have been distinguished for their piety; among whom we find the names of Harvey, Sydenham, Locke, Boerhaave, Arbuthnot, Winslow, Haller, Hoffman, Stahl, Baglivi, Steno, Helmont, Riverius, De Hean, Vesalius, Ruysch, Lancisi, Gaubius; and in our own day, Gregory, Baillie, Bateman, Davy; and many other illustrious men now in existence. They studied the sacred volume, which to use the language of Cicero when speaking of the twelve tables, is, "A little book that alone exceeds the libraries of all the philosophers in the weight of its authority, and in the extent of its utility." In it is to be learned the only complete history of the universe, the divine precepts of religion, which refer all honour and glory to the Author of all [11/12] things, and good will to mankind.

There is laid the foundation of all ethics, "do unto others as you would they should do unto you." This is the great and first principle of human conduct, it commands man to observe justice and benevolence towards his fellow man; duties which by their reciprocity lead to his own and the general good. Of all the calumnies launched against medical practitioners, there is none so diffused and so odious as that of irreligion. It is broadly asserted that medical men are Deists or Atheists. This is a popular error. A medical practitioner cannot be an anatomist and an Atheist, as nothing presents such strong proof of the existence of God, as the wonders of our organization and its functions. The wonderful relation between the structure and functions of all parts of the human body; the marvellous disposition of the bones and muscles; the distribution of the arteries and veins, and their anastomoses; the functions of the brain, heart, lungs, &c., every thing in the study of anthropology attests a superior intelligence. The origin and cessation of life manifest the peculiar power of the Deity. The reproduction of the species, the endowment of life and of mind, can only be ascribed to the same Being. The anatomy of a hair is sufficient to overthrow all the reasonings of materialists.

All the universities, schools, and colleges, inculcate and enforce the cultivation of religion. "Such is the connection between the Deity, religion, and a physician, that without God and religion, no physician can be successful."[4] This reflection of Broesiche's is a great moral truth. A religious physician will not arrogate to himself an absolute empire over the lives and health of men; he will not pretend to govern the progress of diseases; he will not consider himself the God of nature; but he will refer all things to the Supreme Being. "It is from Him he derives his light – it is on Him he calls for succour."

The practice of medicine requires the exercise of all the virtues. Whoever fulfils all the duties it imposes on him, [12/13] obeys the most rigid moral laws. To cultivate religion, to love his country, and to be a truly good man, form the true character of the medical practitioner. It may be urged that there are exceptions; but they are few in number.

In observing the different phenomena of life, physicians are continually obliged to recognize the omnipotence, goodness, and other attributes of the Deity; they cannot consider their art a sovereign power; because they too often find it to fail. It is to the Divinity they must attribute their success or failure. Such is true medical wisdom. Who knows so well as the medical practitioner the miseries of man, his infirmities, and the dangers which every moment threaten his life? Who so well knows the suddenness of death in the most robust constitutions? The whole history of man, and of the healing art, recalls to his mind the idea of a Supreme Being. He discovers in religion the powers opposed to all the annoyances inseparable from his ministry, and the real consolations against the ingratitude of mankind. It enables him to bear all the weaknesses and follies of the sick, to perform the beneficent powers of his art in the most disgusting instances, and to be the instrument in the hands of the Author of all, of affording relief, and diminishing the sufferings of humanity. Hoffman was right when he said, a physician ought to be a Christian – "medicus sit Christianus."

There are intimate relations between religion, morals, philosophy, and medicine. It is well observed by the ancients, that there is a relation between certain physical states, certain characters of the intellectual faculties, and certain passions; that is to say, that according to the habits of the body, the proportion of the members, the colour of the skin, the disposition of blood vessels, lymphatics, nerves, &c., will thought and the train of ideas correspond. Many of the ancient sages found in the organization of man, compared with the functions of life, the solution of the most important moral phenomena.

Many physicians have written esteemed works on philosophy, and on the human mind. It is scarcely necessary to mention the names of Locke, Dugald Stewart, Thomas Brown, &c., nor can I omit the splendid work of Cabanis, on the relations of the Physical and Moral States of Man. He has [13/14] given a most luminous view of the relations between the physical and moral conditions, and followed Locke and Condillac.

Moral philosophy, medicine and philosophy, have numerous relations. Medical philosophy informs us of the formation of ideas, the rules which ought to regulate life, the course that leads to happiness, the influence of the different climates on the physical state of man, and on the institutions of society, that of regimen upon the intellectual faculties and passions, and that of diseases on the operations of the judgment. It embraces the operations which constitute the functions of intelligence, and determine the power of volition; it comprises the different characters of the passions, and supplies the basis of the moral condition. The faculties of the mind are intimately connected with the organization of man, and nothing can separate the study of his physical and moral states. I may briefly illustrate this position.

Vain attempts have been lately made by materialists, among whom are certain phrenologists, to prove that the mind is a mere function of the brain, and consequently dies with it; that the embryo has no mind or soul either in the uterus or at birth, but that its mind is built up by the five external senses, and is annihilated by death. Here is mere assertion without the slightest proof. According to this doctrine, life and mind are distinct, as the living infant, before or at birth, has no mind; but the materialists have not proved that the living brain of the infant in the womb is unconnected with soul: they merely assert it. If their conclusion were admitted, it would be no crime to destroy the fœtus in the womb, or even at birth, no more than the young of any other animal; and fœticide or infanticide might be expunged from the catalogue of crime. These philosophers forget that the first of the human species was perfectly formed and organized before the breath of life and soul were breathed into his nostrils; and it is manifest that life and soul were added to his organization. It, therefore, follows, that life and soul, and matter or brain, were different in the first man, and were not made up by the five external senses. I cannot agree with a recent phrenological physiologist, who maintains "to call the human mind positively a ray of the divinity – *Divinæ particula auræ ex* [14/15] *ipso Deo decerptus, ex universa mente delibatus,*[5] appears to me to be absolute nonsense." Neither can I assent to another absurd position, which denies the interposition of an omnipresent Deity at the time of generation, as if the divine power was not in constant operation in every atom of the universe. Who knows how life is communicated, or what becomes of it, when the body is deprived of it? "That life, or the assemblage of the functions, is immediately dependant on organization, appears to me, physiologically speaking, as clear as that the presence of the sun above the horizon causes the light of the day. Mind is the functional power of the living brain. As I cannot conceive *life* any more than the power of attraction, unless possessed by matter, so I cannot conceive *mind* unless possessed by a brain, or by some nervous organ, whatever name we may choose to give it, endowed with life. I speak of terrestrial or animal mind; with angelic and divine nature we have nothing to do, and of them we know, in the same respects, nothing." I ask, was life and mind distinct from organization in the first of the human species? Is the life and soul of the Deity connected with organization? Nevertheless the terrestrial mind, which we presume differs from the angelic or celestial, is material; and thus the human mind is a compound of terrestrial and celestial, it is material and immaterial according to the doctrine of Dr. Elliotson. Such is the illogical and unphilosophical conclusion of some materialists and phrenologists.[6] The pseudo-science of phrenology or materialism, has led to the too prevalent opinion of the materiality of the mind in this age, though some of the disciples of the newly-revived and fantastic hypothesis of moral philosophy stoutly deny it. There are some phrenologists who disavow materialism: there are more who avow it. I agree with those who consider phrenology and materialism synonymous; and as I hold both one and the other to be totally unworthy of belief, which is the opinion of the greatest physiologists in existence, in this and every other country, I shall repeat my denunciation of this false philosophy as published in the former edition of this work, [15/16] and advise certain commentators not again to misquote and misrepresent my language, as on a former occasion.

The cultivators of medicine have been generally, and most unjustly accused of favouring infidelity and a contempt of religion; though of late it must be acknowledged, with pain, that an abortive attempt had been made which has given too much proof of the justness of the accusation; – this was the revival of materialism. The attempt, however, was as impotent as it was wicked – it was attacking youth on the weak side; and it was extinguished by the universal voice of the most eminent members of the faculty; for it was observed to have unhinged all the bonds of society in a neighbouring nation, and to have produced a degree of anarchy, confusion, and atrocity unequalled in the annals of mankind. It contained not a single argument that had not been urged and refuted a thousand times before; and after all the schemes of a reluctant and erring philosophy, the indispensable resort must be to a Deity and his ordinances. How appropriate the following distich to such philosophers:

Know thou thyself, presume not God to scan,
The proper study of mankind is *man*.

The doctrine is repudiated by the medical profession throughout the civilized world; and out of the whole, amounting to many thousands, are not ten men of any eminence, either phrenologists or materialists. That philosophy which assails the attributes of the Supreme Being is fallacious: that wild hypothesis which is in direct opposition to the principles of revealed religion can have few, if any disciples, in an age so enlightened as the present. The really learned in the medical profession, have never been infected by the poison of infidelity.

The doctrine of the materiality and mortality of the soul, which is that of materialism and phrenology, "should for ever be exploded as totally false, and unworthy of all regard, as subversive of the fundamental principles of all religions, as introducing civil anarchy into the political economy of legislation, as substituting disorder for harmony, despair for hope, and eternal darkness for everlasting light."

If materialism tended to promote the happiness of society, [16/17] to assist our hopes, to subdue our passions, to instruct man in the happy science of purifying the polluted recesses of a vitiated heart, to confirm him in his exalted notion of the dignity of his nature, and thereby to inspire him with sentiments averse to whatever may debase the excellence of his origin, the public and the medical profession would be deeply indebted to the phrenologists. But the tendency of phrenology, however disguised, is to make mind a secretion or function of the brain, and thus to deny the immortality of the soul. Unbelievers, in general, wish to conceal their sentiments; they have a decent respect for public opinion, are cautious of affronting the religion of their country, and fearful of undermining the foundations of civil society. Some few have been more daring, but less judicious; and have, without disguise, professed their unbelief and again retracted their opinions. In denying the immortality of the soul, they deny the authenticity of the Bible; they sap the foundations of all religions; they cut off at one blow the merit of our faith, the comfort of our hope, and the motives of our charity. In denying the immortality of the soul, they degrade human nature, and confound man with the vile and perishable insect, and overturn the whole systems of religion, whether natural or revealed. In denying religion, they deprive the poor of the only comfort which supports them under their distresses and afflictions, and wrest from the hands of the powerful and rich, the only bridle to their injustice and passions; and pluck from the hearts of the guilty the greatest check to their crimes – that remorse of conscience which can never be the result of a handful of organized matter – that interior monitor which makes us blush in the morning at the disorders of the foregoing night, and which erects in the breast of the tyrant a tribunal superior to his power. Such are the consequences naturally resulting from the principles laid down in some of the late phrenological writings. It is no intention of mine to fasten the odium of infidelity on any members of my profession; but it surprises me that men, whose understandings have been enlightened by the Christian revelation, and enlarged by the study of medicine (the most extensive and varied of all human sciences), should broach tenets which [17/18] equally militate against the first principles of reason and the oracles of the Divinity; and which, if true, would be of no service to mankind. Of what benefit to humanity would be the establishment of phrenology or materialism? I answer none; but, on the contrary, the greatest injury. If any man be so unhappy as to work himself into the conviction that his soul is a function or a secretion of the brain, and of course must perish with it, he would still do well to conceal his horrid belief with more secrecy than the Druids concealed their mysteries. In doing otherwise, he only brings disgrace upon himself, for the notion of religion is so deeply impressed on our minds, that the bold champions, who would fain destroy it, are considered by the generality of mankind, and our profession, as public pests, spreading disorder and mortality wherever they appear; and in our feelings we discover the delusions of a cheating and unmedical philosophy, which can never introduce a religion more pure than that of the Christians', nor confer a more glorious privilege on man than that of an immortal soul. In a word, if it be a crime to entertain such a doctrine, it is consummate folly to boast of it. Whence this eagerness to propagate systems, the tendency whereof is to slacken the reins that curb the irregularity of our desires, and restrain the impetuosity of our passions? It must proceed from a corruption of the human heart, averse to restraint, or from the vanity of the mind, which glories in striking from the common path, and not thinking with the multitude. In vain are the phrenological materialists informed by the anatomist, that he can find bile in the liver, urine in the kidneys, but none of the faculties of the mind in the brain. In vain are they told that after death, when volition has ceased, the motions of the muscles can be excited by galvanism, and that though muscular motion be restored, we cannot recal volition, or the other mental faculties; rather a strong proof that motion and volition are not exactly the same thing. In vain have materialists examined dead bodies to explain the most important phenomena of life – their imagination only has answered; in vain have they mutilated the brain, in a hundred different ways, to discover the seat of the intellectual faculties – the vainest hypotheses are the result of [18/19] their researches. What do they know about life, about the astonishing phenomena ascribed to sympathy, or the impenetrable mystery of generation? God reserves these secrets to himself. Facts are observed every instant in practice, which science does not explain; and it was this that made the father of medicine declare, there was something divine in diseases – that is to say, incomprehensible to man. These and ten thousand other proofs are lost on the phrenologists. They set up the proud idols of their own fancies in opposition to the received opinions of their profession, and in opposition to the oracles of the Divinity; and in endeavouring to display absurdities in the Christian religion, fall into much greater. To them we can, with due deference, and without disclaiming our title to good manners, apply the words of St. Paul to the philosophers of his time – "They became vain in their imaginations; professing themselves wise they became fools."

Let the patrons of the revived and long refuted philosophy persuade their wives that their souls die with their bodies – let them instil the same doctrine into the minds of their children – let the doctrine become generally received – unfaithful wives, unchaste daughters, rebellious sons, and general confusion and anarchy will be the blessed fruits of their philosophy. To those philosophers, the words of an able and learned prelate very forcibly apply: "The Bible has withstood the learning of Porphyry and the power of Julian, to say nothing of the manichean Faustus; it has resisted the genius of Bolingbroke and the wit of Voltaire, to say nothing of a numerous herd of inferior assailants; and it will not fall by your force. You have barbed anew the blunted arrows of former adversaries; you have feathered them with blasphemy and ridicule; dipped them into your deadliest poison, aimed them with your utmost skill; shot them against the shield of faith with your utmost vigour; but like the feeble javelin of the aged Priam, they will scarcely reach the mark, will fall to the ground without a stroke [.]"[7]

The doctrine of materialism is not more discordant with [19/20] the principles of revealed religion, than with the opinions of the greatest men who have ever adorned the science of medicine; men who were, and still are, as great ornaments to the literary world in general, and medical literature in particular, as they are useful to mankind. On the other hand, the soul shrinks within itself at the horrors committed by materialists, as they pretended to disbelieve a future state of rewards and punishments, and assented to every evil suggestion of their unhappy minds.

"The first duty," says Napoleon, "of a medical man is to his God, the next to his king and country, and the next to his patients."[8] This was the rule of duty of the greatest ornament of the medical profession, and comprises the whole institutes of professional conduct. The man who pays due homage to his Creator, and obeys the laws laid down for his guidance, must act honourably and justly towards his fellow-creatures. A religious man cultivates all the virtues that adorn the human character, and justly place man the lord of the creation. He will be distinguished for humanity, sympathy, politeness, a large share of good sense, and knowledge of the world. Dr. Gregory considered the chief of the moral qualities peculiarly required in the character of a physician was humanity, "that sensibility of the- heart which makes us feel for the distresses of our fellow-creatures, and which of conse- quence, excites us in the most powerful manner to relieve them; sympathy produces an anxious attention to a thousand little circumstances that may tend to relieve the patient – an attention which money can never purchase; hence the inexpressible comfort of having a friend a physician. Sympathy naturally engages the affection and confidence of a patient, which, in many cases, is of the utmost consequence to his recovery. The patient feels the approach of a man who possesses it, like that of a guardian angel ministering to his relief, while the approach of a rough unfeeling man is like that of an executioner. A certain command of the temper and passions must be natural or acquired to medical men, as sudden emergencies frequently occur in practice, which may flutter [20/21] the spirits and judgment of the best practitioner; and the weakness and bad behaviour of patients and their attendants are well calculated to ruffle the temper of the mildest individual, and cloud his judgment, and make him forget propriety and decency of behaviour; hence the necessity of presence of mind, composure, and steadiness." Let him be remarkable for the humane and liberal exercise of compassionate philosophy, Dr. Gregory has lucidly described the genius, understanding, temper, and qualifications, which are required for the duties and office of a physician, which will be inserted hereafter. He observes, "to excel in this profession requires a greater compass of learning than is necessary in any other." This assertion is borne out by the most exalted testimony. Judge Blackstone gave physicians pre-eminence for "general and extensive knowl- edge." Dr. Johnson was scarcely less favourable in his estimation. "Whether," he observes, "what Temple says be true, that physicians have had more learning than any other faculties, I will not stay to inquire; but I believe every man has found in physicians, great liberality and dignity of sentiment, very prompt effusions of beneficence, and willingness to exert a lucrative art where there was no hope of lucre."[9] The late Dr. Parr, the justly celebrated philologist, remarked, "while I allow that peculiar and important advantages arise from the appropriate studies of the three liberal professions, I must confess, that in erudition and science, and in habits of deep and comprehensive thinking, the pre-eminence, in some degree, must be assigned to physicians."[10] Rousseau spoke as follows of the faculty: "Il n' y a pas d'etat qui exige plus d'etudes que leur: par tous les pays ces sort les gens les plus veritablement utiles et savans."[11] "In addition to all other means of augmenting your true fame," said Professor Godman,

"the observance of one circumstance will be of great importance; this is the unremitted exercise of humanity towards those who seek your professional [21/22] aid, whatever be their conditions. The character of a truly good physician, is one of surpassing excellence, and his reputation is the most exalted we can hope for. He is the friend of the wretched and woe-worn; the cheerer of the despondent; the solacer of the broken-hearted. His soul is the empire of benevolence; his actions the result of a principled charity, and unaffected good-will. He is the blessing conferred on the society in which he lives, and an honour to the human race. Wherever the afflicted dwell – wherever the voice of suffering is heard, he is to be found. The diseased find cheering and consolation from his presence, and the sounds of sorrow are stilled. Even when hopes of life can no longer be given, he calms the tumultuous grief of relatives, by recalling their thought to that better world, where – sickness and sorrow are to be no more – 'where the wicked cease from troubling, and the weary are at rest.'

"Such are the common offices, and frequent exercise, in which he is engaged. His character, even under ordinary circumstances, may be contemplated with gratifying emotions. But there are conditions in which he is presented in a more sublime aspect. It is when the lurid breath of pestilence scatters destruction, desolation, and dismay, throughout the land, and death tramples with indiscriminate fury over the people – when the ties of relationship and affection are sundered by the violence of fear, and utter selfishness seizes on the hearts of men; then the good physician, unmoved by such examples, untouched by terror, regardless of himself, is seen actively discharging every duty. Then he becomes the father, the brother, the friend of the destitute; his steadfast attention smooths the pillow of the dying; he inspires the desolate with hope; and, like a beneficent angel, wherever he goes, is a dispenser of good. Who can estimate the feelings, or measure the fame of such a man? Who would not imitate his example for such a reward? What is there in death's most frightful forms that could withhold us from attempting to deserve it? It is a glorious privilege which our profession confers, of inscribing our memories, not on perishable marble, but in the living affections of our fellow-men, to be cherished as long as [22/23] our race shall endure."[12]

It is unnecessary to illustrate this truism by other citations, as it is universally admitted by those capable of judging. As to the observance of ethics, it may be fairly stated, that there is much room for improvement in this section of the empire.

The following graphic and elegant description of the moral conduct of medical practitioners, is so accurate and so just, that I quote it with much pleasure. The author defended the study of anatomy before the passing of Mr. Warburton's Act, which legalized it.

"Nevertheless, anatomical pursuits are neither criminal in themselves, nor yet fraught with dishonour or disrespect to the dead; they outrage no feelings but such as are of a superstitious nature; neither do they in any way deteriorate or brutalize the character of those who pursue them. Insinuations of the latter kind are, indeed, frequently thrown out, but we repel them with meet and honest indignation. We appeal to observation for the truth of our statement, when we assert, that society does not present another class of individuals more numerous and respectable than that of the medical profession; and one, at the same time, against the general moral conduct of which so little reproach can be made. There is none, too, possessed of more varied and valuable information. In these respects physic has no occasion to shrink even from a comparison with divinity. With temptations infinitely stronger and more diversified, practitioners in medicine do in no wise cede to the clergy, taken generally, in the morality of their conduct. It ought moreover to be borne in mind, that the very vocation of the latter, by abstracting them from temptation, diminishes the merit attached to the rectitude of their walk. At the same time we need not hesitate to affirm, that more benevolent, more truly kind and charitable individuals, than those which have adorned the profession of physic can no where be found. No description of men, whatever their calling or station in life, render such valuable services to the poor and [23/24] needy sick; none expose themselves to dangers equally numerous and great, without the remotest prospects of pecuniary remuneration. They work silently, yet not the less effectually. They make use of no ostentatious preconization of their good deeds, which are of unsolicited and spontaneous origin; and whilst others are idly preaching the duty of charity, they exemplify it in their daily converse with man. No one is better acquainted with the distresses of poverty and sickness than the physician; and no one, therefore, can more fully and deeply sympathize with the afflicted. What a bright galaxy of medical philanthropists does history exhibit to us! Of men who have conferred lasting and invaluable blessings on society; who have laboured through evil and through good report, for the benefit of their fellow-creatures! And do they not still labour in the same cause? Do they not pursue the same undeviating path of benevolence; gratuitously devoting their time and talents to the indigent sick? We will say nothing of what is privately wrought in this respect; let our public hospitals, our dispensaries, and asylums be consulted: let them speak. There are very few such institutions, in which those who have the care of the soul are not adequately remunerated for their trouble; whilst, universally, those who cure the body bestow their time and ability gratuitously. And yet, time is infinitely more precious to the latter than to the former.

"Again, no body of men in the community can boast of brighter ornaments to science than are to be found amongst the members of the profession of physic. Where shall we find more truly liberal and enlightened philosophers? Individuals, that have more effectually contributed to dissipate error and superstition, or more zealously promoted the general good of mankind? Where, in fine, shall we meet with men who have united higher cultivation of mind with a more truly virtuous nobility of character?

"We repeat, then, neither in mental nor moral attributes does the profession of physic yield to that of theology. Let no one imagine we are instituting an invidious comparison, with the intention of exalting the merits of one body in the community, by depreciating those of another. We have no [24/25] aim but that of evincing the general worth, industry, and acquirements of medical practitioners. We wish to show, that the study of anatomy does not exert any baneful influence on their characters; that it does not deprive them of the distinguishing sensibilities of humanity, and thus render them callous to the sufferings of their fellow-men. No! they pursue an honourable and dignified vocation, and are urged on in their career by the noble ambition of achieving the utmost possible good. It would be difficult, indeed, to point out in society, individuals of a more laborious, persevering, and indefatigable character. At all hours, at the table of repast, on the couch of repose, amid the inclemency of weather, the harass of an anxious mind, and the oppression of bodily fatigue, they must be ready to obey each capricious call! And yet how ill-treated and ill-requited! Patients rigorously exact an assiduous attention; whilst with all latitude which may suit their fancies, they will follow the advice of a medical attendant, yet immediately suspect the extent of his skill, should the amelioration demanded not ensue. But this is not all; they even seek at the hands of the law to obtain compensation for any supposed deficiency of skill, to the attainment of which, nevertheless, both themselves and the law are equally opposed!"[13]

They cease to suit their own convenience and to attend to their private affairs, as they are always ready to wait on suffering humanity; they allow nothing whatever to prevent them from attending to the sick, they are always ready; they bear with the injustice, the caprices, and ingratitude of men; they expose their lives in the most dangerous circumstances – when pestilence devastates the earth, in all times, in all places; they possess courage, exemplary patience, and an entire abnegation of self. Such are the virtues of medical practitioners.

They are, however, *amply rewarded* for all annoyances and ill-treatment. They have numberless opportunities of giving that relief to distress which is not to be purchased with the wealth of India. This, to a benevolent mind, is one of the [25/26] greatest pleasures. "Is there any thing in the world more estimable," asks Voltaire, "than a physician; who, having studied nature in his youth, knows the relations of the human body; the diseases that torment it, the remedies that may assuage it, exercises his art in defiance of it, takes equal care of the rich and the poor, who does not receive remuneration but with regret, and employs it to the succour of the indigent? Men," continues he, "who are occupied in affording health to other men, exercise the only principles of beneficence, are far above all the great ones of the earth, they partake of the divinity."

They are now duly appreciated by kings, nobles, poets, and literary and philosophical characters. All the virtues are displayed in the exercise of the functions of a physician – his ministry commands the respect of men, and the admiration of sages.

As there is no perfect code of ethics to guide the profession in this country, a succinct detail of the rules of conduct prescribed in the different works of authority may be given, and then the leading points can be fully discussed under separate heads. The following is an imperfect epitome of the ethics of the last century, collected from various sources. To this will be appended the ethical code of the present period.

Notes

1. *Author's note*: He flourished about 460 years before the present era.
2. *Author's note*: The Edinburgh University requires this oath on conferring the degree of M.D. and is the only university or college in this empire that binds its members by so serious and necessary an obligation.
3. *Author's note*: De Medico, de decenti Habitu aut decoro, Prenotiones.
4. *Author's note*: "Tanta est inter Deum, religionem, et medicum connexio, ut sine Deo et religione nullus exactus medicus esse queat" (*Macoppe op. cit. Editors' note:* See Footnote 1, Chapter 1).
5. *Editors' note*: "a divine ray of light derived from God alone extracted from a universal mind."
6. *Author's note*: See Mr. Roberton's Strictures on Dr. Elliotson's *Physiology. – Medical Gazette*, November, 1835.
7. *Author's note*: Bishop Watson's Apology for the Bible.
8. *Author's note*: O'Meara's Voice from St. Helena.
9. *Author's note*: Lives of the Poets, Garth.
10. *Author's note*: Remarks on the Statement of Dr. Charles Coombe, pp. 82, 83.

11. *Author's note:* Letters.

12. *Author's note:* Address delivered on Professional Reputation, before the Philadelphia Medical Society, 1826, by John D. Godman, M.D., Professor of Anatomy and Physiology in Rutger's Medical College.

13. *Author's note*: Dr. Corden Thompson's Letter on the Necessity of Anatomical Pursuits, 1830.

Chapter III
Medical Ethics of the Middle Ages[1]

First of all things, a medical man ought to exercise piety, and give due honour to the Supreme Being. Next, he ought to render to every one his due; obedience to his superiors, concord to his equals, and equity to his inferiors. He ought to preserve a clean heart, and silent tongue, and cultivate every virtue. The whole praise of virtue consists in action. He is to avoid anger, and suppress all its perturbations, intemperance and insolence, having always before his eyes the great deformity of mind produced in those who give way to them, and the amiableness and gracefulness of those who avoid them. Sensuality, intemperance, and dissipation, produce concupiscence and carnal gratification, which increase rapidly, and would eventually ruin a medical practitioner. These are to be strenuously avoided, as well as every luxury. Continence consists in moderating pleasure; gluttony, debauchery, and ignominy, in abusing it. An incontinent, or an intemperate man never rose to eminence, and is completely unfit for medical practice. Men of loose and dissolute habits, and of excursive amours, debase themselves to the rank of the brute creation, and render the mind stupid and inert, and totally unfit for the pursuits of science. Such profligate and abandoned characters cannot be found in the history of the medical profession – in truth, men so vitiated could not long pursue the practice of medicine. What man would commit the care of his wife, daughters, or female relatives to a medical practitioner, if such could be found, of so debased and brutal character – to a man burning with desire of violating the conjugal and vestal honours of his neighbour's family. Hence the necessity of practising chastely and honourably; and hence the preference which is given to those members of the profession who have entered into the sacred bonds of matrimony, especially in obstetric practice. Every one is [27/28] bound to support his own and the professional dignity, with noble sentiments, probity, and humanity. Sadness and fear depress the mind and body, and unman the practitioner. Fortitude is opposed to sadness and fear, and is often necessary to enable us to bear patiently the calumny and contumely, to which no class of men are more exposed than the professors of medicine. All these passions should be expunged from the medical character, and an ardent desire of fame and glory be substituted in their place. Avarice, pride, and envy, are evils which must be carefully avoided. Avarice was considered the chief of all improbity by the ancients; and it is highly cruel in medical men, when it precludes aid from the sick. But those do not consider its cruelty whose sole object is the accumulation of riches. Well might the poet exclaim

Quid non mortalia pectora cogis
    Auri sacra fames?[2]

The sick should never want aid on account of pecuniary consideration; and the practitioner ought to be satisfied according to the affluence of the patient. Above all things, pride is to be avoided. It is odious in the sight of God and man, as it excites an inordinate desire of excellence, and induces one to think, that he enjoys from himself all the gifts of nature, talent, intellect, memory, power, and science, which are bestowed on him by the Deity: he despises others, and thinks they are to submit to him, although his superiors; and hence follows his insatiable desire of praise, fame, honour, glory, and reverence, which is but vain glory. Physicians ought never to be guilty of such an error, nor of presumption, ambition, nor curiosity. On the contrary, let them display humility without sordidness. Envy at the prosperity or success of another, is an evil which ought not to be named among the profession. An envious man is pusillanimous, of a narrow mind and abject talent, for he shows by envying others, that he is inferior to them; he envies what he does not possess, but vehemently desires. Envy is a compound of hatred, dissimulation, avarice, pusilla-nimity, mendacity, and ambition; and is opposed to friendship, liberality, truth, mag-[28/29] nanimity, and prudence. Medical men must avoid this most pernicious evil. All ought to enjoy fortune happily, and no man should be sad at the prosperity of another. Having thus pointed out some of the vices which are to be avoided by medical men, it is right to enumerate the virtues which they ought to cultivate. These are prudence, circumspection, foresight, caution, perspicacity, continence, sobriety, mildness, modesty, taciturnity, veracity, gravity, magnanimity, liberality, and honesty – friendship towards acquaintances, affability and civility to strangers, and decorum according to age, sex, and condition. The contrary vices are impru-dence, stupidity, precipitation, enmity, cunning, curiosity, and all the excesses of the will and desire, as irascibility, concupiscence, which are to be avoided by medi-cal practitioners as the most mortal pestilence. About which let them consult the writers on morality, especially Plato, Aristotle, Plutarch, and Seneca; and, above all, let them peruse the Sacred Volume, which will lead to virtue and prudence, from which they will learn how to be good men and prudent practitioners, and to think piously. The conversation is to be moderate and veracious; the morals must be grave, benign, and cheerful; the diet temperate and frugal; the apparel respect-able and professional; witticisms to be free but few, and to these may be added the ancient precept –

Mens humilis, studium quærendi, vita quieta,
    Scrutinium tacitum, paupertas, terra aliena.[3]

A medical man should be affable and hospitable; friendly to his relations and neighbours: polite to all without any moroseness; he is to relieve the sadness of the suffering patient with placid and mild discourse. Let him not be peevish, but ingenuous, affable, familiar, and enforce his authority without any disdainful gesture. Let nothing fictitious, nothing simulated be in him, nothing low or base, but let his mind soar with sublimity above all the vicissitudes of fortune. Let the studies of his life be the meditation on the delight and riches of science, virtue, and honour. Let him shun litigation and vain popular applause: he ought candidly to praise the good, always avoid detraction, tolerate the bad, [29/30] indulge the inferior, agree with his equals, and obey his superiors; he is to injure no one; he is to live in the greatest harmony with his family, for this tends to support dignity, and a good reputation; he is to live on the best terms with all, and endeavour to obtain universal esteem. Let him maintain his opinions with modesty and eloquence, and always with veracity, but without obstinacy. He is not to be proud or haughty, but cure rich and poor, slaves and free, of whatever nation, for medicine is the same to all. He must be careful to observe, that his remedies and directions shall be faithfully exhibited and attended to. *He is to esteem as hidden mysteries whatever is said or done in the house of his patient,* and thus he and his art will acquire more praise. By the observance of these institutes a medical practitioner will obtain a distinguished place among the wise and good.

The display of a diploma is to be left to jugglers.[4] The greatest cleanliness is necessary; and particular attention to the hand and fingers is requisite. In visiting the sick, let the countenance be meditative, not melancholy or peevish, which is odious to all. Risibility and hilarity are deemed intolerable, from whence the axiom, "medicus garrulus ægrotanti alter morbus."[5] His whole gesture is not to be so humble as to excite contempt, nor so proud and arrogant as to excite hatred for him. No perfumery is to be used, but if there be any unpleasant odour in the breath, it is to be corrected, lest the patient have reason to exclaim, "cure thyself." The visits to the sick should not be too frequent or too rare, twice a day, except in extreme cases, is quite sufficient. Dr. Gregory maintains an opposite opinion, but reason and common sense are against him. Frequent visits are deemed troublesome by some, while they are esteemed angelic visitations by others. They are to be regulated according to the wish of the patient, or danger of his disease. By gravity and affability we can best learn the nature of disease.

In visiting the sick we should recollect a proper approach, [30/31] authority, silence, and answers to all inquiries, promptitude in prescribing, that nothing be done precipitately. It is proper to enter the chamber calmly, and not boisterously, not with petulance, noise, garrulity, nor an elevated voice, so that in approach, aspect, or any other manner, there can be nothing indecorous. In conversation with the sick, our questions must be grave, plain, and intelligible, without solecisms, pedantry, harshness, not foolishly advising them to be of good hope; neither are we to condemn the former errors of life too severely, which might lead the patient to suppose his case hopeless. The sick are generally pusillanimous and suspicious; and many medical men are so imprudent and loquacious, that they express their opinion openly, rendering the patients melancholy and timorous, who fear that which is related is their own condition. We should be extremely cautious in our expressions, and duly consider what is proper to be said to the sick and the attendants; and enforce with mildness and confidence the necessity of obeying all directions and injunctions, and of faithfully administering the remedies prescribed. In this way the sick can never be alarmed, and we have the best chance of success from the due exhibition of medicines. Some men are harsh and overbearing to the sick, and others too accommodating and flattering; both err exceedingly. Humanity, moderation, and suavity are indispensable, because, unless the patient looks on his physician almost in the light of a deity, he will never obey his precepts. Let him therefore be careful of his appearance, voice, manners, and actions, if he wish compliance with his precepts. These requisites captivate society in every rank and condition.

On entering the patient's chamber, a medical man is recommended to fix his eyes on the floor, or prudently on the countenance of the sufferer, to salute the attendants and patient politely; and then he is patiently to inquire of all, the age, temperament, habit, constitution, symptoms, and causes of the present illness; and he is also to recollect the season and state of the weather. All these things were comprehended in a single distich by the ancients—

> Ars, ætas, forma, complexio, virtus,
>   Mos et symptoma, repletio, tempus et usus.

[31/32] After he shall have patiently heard the complaints and narration of the patient, and however tedious, of the attendants, he should consider the state of the apartment, the ventilation, warmth, moisture, and so on. He is next to learn the order of symptoms, and examine the state of the cerebral, respiratory, circulatory, and assimilatory functions. The countenance, tongue, respiration, and pulse, are to be examined; the condition of the bowels and urinary secretion ascertained; and the state of the uterine evacuation in women. He is to inspect all the egesta, expectoration, alvine, uterine, and urinary discharges. If these things be well observed, you will explain the feelings and state of the patient better than he can do himself; and, therefore, he will admire and proclaim your skill and excellence, and will repose more confidence in your opinion. The confidence of the patient is often a more certain cure, than can be accomplished by the practitioner and his medicines. In prognosticating the event to attendants and friends of the sick, we must duly consider his former and present condition; and only predict those things which we know must eventually happen. Thus, they will perceive, that you are not the cause of death, should it occur, but of health, should it be restored. It is from prognosis the world extols a physician, and celebrates his name with utility and honour.

In prescribing remedies, it is necessary to inquire the form which the patient prefers. In this, as in all other desires which are not injurious, he is to be gratified; and all medicines are to be made as palatable as possible. In some cases, the parents must persuade the child, and the child the parents, or one friend another, to take certain remedies. There is no objection to this, on the contrary, we must avail ourselves of it as an auxiliary, or life may be sacrificed. In many diseases the sick must be kept tranquil, and all visitors excluded. In this injunction we must be peremptory, and no one can be so unreasonable as to require admittance, at the risk of destroying the life of the patient. Some patients are exceedingly sad and melancholy, and these must be preserved from the visits of all who would further afflict them. The intemperate zeal of the clergy, and even when there is no necessity for their interference, is often fatal to the patient. Every remedy sanc- [32/33] tioned by the major part of the faculty ought to be faithfully employed. Too much physic is highly injurious to the constitution; and disease can often be cured as well without, as with medicines.

Those whom nature cures, are injured by medicines. Administer nothing timorously or rashly, and never change medicine without having given it a fair trial. A man who is constantly changing his remedies will not possess the confidence of his patient, because he affords strong evidence that he is ignorant of the means of cure.

Cleanliness, ventilation, and regulation of temperature in the sick chamber, are as valuable as medicines. Should the disease be incurable, and prove fatal, the medical attendant may pay a visit of condolence to the relatives which always looked on as a proof of his sincerity and friendship. In some cases he may opportunely mention patients whom he had cured of a disease similar to that under treatment; if these were the friends or acquaintances of the sufferer, so much the better; he serves the sick by inspiring confidence, and also increases his own reputation. If spoken to about compensation for his attendance, he should declare that the recovery of the patient was his chief concern; but in many cases he must stipulate for his fees, or he will be cheated. This may be done through the friends and attendants, but not with the patient when dangerously ill, unless required by himself. A medical man should be affable to all, but not too familiar with any, especially with the illiterate; and his conversation should be always dignified and reserved.

A physician should be always ambiguous in his prognosis, unless the most certain and infallible indications of health, or death be manifest. "Ambigue futura pande," says Macoppe, "predict the future ambiguously." The sick, their relations, domestics, and neighbours, are too desirous of knowing the result of disease. They wish us to be oracles and demi-gods instead of men. He who expects inheritance or title; he who loves a father, mother, wife, child, friend, or acquaintance; and he who hates an enemy, are all anxious to learn our opinion on the issue of disease. It is extremely difficult to form an accurate prognosis; and in a vast majority of cases it must be doubtful. Professional fame is often [33/34] injured by a prognosis. When it is doubtful, if the patient perishes, it is the fault of the disease and not of the practitioner; if he recovers, a new Esculapius is lauded to the skies. If, on the contrary, the patient dies after his recovery had been predicted, the medical attendant will be censured, and it will be said, he mistook the nature of the disease and caused death. He is to be moderate in his promises, and always rather pronounce a hope of recovery, than certain death. If a patient is deserted, and afterwards recovers by nature or chance, which often happens, the practitioner will incur much blame. But if he had afforded a hope of health, and death occurs, the ignominy is not so great, because many errors and excesses may take place, or even a new disease; and the commutation is much easier from health to death, than from death to health, which, in the course of nature, is almost impossible. A medical man should not interfere in recommending the patient to make a will, it is the strongest proof of despair, and a matter with which he has no concern. When he sees it necessary, he ought to apprize the friends or attendants, and let them advise the patient to arrange his affairs, in order to prevent disputes and litigation among his heirs. Even this requires great caution, for most probably other aid will be had, and should the patient ultimately recover, it will be to the injury of the first attendant. The friends, relatives, or clergy, are the most proper persons to advise the patient to arrange his affairs; but if the medical attendant be present, he, with a composed countenance, mixed with sincere hope, is to console the patient, and tell him, that the making of a will cannot affect his recovery, for many have done so, who have recovered and lived for many years, or are now in existence. When the patient is labouring under a disease that may speedily suspend the mental faculties, or endanger life, it is right to apprize his relatives or attendants, and advise them to propose sending for a clergyman, to afford the consolations of religion. In Catholic countries the law is, that the medical attendants should recommend a clergyman at the fourth visit, when the disease is acute. I feel convinced, that I have known some individuals whose lives were destroyed by the premature recommendation to see a clergyman; [34/35] and others who shared a similar fate when advised to settle their affairs by making their last will and testament. Others have fallen victims in consequence of the indiscreet zeal of the clergyman, more especially when he professes, as is sometimes the case, without any real pretension, a knowledge of medicine. Again, a decided opinion delivered by a medical practitioner, in whom great confidence is placed, may turn the balance in favour of life or death. The greatest caution is therefore necessary in forming a prognosis. Health or recovery ought never to be promised, lest it offend the Deity, who alone can decide the fate of the patient: if the healthy and robust are uncertain of tomorrow, how much more are the diseased and infirm. In making promises, many natural and auxiliary causes are to be considered, the compliance and obedience of the sick, the diligent application of remedies, and the sudden and unaccountable changes of diseases. We may state, there is or is not danger at present, that there are many changes in diseases, and that the case is in the hands of Divine Providence. Lastly; that every thing which art affords shall be carefully tried. How powerful the first aphorism of Hippocrates, "judicium difficile;" and also that of Macoppe, "ambigue futura pande."

Plato, Hippocrates, and Galen, maintained, that "medical men were justified in deceiving the sick for the cause of health." They held statesmen and physicians excusable for this breach of ethics, but no other class of society. The Christian religion, however, is opposed to mendacity in any circumstance. To the physician deception is however almost necessary, because his patients are often timid, suspicious, and observant of his words and aspect, and drawing unfavourable conclusions from his very appearance. David, in order to escape from king Lachish, changed his dress, and simulated insanity. It is a fact, that a tone, a word, a look, will destroy life in delicate and dangerous cases. How often must we disguise medicine, in order to conquer the aversion and idiosyncrasy of the sick. But mendacity is an evil by which we misrepresent the order of the Creator and of nature. We express not that which we think. We are not to do evil, that good may follow. Nevertheless, Abraham said he was his [35/36] wife's brother; Jacob said to his father, he was Esau, his eldest son; and the midwives of Egypt said they could not kill the Hebrew infants, a murderous injunction, for the breach of which they were justly commended. Official mendacity is less pernicious than that which is malicious. We must never promise recovery, without premising that it depends on the Divine will, on the due performance of all medical precepts, both by the patient and the ordinary attendants; and even then we must pronounce our prognosis with the greatest caution. The learned physician will bear in mind the great influence of change of weather on the human body; and the many changes a disease of the most favourable aspect may undergo. Presumption must be avoided. The presumptuous practitioner knows every thing, has read all works, and seen all diseases; the most difficult cases do not alarm him, the most delicate operations are undertaken by him with alacrity; in fact, nothing embarrasses him. He speaks of himself in flattering terms, he knows more of the profession than the whole body, he can cure consumption, cancer, hydrophobia; the first aphorism of Hippocrates does not apply to him; in fine, he believes he possesses the genius and power of Esculapius. Such is the presumptuous practitioner, though he is, in truth, very superficially informed, and generally an ignorant individual. He is an imitator of the uneducated and daring empiric, who professes to cure every one, and proclaims "all diseases incurable to the faculty cured here." No well-informed practitioner can be accused of such ridiculous vanity; he duly appreciates the difficulties and dangers in practice, the mutations and incurability of many diseases, and he is cautious in his prognosis.

Timidity, on the contrary, is a bad feature in the character of a medical practitioner. Some are frightened at every case that comes under their care, they have not courage to employ active or powerful remedies, they do not kill their patients, but allow them to die. They accuse medicine of the unfavourable issue, and not their own ignorance and culpable neglect, in not applying proper means of treatment.

There are others who may be considered fanatics in medicine, the partizans of this or that doctrine. In vain have [36/37] new discoveries changed the face of science, they are and remain incredulous. Such are the disciples of Abernethy, of Broussais, and of Hanhemann. They forget that diseases are as numerous as the sands of the ocean; that they depend upon a variety of morbid changes, and that they are successfully treated by the most opposite remedies. Thus the inflammations and fevers are cured by antiphlogistic measures, and sometimes by powerful stimulants; while the painful, spasmodic and nervous affections are best relieved by sedatives, tonics, and improvement of the general health.

A few words may be said on the duties of the sick, and their ordinary attendants during the cure of disease.

Hippocrates laid down, in his first aphorism, that the physician should not only excel in the due exercise of the best of his skill, but also direct his attention to the conduct of the patient and his attendants, as well as to all extraneous affairs. He said, the medical art consisted in three things, in disease, in the sick, and in the physician, the minister of nature and art. We cannot cure disease, unless the patient assists us. If he refuses remedies, the disease must in general remain, or prove fatal. Hence, the mortality of violent maniacal patients. The assistants are the minor order of medical men and domestics; and the physician must be acquainted with all their duties. Thus, Galen was of opinion, that he must be conversant with pharmacy, venesection, obstetricy, culinary affairs, and even menial servitude. Thus qualified, he can discover artifice, amend, and avoid it. The ordinary attendants should be chosen from those who are acquainted with the patient, and are most likely to pay implicit obedience to the medical directions, and who are accustomed to wait on the sick. They are to report all the symptoms at each visit, and preserve the apartment in the best order. They are to amuse the sick with cheerful conversation, and never indulge in frightful or ominous narrations; they are to preserve all the egesta, sputa, alvine, and urinary discharges, and whatever is rejected from the stomach, or is passed by hæmorrhage. Of all things they must avoid quackery, and never administer either diet or medicine, unless that which has been prescribed. They must not indulge in the use of ardent or inebriating liquors; they [37/38] must avoid informing the patient of any thing that may distress him; they must not exhibit any of their own food or drink to the sick; and, in order to compel them to comply with their duties, a faithful servant ought to be placed as a watch over them. They should not divulge any thing they may see or hear, but adopt the old adage – "Hear and see, and say nothing." Let the young physician duly consider, how much the event of the disease depends on the proper performance of the duties of nurses; and he will be very cautious in his prognosis.

This precept is to be chiefly observed by a medical man; never to visit a patient unless requested to do so. Should he volunteer his services, he renders himself liable to be suspected of ambition or of avarice. When one goes unsolicited, he may be certain he will not have the confidence of the sick, and will be deemed indigent or necessitous; and he cannot expect any reward for his spontaneous visitation. If known to the sick, he will be despised for his intrusion; if unknown, his conduct will be suspected. A visit unsolicited is not only useless and indecent, but suspicious. Another medical practitioner may be already in attendance, and will view any intrusion as highly unprofessional. The only exception to this conduct, is when we visit a friend; but even then we are to come as friends, and not in a professional capacity. Again, a patient might be alarmed at seeing a strange practitioner, having one in whom he reposes all confidence already in attendance. It is also derogatory to the dignity of the profession to be called in, through the recommendation or persuasion of nurses and others. Attendance even on friends must not be given, unless regularly requested. It is also improper to visit a patient at the request of a relative, unless the person affected, if rational, concurs in the request.

There are some avaricious and eccentric men who refuse medical aid, because it is expensive, but their friends solicit attendance in the usual way. It may be given, provided due compensation be made to the practitioner, but not otherwise. It is monstrously improper for eminent professional men to give advice gratuitously to the affluent at hospitals; it is a direct injury to the rising members of the profession. [38/39].

If a relation or contemporary who has quarrelled with a medical man be ill, ought we to visit him? Most certainly, even if he has called in another. A good and generous man will conquer evil with good, not considering the deserts of the patient, but what is right and meet for himself, and to promote the honour and glory of the Deity.

A patient is to be visited once a day in chronic cases, twice in acute, and oftener in the most acute. After a lapse of 40 days, two visits in the week may be sufficient; but in all cases they must be regulated by the patient or his friends. In acute and dangerous cases the friends are to be informed of the perilous condition of the sufferer, and should they not speak of the next visit, they are to be apprized of its necessity, however painful, to the practitioner. After convalescence attendance may be occasionally given, and a gratuity is not to be refused; but must not be accepted, if the patient declare his restoration to health. In such cases some complimentary visits ought to be made. Medical aid is to be afforded to every human being, friend or foe, native or stranger. A physician should not undertake the cure of himself, his wife, or his children in dangerous diseases, as his mind and reason will be perturbed, and unfit him for the arduous and difficult treatment of violent cases. A lawyer who pleads his own cause is said "to have a fool for his client."

Some men suppose if they once pay a physician he is to attend them through a succession of diseases; but who will be imposed on in this manner? Others who pretend friendship, will not call a practitioner, but should he visit them, will enjoy his labour without recompence. This is very often the conduct of relations, and of many towards young practitioners; in such cases attendance ought to be withheld, unless in the regular manner. What other class of society will act without reward? Will the clergy, or lawyers, or military, or kings, or any other class of society act so generously? Surely the labourer is worthy of his wages.

A patient is not to be deserted in the most hopeless condition, for recovery may still happen. Let the practitioner ever recollect the adage; "dum anima est, spes est," while there is life, there is hope. With respect to consultations, [39/40] those who form them should be regularly qualified, and only look forward to the welfare of the patient, and in preference to fame and lucre. The sick have a right to select those they please. No medical practitioner is justified in refusing to meet another in consultation, if both be duly educated and legally qualified, whether in the same university or another. The patient is not to suffer by their disputes, with which he can have little or no concern; the great object of a consultation is the cure or relief of the patient, and not disputation. Consultations are highly necessary in dangerous cases; and no man should ever object to them. A remedy may occur to one, which has escaped the recollection of another, "Et quod tu nescis fortassis novit Ofellus." The physician called in should be older and more eminent in the art than the former attendant. It is the duty of the attendant to communicate the history and treatment of the case, that both may consult what may be added to the cure. If a new remedy be proposed, it is to be tried if either of the attendants would use it himself, if afflicted in a similar manner to the patient. A few words may be said on the manner in which a medical man ought to behave on receiving compensation.

It is unnecessary to enumerate the arguments in proof of the justice of rewarding medical practitioners for their labours. An art that is not purchased, is disesteemed, therefore the correctness of the axiom,

Exige dum dolor est, nam postquam pœna recessit,
    Audebit sanus dicere, multa dedi.[6]

And of another,

Accipe, dum dolet.[7]

What will not a man give, to one, who liberates him from the greatest danger? Or is there any one so insane, as to spare his riches, when life is in danger? All that a man possesses, says Job, he will give for his life. Thus Philip, the father of

Alexander, said to his physician, after he had reduced his dislocated clavicle, "Accipe omnia quae voles, [40/41] quando quidem clavem habes."[8] Many similar examples might be quoted; for,

> Medicis in morbis totus promittitur orbis,
>   Mox fugit a mente medicus, morbo recedente: [9]

A physician was said to possess three casts of countenance. When he converses with the healthy, he displays his ordinary one; when he approaches the sick, labouring under acute and painful disease, he is said to display his angelic one; and when he visits his patient after cure, and seeks compensation, then his aspect is satanic,

> Dum præmia poscit medicus, Satan est. [10]

As health is above all temporal blessings, no one can give an adequate reward for its recovery. Nor has a medical man ever received sufficient compensation, for the labours and troubles which he experiences for the calamities and miseries of others, which he makes his own. What can compensate him for the continued anxiety, inconvenience, loss of rest, deprivation of ordinary pleasures and comforts, which he always experiences? The members of the church and law have time for recreation and amusement; they are not always employed, but the medical man must be ever at his post, his motto is "semper paratus;"[11] and it was for this reason, Dr. Johnson defined the duties of a medical man thus, "a truly melancholy attendance on misery, a mean submission to peevishness, and a continual interruption to rest and pleasure."[12] Soranus said, "if rewards be given, let them be accepted and not refused if they be not given, let them not be required; because however much any one can give, is inadequate value for the benefits conferred by medicine." The remuneration of medical practitioners ought to be regulated according to custom and to the rules of the profession. It ought, however, to be proportionate to the circumstances or pecuniary means of the sick, and to the standing and eminence of the practitioner. The fee of a guinea a visit is much too high for a [41/42] preponderating majority of society in most countries, and it has led even the most eminent physicians and surgeons in London, to pay two visits to those in the middle ranks in life for one *honorarium* or fee. It is unreasonable to expect, that a junior will be remunerated on the same scale as a senior; nor is this done in the church, at the bar, nor in the army or navy. The custom in France, Germany, Italy, and most foreign countries, of regulating fees according to the standing of the practitioner and means of the sick, is, in my opinion, much wiser and more beneficial to the junior members of the profession, than our system of guinea fees. In the countries now mentioned, the fee of a junior practitioner, a doctor in medicine and surgery, is, at first, about a shilling a visit; but, as practice increases, the fee is gradually raised to a sum equal to a guinea. The system of fees in this country has originated general practitioners, or surgeon-apothecaries, who generally receive no fee, but are remunerated by their medicines. This system was declared to be objectionable, by the heads of that body before Mr. Warburton's Parliamentary Committee, 1834, and all were in favour of small fees, which will most probably be enforced by law. Many eminent physicians and surgeons in this metropolis are in the habit of visiting a patient in the middle ranks of life, as often as three times for one fee; and they also cheerfully receive half a sovereign for a consultation at their own habitations. This being a fact, what chance has a junior physician or surgeon, of competing with them in practice, if even he was satisfied with the same remuneration? For these and many other reasons, it is expedient and necessary to regulate and proportion the fees of medical practitioners, so as to protect the junior members and surgeon-apothecaries. This subject will come under full consideration in the section following the chapter on education, and in that relating to the laws of the medical profession.

It is contrary to medical morals to bargain for a certain fee before a cure is effected. This is the system of empirics who make a contract, "no cure no pay." But physicians and surgeons in extensive practice, will not go several miles distant, or perform capital operations, without specifying the exact remuneration they require. They do not act in the [42/43] above manner, nor violate the ethical rule – "de mercede ne pasciscaris. Hoc impostoribus et circulatoribus relinque."[13]

Many presents are given to medical practitioners in addition to fees, and sometimes a large amount of property both in money and estates. Others bestow books, carriages, &c. In former times, it was considered derogatory to the profession to accept gifts of this kind in place of fees. Many instances of ample and magnificent fees and gifts to medical practitioners in past and modern times might be quoted.

A question has been discussed in all ages, namely, whether a medical practitioner ought to be punished for bad practice? The most eminent philosophers and legislators of old, were unanimously of opinion, that the errors of a medical man, if involuntary, should be forgiven. The father of medicine maintained, the only punishment should be his own ignominy. Plato de Republica says, "quivis medicus, si is, qui ab eo curatur, moritur, invito ipso purus sit secundum legem."[14] This opinion obtains in some civilized countries.[15] Pliny, lib. 29. is of the same opinion – "Nulla lex est, quæ puniat inscitiam capitalem medicorum, nullum exemplum vindictæ, discunt periculis nostris, et experimenta per mortes agunt, medicoque tantum hominem occidisse summa impunitas est, quando in hac artium sola eveniat, ut unicuique medicum se profitenti statim credatur."[16] Barbarians held an opposite opinion, and punished medical men with death, if they had failed in curing the sick. Manes promised to cure the son of the king of Persia, and having failed, he was ordered to be flayed, which sentence was executed. Zerbus having failed to cure Bassa, the emperor of the Turks, was instantly sabred by the soldiers. The experienced and talented Avenzoar, was thrown into chains for a similar failure.[17] But I must proceed to consider the other duties of medical practitioners. [43/44]

Notes

1. Editors' note: Only in the next chapter, p. 46 in the original, does Ryan acknowledge that all the ethical reflections in this chapter "were compiled from a work by Roderic a Castro, entitled 'Medicus-Politicus sive de officiis Medico-Politicis tractatus, quatuor distinctus libris: in quibus non solum bonorum medicorum mores ac virtutes exprimuntur, malorum vero fraudes et impostures deteguntur: verum etiam pleraque alia circa novum hoc argumentum utilia atque

jucunda exactissime proponuntur. (Hamburg. anno 1614.)'" As noted in the Introduction, a full translation of this title might be rendered as: "The State Physician: Treatise concerning the office of State Physician in four separate books in which not only are good, medical mores and virtues described, but even deceptions of evil and deceit are revealed: truly also a new argument is most precisely proposed by another method both useful and agreeable. The full work being fit for a physician, student of illness, attendant upon the ill, and by all others of letters, and also by supporters of state education."

2. Editors' note: "Why, sacred longing for gold, do you not constrain the hearts of mortals?" (Virgil, Aeneid, 3.57–8).

3. Editors' note: "A humble mind, zeal for learning, a quiet life, silent investigation, poverty, a foreign land" (From Bernard of Chartres, Policraticus).

4. Author's note: It is a common custom in England to frame and glaze diplomas.

5. Editors' note: "a garrulous doctor causes another sickness."

6. Editors' note: "Collect while the pain still persists, for after it recedes, he will dare to say, once healthy, 'I suffered much.' "

7. Editors' note: "Accept [the offered fee] while [the patient] still suffers [from the throes of pain]."

8. Editors' note: "Take anything you wish, once you have indeed reduced the clavicle."

9. Editors' note: "In sickness the patient will promise the whole universe to the physician; but the physician soon slips from the mind when the illness subsides."

10. Editors' note: "When the physician demands recompense, he is Satan."

11. Editors' note: "Always ready."

12. Author's note: Thus the axiom, Medicè vivere est miserè vivere.

13. Editors' note: "Concerning price, do not grow fat on high fees. Leave such practices to imposters and peddlers."

14. Editors' note: This is a paraphrase rather than an actual quotation from Plato's Republic: "I argue that any physician is blameless according to the law if he attempts to cure someone who later dies."

15. Author's note: Not in Great Britain and Ireland, a civil, and even a criminal action, may be instituted for mala praxis, if death ensues.

16. Editors' note: "There is no law in existence whereby to punish the ignorance of physicians, no instance before us of capital punishment inflicted. It is at the expense of our perils that they learn, and they experimentalize by putting us to death, a physician being the only person that can kill another with sovereign impunity. To all this, however, we tend to pay no attention, [due to the seductive hope of recovery each person feels when ill]." This is a rather fractured, out-of-order quotation from Pliny the Elder, Natural History, Chap. 8, trans. John Bostock and H.T. Riley.

17. Author's note: This horrible and barbarous law is still in force in Turkey.

Chapter IV
Ethics of the Present Age

I will never set politics against ethics, for true ethics are but as a handmaid to divinity and religion.

– Bacon

It is not easy to conceive the reason why the cultivation of ethics, a matter of primary importance to the success of medical practitioners, in the commencement of their career, should be almost totally neglected in the medical schools of an age so enlightened as the present. The fact is so, however incomprehensible it may appear. It is now the custom to initiate men into the mysteries of medicine, without the slightest allusion to the duties they owe each other or the public; or to the difficulties to be encountered on the commencement of their practice. Hence arise the frequent misunderstandings, disputes, and improper behaviour between medical practitioners, which are so disreputable and injurious to the dignity and interests of science. From some cause which remains to be explained, the majority of medical professors have excluded the discussion of ethics from their instructions; the faculties of physic and surgery have acted in like manner, so that there is no code of ethical institutes to be referred to, in the daily violations of those high moral principles, which have always characterized the true cultivators of medicine. The moral statutes and obligations which are required by some of our colleges, are so few, and so little known, that they are nearly useless; they are seldom observed, obeyed, or enforced. Indeed, the only works we have on the subject are those of Dr. John Gregory and Dr. Percival; but these are not deemed authority, nor are they perused by medical students. The former was published above half a century since; the latter previous to the changes made in the constitution of the profession by recent legislation – and both unsuited to the state of the profession at the present time. [44/45] There is, therefore, a fair field for further observations upon the subject.

In the subsequent remarks it will be necessary to describe the present state of the profession, and necessarily comprehend the condition of every class of practitioners. To execute this task in a satisfactory manner to all parties, is not to be contemplated, as the conduct of all richly deserves animadversion and censure. Truth, justice, impartiality, and an ardent desire to promote the dignity of my profession, shall guide me in the execution of my subject; and I am confident that I cannot be accused of partiality towards any denomination of the profession. My motto is, "amicus Socrates, amicus Plato, sed magis amica veritas." It is an old adage, that "it is impossible to please all parties;" neither shall I attempt it, nor endeavour to please one party more than another. Here I follow the steps of Dr. Gregory, who fearlessly avowed his determination to expose abuses, whatever might be the animadversions of his contemporaries. "Whatever opposition," says the editor of the Duties of a Physician, in 1770,

> this part of the work may meet with, from those who find their own foibles, or rather vices, censured with a just severity, the ingenuous part of mankind, however, will not fail in bestowing that degree of applause so justly due to its merit. At present, there seems to be a general disposition in mankind to expose to their deserved contempt *those quackish, low, and illiberal artifices,* which have too long disgraced the profession of medicine. It is therefore hoped, that the general spirit will have a remarkable tendency to promote this laudable end; and that it will excite men of influence and of abilities to exert themselves in crushing that arrogance which has frequently served to cover the ignorance of many practitioners of medicine, by means of which alone they acquire such a share of practice as they are by no means entitled to.

The same defects and abuses on which Dr. Gregory animadverted still remain uncorrected, and consequently are deserving of further exposure; in fact, they are too glaring to be defended, except by those whose personal interests render them insensible to the important advantages of reformation and improvement. A faithful picture of modern medical [45/46] ethics is certainly much wanted; a comparison of what the profession is and was, may be entertaining and instructive to a large majority of my readers.

The only works on medical ethics which can be cited, are those of Drs. Gregory and Percival; and to these will be added the oaths required by one or two of our colleges, and the moral statutes sanctioned by them. These must be laid under free contribution. The first entitled, "Lectures on the Duties and Qualifications of a Physician," has justly received the universal approbation of the profession for more than half a century; and is an excellent abridgement of the maxims of preceding writers, and on many occasions their language is quoted *ipsissimis verbis.* In proof of this assertion, it is only necessary to compare the author's observations with the ethics in the last article, which were compiled from a work by Roderic a Castro, entitled "Medicus-Politicus sive de officiis Medico-Politicis tractatus, quatuor distinctus libris: in quibus non solum bonorum medicorum mores ac virtutes exprimuntur, malorum vero fraudes et imposturæ deteguntur: verum etiam pleraque alia circa novum hoc argumentum utilia atque jucunda exactissime proponuntur. Hamburg. anno [...]1614."[1] The comments and original views of Dr. G. are, however, numerous and valuable, and will be esteemed so long as medicine will be cultivated. As his work is almost obsolete, I trust that an analysis of its contents may be presented in a form which cannot fail to meet the eye of the industrious student. The author's remarks on the dignity and importance of medicine, and on the genius and education required for its proper cultivation, need not be inserted, as these points have been already discussed in a former section. The extracts I shall give do not obviate the necessity of referring to the original work – a production that ought to be in the hands of every practitioner.

Dr. Percival's valuable and truly classic work was re-published in 1803, and completed what Dr. Gregory had omitted. It consisted of the following chapters:

(I) On Professional Conduct in Hospital Practice; (II) In Private Practice; (III) In relation to Apothecaries; and (IV) In Cases Which Require a Knowledge of the Laws; to which were added, a Discourse on Hospital Duties, being the substance of a Sermon preached [46/47] by the Rev. T. B. Percival, and some valuable notes and illustrations by the author himself. It was his intention to have treated of the powers, privileges, honours, and emoluments of the faculty, an object he did not accomplish. The work was arranged in 1792, as a code of institutes and precepts for the professional conduct of the physicians and surgeons of the Royal Manchester Infirmary, and afterwards extended into a system of Medical Ethics, "in which," says his biographer, "he has drawn a portrait of himself, by tracing, with his own hand, what sort of character a physician ought to be." The work consists of several aphorisms, on which very inappropriate comments have been made by the anonymous editor of the last edition. Of the original opinions of Dr. Percival I shall avail myself; and therefore cannot be considered to interfere with the copyright of the last and worst edition of his work.

In citing the opinions of Dr. Percival and others, I claim the usual privilege of all writers, to collect and arrange materials from all available sources, which are public property. This privilege is sanctioned by the laws of the country. In a word, I shall endeavour to prove that the most eminent members of the profession are the strongest advocates of a branch of education which has been most preposterously overlooked and neglected. Want of leisure precludes me from making sufficient research for the compilation of a complete system of ethics; but I trust I shall be able to accumulate a mass of facts, which will perfect the subject they so ably and laudably commenced.

The following is a condensed detail of Dr. Gregory's opinions on the duties of medical practitioners:

Physicians, considered as a body of men who live by medicine as a profession, have an interest separate and distinct from the honour of the science. In pursuit of this interest, some have acted with candour, with honour, with the ingenuous and liberal manners of gentlemen. Conscious of their own worth, they disdained every artifice, and depended for success on their real merit. But such men are not the most numerous in any profession. Some impelled by necessity, some stimulated by vanity, and others anxious to conceal [47/48] ignorance, have had recourse to various mean and unworthy arts to raise their importance among the ignorant, who are always the most numerous part of mankind. Some of these arts have been an affectation of mystery in all their writings and conversations relating to their profession; an affectation of knowledge, inscrutable to all, except the adepts of the science; an air of perfect confidence in their own skill and abilities; and a demeanour solemn, contemptuous, and highly expressive of self-sufficiency. These arts, however well they might succeed with the rest of mankind, could not escape the censure of the more judicious, nor elude the ridicule of men of wit and humour. The stage, in particular, has used freedom with the professors of the salutary art; but it is evident, that most of the satire is levelled against the particular notions or manners of individuals, and not against the science itself.

The practice of the healing art affords a vast field for the exercise of humanity. A physician has numberless opportunities of giving that relief to distress, which is not to be purchased with the wealth of India. This, to a benevolent mind, must be one of the greatest pleasures. But, besides the good which a physician has it often in his power to do, in consequence of skill in his profession, there are many occasions that call for his assistance as a man; – as one who feels for the misfortunes of his fellow-creatures. In this respect he has many opportunities of displaying patience, good nature, generosity, compassion, and all the gentler virtues that do honour to human nature. A physician endowed with this virtue diffuses consolation and comfort; he employs his talents, his time, and his fortune, in removing misery. He is the friend of the poor and unfortunate. The victims of misery, those of disease and of death, offer a distressing, a shocking picture. Here is an opportunity of doing good; in such cases man can assist man without any witness; it is here that generosity, benevolence, and tender pity, may be exerted to dry up tears and wailings. Is there any other class of citizens that perform these duties with so much zeal and courage as medical practitioners? These labours, these pleasures, are those of almost all the practitioners of medicine; they have their first lessons from the poor and miserable, who feel their [48/49] benevolence and virtue. The disinterested cares given to the sick are always recompensed, the practitioner invariably finds, that his benevolence and charity are the foundation of his celebrity. When he acquires a high reputation, he must not forget the means by which he has acquired it; the poor and miserable are the base of his fame and prosperity. When he acquires fame and wealth, he still gives his aid to the poor at hospitals, dispensaries, or at his own residence. The faculty has often been reproached with hardness of heart, occasioned, as is supposed, by their being so much conversant with human misery. I hope and believe the charge is unjust; for habit may beget a command of temper, and a seeming composure which is often mistaken for absolute insensibility. But, by the way, I must observe, that, when this insensibility is real, it is a misfortune for a physician, as it deprives him of one of the most natural and powerful incitements to exert himself for the relief of his patient. On the other hand, a physician of too much sensibility may be rendered incapable of doing his duty, from anxiety, and excess of sympathy; which cloud his understanding, depress his mind, and prevent him from acting with that steadiness and vigour, upon which perhaps the life of his patient in a great measure depends.

Though a physician possess that enlarged medical genius already described, yet talents of another kind are necessary. He has not only for an object the improvement of his own mind, but he must study the temper, and struggle with the prejudices of his patient, of his relations, and of the world in general; nay, he must guard himself against the ill offices of those, whose interest interferes with him; and it unfortunately happens, that the only judges of his medical merit are those who have sinister views in concealing or depreciating it. Hence appears the necessity of a physician's having a large share of good sense, and knowledge of the world, as well as medical genius and learning.

Such are the genius and talent required in a physician; but a certain command of the temper and passions, either natural or acquired, must be added, in order to give them their full advantage. Sudden emergencies often occur in practice, and diseases often take unexpected turns, which are [49/50] apt to flutter the spirits of a man of lively parts and of a warm temper.

Accidents of this kind may affect his judgment in such a manner as to unfit him for discerning what is proper to be done; or, if he does perceive it, may nevertheless render him irresolute. Yet such occasions call for the quickest discernment, and the steadiest and most resolute conduct; and the more, as the sick so readily take the alarm, when they discover any diffidence in their physician. The weaknesses too and bad behaviour of patients, and a number of little difficulties and contradictions which every physician must encounter in his practice, are apt to ruffle his temper, and consequently to cloud his judgment, and make him forget propriety and decency of behaviour. Hence appears the advantage of a physician's possessing presence of mind, composure, steadiness, and an appearance of resolution, even in cases where, in his own judgment, he is fully sensible of the difficulty.

I come now to mention the moral qualities peculiarly required in the character of a physician. The chief of these is humanity; that sensibility of heart which makes us feel for the distresses of our fellow-creatures, and which of consequence incites us in the most powerful manner to relieve them. Sympathy produces an anxious attention to a thousand little circumstances that may tend to relieve the patient; an attention which money can never purchase: hence the inexpressible comfort of having a friend for a physician. Sympathy naturally engages the affection and confidence of a patient, which in many cases is of the utmost consequence to his recovery. If the physician possesses gentleness of manners, and a compassionate heart, and what Shakspeare so emphatically calls "the milk of human kindness," the patient feels his approach like that of a guardian angel ministering to his relief; while every visit of a physician who is unfeeling, and rough in his manners, makes his heart sick within him, as at the presence of one who comes to pronounce his doom. Men of the most compassionate tempers, by being daily conversant with scenes of distress, acquire, in process of time, that composure and firmness of mind so necessary in the practice of medicine. They can feel whatever is amiable in pity, without [50/51] suffering it to enervate or unman them. Such physicians as are callous to sentiments of humanity, treat this sympathy with ridicule, and represent it either as hypocrisy, or as the indication of a feeble mind. That sympathy is often affected, I am afraid, is too true; but this affectation is easily seen through. Real sympathy is never ostentatious, on the contrary, it rather strives to conceal itself. But what most effectually detects this hypocrisy is a physician's different manner of behaving to people in high and people in low life; to those who reward him handsomely, and those who have not the means to do it. A generous and elevated mind is even more shy in expressing sympathy with those of high rank, than with those in humbler life; being jealous of the unworthy construction so usually annexed to it. The insinuation that a compassionate and feeling heart is commonly accompanied with a weak understanding and a feeble mind, is malignant and false. Experience demonstrates, that a gentle and humane temper, far from being inconsistent with vigour of mind, is its usual attendant; and that rough and blustering manners generally accompany a weak understanding and a mean soul, and are indeed frequently affected by men void of magnanimity and personal courage, in order to conceal their natural defects.

There is a species of good humour different from the sympathy I have been speaking of, which is likewise amiable in a physician. It consists in a certain gentleness and flexibility, which makes him suffer with patience, and even apparent cheerfulness, the many contradictions and disappointments he is subjected to in his practice. If he is rigid and too minute in his directions about regimen, he may be assured they will not be strictly followed; and if he is severe in his manners, the deviations from his rules will as certainly be concealed from him. The consequence is, that he is kept in ignorance of the true state of his patient; he ascribes to the consequence of the disease what is merely owing to irregularities in diet, and attributes effects to medicines which were perhaps never taken. The errors which in this way he may be led into are sufficiently obvious, and might easily be prevented by a prudent relaxation of rules that cannot well be obeyed. The [51/52] government of a physician over his patient should undoubtedly be absolute; but an absolute government very few patients will submit to. A prudent physician should therefore prescribe such laws, as though not the best, are yet the best that will be observed; of different evils, he should choose the least; and, at no rate, lose the confidence of his patient, so as to be deceived by him as to his true situation. This indulgence, however, which I am pleading for, must be managed with judgment and discretion; as it is very necessary that a physician should support a proper dignity and authority with his patients, for their sakes as well as his own.

There is a numerous class of patients who put a physician's good nature and patience to a severe trial; those I mean who suffer under nervous ailments. Although the fears of these patients are generally groundless, yet their sufferings are real; and the disease is as much seated in the constitution as a rheumatism or a dropsy. To treat their complaints with ridicule or neglect, from supposing them the effect of a crazy imagination, is equally cruel and absurd.[2] They generally arise from, or are attended with, bodily disorders obvious enough; but supposing them otherwise, still it is the physician's duty to do every thing in his power for the relief of the distressed. Disorders of the imagination may be as properly the object of a physician's attention as those of the body; and surely they are, frequently, of all distresses the greatest, and demand the most tender sympathy; but it requires address and good sense in a physician to manage them properly. If he seems to treat them slightly, or with unseasonable mirth, the patient is hurt beyond measure; if he is too anxiously attentive to every little circumstance, he feeds the disease. For the patient's sake therefore, as well as his own, he must endeavour to strike the medium between negligence and ridicule on, the one hand, and too much solicitude about every trifling symptom on the other. He may sometimes divert the mind, without seeming to intend it, from its present sufferings, and from its melancholy prospects of the future, by insensibly [52/53] introducing subjects that are amusing or interesting; and sometimes he may successfully employ a delicate and good-natured pleasantry.

This class of patients are in general extremely unreasonable. They are constantly complaining, nothing does them good, and every thing injures them. They suppose that they labour under one or many incurable diseases. They complain of every disease, and more than every disease; or, as it is well expressed, "de omnibus rebus, et quibusdam aliis."[3] They do not know, nor can they understand, that disease in any part of the body may derange the whole organs, in consequence of the universal nervous connection; or that all their unpleasant, and to them unaccountable sensations, are as obvious as the noon-day sun to every well-educated medical practitioner. They are excessively, or morbidly sensitive; they watch every cast of countenance; they put the most unfavourable construction on every sentence uttered by their medical attendant; they fear the worst; and, should he deliver an ambiguous or doubtful prognosis, they fear him much more than they are pleased with him. Their esteem and respect are very variable and inconstant, and very readily converted into hatred and contempt. No matter how much you relieve them, even after the most eminent practitioners have failed, you are suddenly dismissed without the slightest reason, more especially when women are your patients. I have been repeatedly treated in this way, and after many other practitioners have been consulted, again recalled. But we must not forget the adage, "varia et mutabilis fæmina," in plain language, women are variable and changeable; and I must add, when excessively nervous or hypochondriacal, very often most unreasonable. Disease, however, is their excuse; and every humane physician will forget and forgive their inconsistencies, however remarkable. Such patients may imagine the most improbable conditions of constitution, as being made of glass, butter, &c., being a grain of wheat, pregnant of a hen and her chickens, & c.; examples of which have fallen under my own observations. The fact is, that many nervous persons and hypochondriacs do not possess perfect reason; and, therefore, great allowances must be made for them. [53/54]

We sometimes see a remarkable difference between the behaviour of a physician at his first setting out, and afterwards when fully established in reputation and practice. In the beginning, he is affable, polite, humane, and assiduously attentive to his patients; but afterwards; when he has reaped the fruits of such a behaviour, and finds himself independent, he assumes a very different tone; he becomes haughty, rapacious, and careless, and often somewhat brutal in his manners. Conscious of the ascendancy he has acquired, he acts a despotic part, and takes a most ungenerous advantage of the confidence which people have in his abilities.[4] A physician, by the nature of his profession, has many opportunities of knowing the private characters and concerns of the families in which he is employed. Besides, what he may learn from his own observation, he is often admitted to the confidence of those who perhaps think they owe their life to his care. He sees people in the most disadvantageous circumstances, very different from those in which the world views them – oppressed with pain, sickness, and low spirits. In these humiliating situations, instead of wonted cheerfulness, evenness of temper, and vigour of mind, he meets with peevishness, impatience, and timidity. Hence it appears, how much the characters of individuals, and the credit of families, may sometimes depend on the discretion, secrecy, and honour of a physician. Secrecy is particularly requisite where women are concerned. Independently of the peculiar tenderness with which a woman's character should be treated, there are certain circumstances of health which, though in no respect connected with her reputation, every woman, from the natural delicacy of her sex, is anxious to conceal; and, in some cases, the concealment of these circumstances may be of consequence to her health, to her interest, and to her happiness.

Temperance and sobriety are virtues peculiarly required in a physician. In the course of an extensive practice, difficult cases frequently occur, which demand the most vigorous exertion of memory and judgment. I have heard it said of some eminent physicians, that they prescribed as well when [54/55] drunk as when sober. If there was any truth in this report, it contained a severe reflection against their abilities in their profession. It showed that they practised by rote, or prescribed for some of the more obvious symptoms, without attending to those nice peculiar circumstances, a knowledge of which constitutes the great difference between a physician who has genius and one who has none. Drunkenness implies a defect in the memory and judgment; it implies confusion of thought, perplexity, and unsteadiness; and must therefore unfit a man for every business that requires the lively and vigorous use of his understanding.

An obstinate adherence to an unsuccessful method of treating a disease, must be owing to a high degree of self-conceit, and a belief in the infallibility of a system. It has been the cause of the death of thousands. Patients ought to be indulged in every thing consistent with their safety; and if they are determined to try an improper or dangerous medicine, a physician should refuse his sanction, but he has no right to complain of his advice not being followed. A physician is often at a loss in speaking to his patients of their real situation, when it is dangerous. A deviation from truth is in this case both justifiable and necessary. It often happens that a person is extremely ill, but he may recover if he is not informed of his danger. Again, a man may not have settled his affairs, though the future happiness of his family depends on his making a settlement. In such cases the physician may apprize the friends, and occasionally the patient, of the necessity of the arrangement and disposal of his property. In all dangerous cases, the real situation of the patient should be communicated to his nearest relatives, as it gives them an opportunity of calling other assistance, if they think it necessary.

The patient is not to be deserted when his case is despaired of; it is as much the duty of a physician to alleviate pain, and to smooth the avenues of death, when inevitable, as to cure diseases; his presence and assistance as a friend may be both agreeable and useful, where his skill is of no further avail. In some cases we should caution the indiscreet enthusiasts among the clergy against too much zeal, as they often [55/56] terrify the patient, and contribute to shorten a life which might otherwise be saved.

Medical men should never involve their patients in private and professional quarrels, in which the sick can have little or no concern. All personal feelings should be forgotten in consultations, the good of the patient ought to be the chief and only consideration. The quarrels of the faculty, when they end in appeals to the public, generally hurt the contending parties, discredit the profession, and expose it to ridicule and contempt. Nothing can justify the refusal to consult but want of temper, nor can such circumstances as the university where a person has taken a degree, "or whether he had any degree at all, justify the refusal." This assertion, I may observe, is at variance with the usages of the profession, though society has sanctioned it. Fellows of the College of Physicians refuse to meet graduates of all the British and Foreign universities in consultation, until admitted into the College. But of this hereafter. It becomes young practitioners to be particularly attentive to the propriety of their behaviour when consulting with their seniors. Besides the respect due to age, these are entitled to a particular deference from their longer and more extensive experience, provided it be scientific.

The revolutions indeed of medical hypotheses and systems are so quick, that an old and a young physician seldom reason in the same way on subjects of their profession; although the difference is sometimes rather apparent than real, when they use only a different language to express sentiments essentially the same. But it generally happens, that the speculations which principally engage the attention of young physicians seldom in any degree affect their practice; and therefore, as they are in a great measure foreign to the business, they should never introduce them in medical consultations. They show equal want of sense and good manners, when they wantonly take opportunities of expressing contempt for opinions as antiquated and exploded, in which their seniors have been educated, and which they hold as firmly established. A little reflection might teach them, that it is not impossible but in the course of a few years, their own most favourite theories [56/57] may be discovered to be as weak and delusive as those which have gone before them: and this should lead them to consider how sensibly they may be hurt themselves, when they find those idols of their youth attacked by the petulant ridicule of the next generation; when, perhaps, they are arrived at a time of life when they have neither abilities nor temper to defend them.

There are, however, many old practitioners who have not kept pace with science, and who despise all new remedies. There are many who oppose and ridicule the stethoscope, the use of the alkaloid medicines, such as iodine, strychnine, &c.; as if these discoveries were not made known by Divine Providence, and as if medicine was a perfect and positive science. If a practitioner is ignorant of the value of new discoveries, he is not justified in abusing them; and, if they have been amply proved to be valuable, they ought to be adopted for the relief of the sick. Macoppe has ably commented on this point: "Si nova inventa, si elegantes recentiorum ignoras eas spurco lividoque non vituperes ore. Tuas liceat extollere, pedibus non alias calcare."[5] It is a just opinion, that a man who is ignorant of new remedies or discoveries ought not to despise them; he may extol his own, but must not condemn those of others. Our predecessors have done much; we have done much; but a vast deal remains to be done by our successors. It is to be recollected, that diseases are cured with remedies, and not by disputation. "When you are an old practitioner, you must not be ashamed to praise the new remedy of a junior, when more efficacious than your own. You must not look on the young practitioner with a jaundiced eye; he has left the schools, and is generally more conversant with the actual state of science than you are. Praise what is just, be silent of what is doubtful or inefficacious. Do not praise first, and vituperate afterwards."

A routine practitioner is ignorant of the progress of science, and is consequently self-opinionated. He is like a machine that always performs the same evolutions. He dispenses with all study, and he is content with what the vulgar consider experience.

This experience is necessarily false; for how can any man [57/58] exercise an art of whose principles he is ignorant ? Dewees has well observed, "that an obstetrician may have thirty years experience, but, without a knowledge of first principles, he is as ignorant and dangerous as he who has had the slightest experience." A question has been asked, whether a senior or junior physician ought to be preferred? It has been answered, that, in common disorders, a senior is preferable; but a junior in difficult and complicated diseases. "I hold," says Kyper, "that a physician cannot be experienced, unless he is learned; because an ignorant man does not know how to make judicious observation, or to deduce correct experience. No one can be a good practitioner unless he has been properly educated in public schools and hospitals, and has kept pace with the progress of science." Students are now better informed than their teachers were 50 years ago; they are taught all the new doctrines, and must excel those who are content with old ones. The ablest physician is he who is of middle age, and possesses real scientific knowledge, and whose judgment is formed on science, after sober and extensive observation on the poor, and at hospitals and dispensaries. He is not less informed than a young practitioner, and has the superior advantage of great experience. He is judicious in consultations, intrepid in dangers, competent to anticipate results, fertile in his resources, and endowed with great sagacity. Knowledge makes a young man old; and ignorance makes an old man a student. Talent compensates for the want of age,

Quid numeras annos, vixi maturior annis;
    Acta senem faciunt, hæc numerandi tibi.[6]

It is not a head adorned with grey hairs that establishes merit; it is superiority in consultation, and at the bed-side of the sick. A young man may be a great physician, but an old man can scarcely be a great surgeon, and decidedly not a great physician. Talent, and not years, makes a good physician or surgeon.

A young man, gifted with genius for medicine, will in a few years be a great practitioner; but a man of 60 years of age, though he has seen 100,000 patients, will never be a good physician or surgeon, if deprived of this valuable gift of nature. [58/59] It is therefore an error to imagine, that the best practitioner is he who has seen the most cases. This is a popular prejudice; but, as Zimmerman observes, they do not inquire whether such a practitioner has received a proper medical education, whether he is a man of penetration and genius, though he is grey-headed. It is for this reason that many prefer an old to a young practitioner; they consider senescence and experience inseparable. They cannot distinguish scientific experience from common routine. It is also a fact, that old practitioners entertain the vulgar notion; in their opinion, a young man, of the greatest talent, is only a young man, and they can never believe that he can be equal to themselves. In their consultations and writings they maintain their superiority, even when the public confidence is being withdrawn from them, and when the world thinks they ought to retire. They forget, or perhaps have never seen the saying of Galen, that men of mere experience, without previous education, may be considered idiots; but the axiom of this venerated author is so apposite, that I must quote it: "Medicos qui solam experientiam sequuntur, non admittimus; quoniam ipsi idiotæ faciunt, quæ vident inspicientes, et rerum quidem eventum continentes, causam autem ignorantes."[7] This remark applies to uneducated practitioners only, and not to the scientific and experienced.

The latter, when advanced in life, are esteemed and cherished by the public and the profession, they are the Mentors of their juniors; their sentiments are received with profound respect; and they are heard with a religious emotion in the lecture-room and in consultations.

But as age advances the intellectual faculties become enfeebled. Horace said,

Multa senem circumveniunt incommoda.[8]

And Virgil expressed the same opinion:

Tarda senectus Debilitat vires animi mutatque vigorem.[9]

On the arrival of senescence the most renowned physicians have lost all their practice. They have often proclaimed their prerogative of experience, and fruitlessly endeavoured to deprecate their juniors. They have said that these had little patience, [59/60] no assiduity, circumspection; their impetuosity confuses them, they cannot coolly observe nature, judge accurately, preserve constancy, or form an accurate opinion of the nature or treatment of diseases.

It must, however, be obvious, that these assertions are groundless, because there is no reason why young practitioners, who are properly educated, who are conversant with the old and new opinions, and who are in the prime of life, when the intellectual faculties are fully developed, should not observe, think, reason, and practise, as well as their seniors. Many proofs might be adduced to support this statement, but a few shall suffice. Harvey discovered the circulation of the blood at an early age; Baglivi, who was the restorer of medicine, died at the age of 39; Bichat died at the age of 31 years.[10]

Indeed, a man who is not a good physician or surgeon at the age of 30, provided he has enjoyed ample opportunities, will never be one; for it is not years, but knowledge, that constitutes such a character.

A young physician of talent, who has received a good and extensive education, who has studied with assiduity all the elementary sciences, as well as the practice of medicine, surgery, obstetricy, and pharmacy; who has read all the best works in the vernacular and other languages, who has rigidly observed the effects of medicines for years, will far surpass his contemporaries in the aggregate, and leave them, whether old or young, immeasurably behind him. It is true, he may learn wisdom and caution from his seniors, when scientific and acquainted with the actual state of the science and practice of the healing art; but this very seldom happens. Those advanced in age, and actively engaged in practice, are generally unacquainted with the rapid progress of medicine; and their opinions are too often erroneous. Nevertheless, the junior is bound to respect his senior, because to honour a master is to honour one's self. Age is honourable, and ought to be always respected.

Dr. John Gregory defended the necessity of medical men being versed in all the branches of the healing art, and concludes by observing

Every department of the profession is respectable, when exercised with capacity and integrity. I only contend for an evident truth, either that the different [60/61] branches should be separately professed, or, if one person will profess all, that he should be regularly educated to, and thoroughly master of all. I am not here adjusting points of precedence, or insinuating the deference due to degrees in medicine. As a doctor's degree can never confer sense, the title alone can never command regard; neither should the want of it deprive any man of the esteem and deference due to real merit. If a surgeon or apothecary has had the education, and acquired the knowledge of a physician, he is a physician to all intents and purposes, whether he has a degree or not, and ought to be respected and treated accordingly. In Great Britain, surgery is a liberal profession. In many parts of it, surgeons or apothecaries are the physicians in ordinary to most families, for which trust they are often well qualified by their education and knowledge; and a physician is only called where a case is difficult, or attended with danger. There are certain limits, however, between the two professions, which ought to be attended to: as they are established by the customs of the country, and by the rules of their several societies. But a physician of a candid and liberal spirit, will never take advantage of what a nominal distinction, and certain privileges, give him over other men who, in point of real merit, are his equals; and will feel no superiority, but what arises from superior learning, superior abilities, and more liberal manners. He will despise those distinctions founded in vanity, self-interest, or caprice; and will be careful that the interests of science and of mankind shall never be hurt, on his part, by a punctilious adherence to formalities.

Much stress has been laid on the formality of a physician's dress, but there is no reason in preferring one garb to another. In some cases there is great impropriety in having any distinguishing formality in dress and manners. Negligence and luxuriousness of dress are extremes, and ought to be avoided. A respectable appearance, propriety, convenience, and elegance without pretension, are the characters which ought to mark the costume of a medical practitioner. Carelessness or foppishness in dress are unworthy of the followers of the healing art. The vulgar attach great importance to [61/62] personal appearance. The professions of theology and law take advantage of this weakness; and still continue to wear peculiar habiliments, while the scarlet robes, velvet caps, and gold-headed canes of the professors of physic are abandoned. The divines and juris-consults are, perhaps, as wise as the Esculapians, and though the latter may dress as they please, I cannot help agreeing in opinion with those who maintain that external appearance is every thing with the world. For this reason the dress of medical men should be grave and elegant, not showy or *à la mode*. A cravat tied in the newest fashion, a coat of the colour and cut of the day, and all the other frivolities of the passing hour, are unworthy of the professors of the noble science of healing. A philosophic physician allows his tailor to equip him. Macoppe makes some happy remarks on moustaches, and contends that knowledge does not depend on hair, which characterizes the head or hope of the flock – vir gregis, ipse caper, but is no proof of philosophical or medical knowledge. Perfumery and odours were also condemned by Hippocrates, as they are often disagreeable and hurtful to delicate patients, especially to nervous and delicate women in whom they may excite fainting, vomiting, spasms, &c. The father of medicine taught, that a medical practitioner ought to avoid all odoriferous apparel, and that he was most perfumed who was not at all perfumed: optime olet medicus quum nihil olet. Septal, Roderic a Castro, Triller, and other medical moralists, inculcate this doctrine. Musk, amber, and other perfumes, are disagreeable to many persons, cause head-ache, vertigo, convulsions, hysteria, &c.; and he who uses them is the author and not the remover of diseases.

The attendance should be in proportion to the urgency and danger of the disease. A patient or his friends have a curiosity to know the nature of the medicine prescribed, which it is often very improper to gratify; but other cases occur in which it may be proper to acquaint the patient with the nature of remedies, as the peculiarities of constitution require great attention, both as to the quantity and quality of certain medicines. Such are the chief of the duties of medical men, according to the amiable and revered Dr. John Gregory; [62/63] the observance of which cannot fail to promote the honour and dignity of the profession. He included many minor topics, which need not be recorded at the present period.

There are certain duties belonging to the learned professions which are supreme, and which no individual and no set of men can, either for themselves or their successors violate, renounce, or neglect, without substantial injustice. These duties, so far as they relate to physicians, are comprised in the oaths required by the Universities, Colleges of Physicians, and in one of the Colleges of Surgeons, in this empire. The substance of these oaths is that proposed by Hippocrates nearly 2000 years ago, and the oath was formerly administered in all Universities in which medicine was taught, to those who were created doctors, and to those who were licensed to practise by the Colleges of Physicians. The oath required by the Edinburgh University is in the following words. After an invocation of the Deity, the graduate pronounces these words:[11] "To practise physic *cautiously, chastely, and honourably*; and faithfully to procure all things conducive to the health of the bodies of the sick; and lastly, never, without great cause, to divulge any thing that ought to be concealed, which may be heard or seen during professional attendance. To this oath let the Deity be witness."[12] I believe no similar oath is required by the Universities of Oxford, Cambridge, Dublin, Glasgow, Aberdeen, or Saint Andrews, or by any of the Colleges of Physicians or Surgeons, except those of London. The Royal College of Physicians requires the following promise (not an oath) from its members, fellows, and licentiates, and prescribes a code of moral statutes: – [13][63/64] "You faithfully promise that you will observe the statutes of the College, and that you will promptly discharge all fines imposed on you for the breach thereof, and that you will do everything in the practice of medicine for the conservation of health, to the honour of the College, and the good of the public." The following are the *Statutes of Morality* of the Royal College of Physicians of London in 1835:[14]

1. No fellow, candidate, or licentiate, shall accuse a fellow, candidate, or licentiate, of ignorance or mala praxis of his art, unless before legitimate judges, or before those concerned. If it be known to the president and censors, or the majority of them, that any person shall so act, he shall pay £4. for the first offence, and the fine will be doubled for the second; but if he transgress a third time, and be convicted in the manner mentioned, if he is a fellow or candidate he [65/66]

shall be expelled from our society, or from the order of candidates; and if he is a licentiate he shall pay £10, and we ordain, that licentiates shall be fined a like sum for every similar transgression.

2. No fellow, candidate, or licentiate, shall afford medical aid or prescribe for a patient whom he knows is under the care of another physician, whether fellow, candidate, or licentiate, and to whom he has not been regularly called.

3. If any one be convicted of this vice, besides the known ignominy which we wish him to suffer, he shall be fined £2. by the president and censors.

4. If any one shall bargain with apothecaries for any percentage on prescriptions, if a fellow or candidate, and if convicted in the manner before mentioned, he shall be expelled from the fellowship, or from the order of candidates.

5. If a licentiate, he shall be fined £10. for each offence.

6. Every physician, whether fellow, candidate, or licentiate, shall inscribe his initials, the date of the prescription, and name of the patient, on every prescription, unless some cause intervenes which shall be approved by the president and censors.

7. If many physicians be called to a patient, they are to consult with great modesty, and in the absence of witnesses or unprofessional persons. Nor shall any one prescribe or insinuate what is to be done to the sick or attendants, before he has stated his method in consultation. But, as medical men have different opinions, so that they may not agree in the plan of treatment, they are to conduct themselves with the greatest prudence and moderation; the ordinary attendant shall signify to the sick and attendants their dissension, so that it may appear as trifling, and as slightly disagreeable to the patient or his friends as possible.

8 Whoever will not obey these rules of consultation, and is convicted by the president and censors, shall be fined £5.

9. Finally, no physician, fellow, candidate, or licentiate, shall consult in the city of London, or within 7 miles thereof, unless with a fellow, candidate, or licentiate, under a [66/67] penalty of £5 as often as convicted by the president and censors, or majority of them.

10. All fines imposed by these statutes must be paid.

It is much to be regretted, that the great bulk of the profession – university graduates in medicine, surgeons, and apothecaries – have no opportunity of being acquainted with these admirable statutes, or have nothing similar to inform them of the etiquette they owe to each other. In printing these statutes, and placing them before the medical public, I hope and trust I may add to the honour and dignity of the profession. The majority of the tenets maintained in them are highly conducive to the fame of every class of medical men; and, if duly observed, would extinguish that base and unprofessional and ungentlemanly behaviour, which of late has characterized too many medical practitioners, and has debased and degraded the profession. The disputes and calumnies of medical men have been so frequent, so violent, so notorious of late, that the character of the profession is lowered in the estimation of the public to a degree unequalled in the history of medicine. Actions against medical men by their contemporaries, or their patients, are now frequent in our courts of justice. This degeneracy of the profession is not confined to this country, it extends throughout Europe, and has even crossed the Atlantic Ocean; and it arises from the exclusion of medical ethics from the prescribed courses of professional education. This malignant spirit pervades every branch of the healing art; the physicians, the surgeons, and the apothecaries, are the most prominent of litigants in our courts of justice. What a falling off is here! If we turn to private practice, we find those uninfluenced by the statutes under consideration, vituperating each other, "by look, gesture, and suspicious silence," and often without any disguise; and the injured individual has no remedy afforded him by the body to which he belongs, and which gravely promises him rights, privileges, immunities, and protection in the discharge of his vocation; his only remedy is an appeal to the laws of his country. But the fact is, our Colleges of Surgeons, and Companies of Apothecaries, have no power to protect their members; nor is there any country in [67/68] the world, in which the laws relative to the practice of the medical profession are so imperfect and defective as in the British empire.

But to return to the subject immediately under consideration: I have to insert the oath required by the Royal College of Surgeons in this city, which is as follows:

> You swear that while you shall be a member of the Royal College of Surgeons in London, you will observe the statutes, bye-laws, ordinances, rules, and constitutions thereof; that you will obey every lawful summons issued by order of the court of assistants and examiners of the said college, or of either of them, having no reasonable excuse to the contrary that you will pay such contributions as shall be legally assessed upon and demanded of you; that you will demean yourself *honourably* in the practice of your profession; and to the utmost of your power maintain the dignity and welfare of the college – So help you God.

It is to be feared that some surgeons forget to demean themselves honourably in the practice of their profession, more especially as their rivals, the general practitioners, are under no such obligation. From the open violation of our laws relative to the practice of medicine, the surgeons act as physicians, and must become apothecaries in self-defence; the apothecaries act as physicians and surgeons, while the chemists and druggists, without any medical education whatever, act as physicians, surgeons, and apothecaries; and as to quacks, they are allowed to flourish to an illimitable extent, and to destroy more than the sword, famine, and pestilence united. Such is a true picture of the medical profession in the greatest nation upon earth – in a country pre-eminent for literature, the sciences and the arts: – such is the state of medical practice in England.

But I leave this part of the subject to another opportunity, and return to the topic more immediately under consideration[.] That medical men should practise *cautiously, chastely*, and *honourably*, and observe strict secrecy in all delicate cases, and in all domestic affairs, which may fall under their notice during professional attendance, is not only consonant to the usage of the profession, but to common sense and justice. It would be highly improper to divulge the nature of certain [68/69] diseases, or expose the affairs of families, to gratify idle curiosity, impertinence, or serve the purposes of an interested knave. The law, however, compels us to violate these principles; and hence the exception in the Edinburgh oath, "not to divulge without weighty reasons." In such cases the violation or renunciation of our moral and professional duties is compulsory.

Chastity and honour are general moral duties, and not peculiarly belonging to any one profession. No profession commands a greater purity of morals than the medical. The intimate confident of the other sex – the adviser in her mental and corporal diseases; but never the abuser of these advantages. Never has a properly educated physician employed his ascendancy to seduce innocence, which places herself in his hands; and scarcely ever has his voice been employed in corrupt discourses towards women who have selected him as their consoler and friend. A medical practitioner is often placed between his duties and vice; his station almost daily exposes him to sacrifice honour to interest; he frequently feels, without danger, the strongest passions; – but he thinks it more glorious and virtuous to conquer them. It is for the good of society that he employs his powerful influence in the most virtuous and honourable manner. Men who confide to him all that is dearest to them – their wives and children – have a just right to require of him a pure heart and untarnished morals. Such is the moral character of the entire profession in all countries, and it is but rarely, scarcely ever tarnished, notwithstanding the immense number of medical men, and the certainty that there must of necessity be a few exceptionable, alias unprincipled individuals, amongst them. Were medical practitioners to act immorally, they would cease to be employed; and society would place the signet of reprobation upon them. I am certain that there have not been half-a-dozen actions against medical practitioners in this country, for immoral conduct, since the first century to the present date. My position is therefore proved by historical records.

The multiplied studies of a truly informed medical practitioner; his various duties; the practice of his profession; the care of his reputation – all prevent him from participating [69/70] in the commotions which occupy societies and empires. He avoids politics and political assemblies, while engaged in the peaceful pursuits of his profession. He takes no interest in the quarrels of sovereigns. There have been a few exceptions, but wise practitioners avoid all political and religious discussions. Their patients may entertain the most opposite opinions on these subjects; and no good ever results from medical practitioners involving themselves in disputes concerning them. In time of war, medical men feel equally bound to aid friends and enemies, according to the law of nations.

The duty of *caution* in practice, means "care not to expose the sick to any unnecessary danger." The best rule of conduct on this important point, is the simple and comprehensive, religious and moral precept, "Do unto others as you would they should do unto you." Whatever the practitioner does or advises to be done for the good of his patient, and what he would do in his own case, or in the case of those who are dearest to him – if he or they were in the same situation – is not only justifiable on his part, but it is his indispensable duty to do. The patient should have the chance, whether it be a hundred to one, or only one in a hundred in his favour. Whatever may be the result, the practitioner has the greatest of all consolation – the consciousness of rectitude – "mens conscia recti;" this will be his solace, should the case terminate unfavourably, when the vulgar, the ignorant, the envious, the malicious, and the interested, will not fail to blame him for the death of his patient. But if he administered a dangerous medicine, merely to gratify his own curiosity, or zeal for science – to ascertain the comparative advantage or disadvantage of some new remedy, either proposed by himself or suggested by others – he is held guilty of a breach of ethics, and of a high misdemeanour, and a great breach of trust towards his patient; and if the patient die, I apprehend, he might be severely punished.

Medical men have tried the most dangerous experiments upon themselves, from their zeal for science, and even sacrificed their lives; but patients, in general, have no such zeal for science – no ambition for such a crown of martyrdom – [70/71] and generally employ and pay their medical attendants for the very opposite purpose. It must be admitted, that men who would try experiments upon themselves, would be very apt to try experiments on their patients. It is a melancholy truth, but cannot be denied. The profession, however, has always reprobated such conduct; and the medical phrase of reproach and contempt for it, "corio humano ludere," to play with the human hide, abundantly testifies in what abomination it has been held by the faculty. It is unnecessary to dwell upon this point in this age, because all experiments are made upon the inferior animals; and the just reproach entertained by the faculty, in former times, is now inapplicable. But every man of common understanding well knows, that neither physic nor surgery can be practised without some danger to the sick. It is universally known, that many surgical operations are dangerous to life, and that all our most powerful remedies are highly dangerous; and more especially when improperly employed, or when they cannot be borne. A safe medicine is often extremely dangerous, from the peculiarity of constitution: and the great and urgent danger, in many diseases, requires the immediate use of dangerous remedies. It is admitted, by the best practitioners, that many remedies are still wanted for the cure of disease, and this want leads us most justifiably, and almost inevitably, to try new remedies on many occasions; and such experiments are not blamable, for they are necessary: sic enim medicina arta; subinde aliorum salute, aliorum interritu perniciosa discernans a salutaribus.[15] From these causes there results much inevitable danger in the practice of physic. From this acknowledged danger, results the important duty of caution in a physician, or care to make the danger as little as possible. Whatever is best for the sick, it is the indispensable duty of a medical man to do for them. It is his duty and obligation, "faithfully to do all things conducive to the health of his patients;" and this is so complete and indefeasible, that it cannot be set aside by any cause whatever.

The last obligation imposed on a medical practitioner, is discretion, or secrecy. The depositary of family secrets – often the possessor of the reputation of those who have [71/72] reposed in him their confidence – to what disgrace would he not expose them, were he to divulge the mysteries he knows, which ought to be hidden from the public! Here, the unfortunate victim of seduction implores his succour and his silence; there, a father or a mother avows to him the unfortunate consequences of youth abandoned to passion. The confidence reposed in him, and revelations made to him, during his professional attendance, are such that honour commands him not to abuse the one, or publish the other, unless in our courts of justice, which have the power to compel him. Thence the phrase, non sine gravi causa, in the doctors' oath; that is, such secrets are not to be divulged without the greatest necessity; but the French medical moralists contend, that they ought not to be divulged even at the risk of liberty and life. Hippocrates inculcated this conduct in the oath he required of his disciples: "Quæ vero inter curandum aut etiam medicinam minime faciens, in communi hominum vitâ, vel videro, vel audiero, quæ minime in vulgus efferi oporteat ea arcana esse ratus, silebo."[16] The Edinburgh University exacts a similar obligation.

Prudence is indispensably necessary in the practice of medicine – not only in the selection of medicines, but in moral conduct. The greatest care, and constant attention are required, to preserve the integrity of reputation. We are consulted in cases of the greatest delicacy and difficulty. We are the advisers of parents, husbands, wives, youths, and children. The most scrupulous morality is required in cases of girls and women. The presence of the mother, some near relation, or intimate female friend, is required in those delicate and frequent cases, in which the most secret charms of nature are subjected to indispensable examination, when the timid and blushing virgin is compelled to place at her feet the last veil of modesty. In these obstetric, and all female cases, let the following rules be rigidly observed. Modestissimus in curandis mulieribus existas; et si pectus, venter imus, aut alinæ arcanæ partes tangendæ, aut pertractandæ sunt, te fidelem, aut ita dicam, marmoreum et gelidum, animo constanti, vel effinge, vel efforma. Pessima, jure merito, tuo nomini fama inuretur, si lubrica manus, impurus animus, corrupti sermonis [72/73] castitatem per solas etiam aures violabunt. Quorundam morborum, præcipue paellas, matronas, principes divexantium, aut eos, quibus ex his nomen aliquo modo periclitatur, labes ne detegas, secreto ac fido pectore, naturæ mortalium errores conde. Caute tamen in viduarum domos, ubi sunt virgines, petilcæ uxores, pulchræ ancillulæ juventutem intrude. Vidi mulierculas puellasque insano practicantium amore captas, ac dulcæ rabidæ tentigini remedium quærentes; vidique ex virginibus factas matres cum summo incauti præceptoris dedecore.[17]

Such is the line of moral conduct which guides all religious and respectable medical practitioners. It is highly creditable to so large a portion of society in all civilized countries, that their moral conduct has been almost invariably so excellent. If medical practitioners were so immoral as to insul[t] feminine dignity by soliciting the chastity of their patients, or of seducing them, they could not escape detection; and their punishment would be public execration, the severe inflictions of the offended law, and the utter ruin of their professional character. They would be instantly compelled to relinquish the profession. Women, when labouring under disease, or in the agonies of child-birth, are objects of sympathy and compassion, just as all mankind are when suffering from painful infirmities. Sensual impulse and severe pain are contrary influences, setting aside all moral feeling; and gallantry and brutality are equally incompatible. Amorous desire is extinguished by pain, and the brute deprived of reason is restrained by this law of nature. It is this law, as well as a moral obligation, that enables medical practitioners to act in the most virtuous manner towards their female patients, and to relieve their weaknesses and infirmities in the same manner as all other human diseases.

Such is the code of ethics which ought to influence medical men, both in public and in private practice; "but it is matter of question," says Dr. Gordon Smith, "whether it has in reality an existence."[18] This is a truism that cannot be doubted; and yet the rising members of the profession are [73/74] expected to support the honor and dignity of the faculty, without any rules to guide them, without having heard a single word upon the subject, during their education. Hence the cause of that improper conduct which has degraded the profession to a degree unparalleled in the annals of British medicine. I shall not prosecute this subject at present, as it will be more properly considered in my account of the laws relative to the practice of every branch of medicine in this country, and of the constitution of the faculty.

I shall now conclude the subject with the Code of Ethics of Dr. Percival, and append some notes.

Dr. Percival's Medical Ethics[19]

Section 1. Of Professional Conduct Relative to Hospitals, or Other Medical Charities

1. Hospital physicians and surgeons should minister to the sick, with due impressions of the importance of their office; reflecting that the ease, the health, and the lives of those committed to their charge depend on their skill, attention, and fidelity. They should study, also, in their deportment, so to unite tenderness with steadiness, and condescension with authority, as to inspire the minds of their patients with gratitude, respect, and confidence.
2. The choice of a physician or surgeon cannot be allowed to hospital patients, consistently with the regular and established succession of medical attendance. Yet personal confidence is not less important to the comfort and relief of the sick-poor, than of the rich under similar circumstances; and it would be equally just and humane, to inquire into and to indulge their partialities, by occasionally calling into consultation the favourite practitioner. The rectitude and wisdom of this conduct will be still more apparent, when it is recollected that patients in hospitals not unfrequently request their discharge, on a deceitful plea of having received relief; and afterwards procure another recommendation, that they may be [74/75] admitted, under the physician or surgeon of their choice. Such practices involve in them a degree of falsehood, produce unnecessary trouble, and may be the occasion of irreparable loss of time in the treatment of diseases.
3. The feelings and emotions of the patients, under critical circumstances, require to be known and attended to, no less than the symptoms of their diseases. Thus, extreme timidity with respect to venesection, contra-indicates its use in certain cases and constitutions. Even the prejudices of the sick are not to be contemned, or opposed with harshness. For though silenced by authority, they will operate secretly and forcibly on the mind, creating fear, anxiety, and watchfulness.
4. As misapprehension may magnify real evils, or create imaginary ones, no discussion concerning the nature of the case should be entered into before the

patients, either with the house-surgeon, the pupils of the hospitals, or any medical visiter [*sic*].

5. In the large wards of an Infirmary, the patients should be interrogated concerning their complaints, in a tone of voice which cannot be overheard. Secrecy, also, when required by peculiar circumstances, should be strictly observed. And females should always be treated with the most scrupulous delicacy. To neglect or sport with their feelings is cruelty; and every wound thus inflicted tends to produce a callousness of mind, a contempt of decorum, and an insensibility to modesty and virtue. Let these considerations be forcibly and repeatedly urged on the hospital pupils.

6. The moral and religious influence of sickness is so favourable to the best interests of men and of society, that it is justly regarded as an important object in the establishment of every hospital. The institutions for promoting it should, therefore, be encouraged by the physicians and surgeons whenever seasonable opportunities occur. And by pointing out these to the officiating clergyman, the sacred offices will be performed with propriety, discrimination, and greater certainty of success. The character of a physician is usually remote either from superstition or enthusiasm; and the aid which he is now exhorted to give, will tend to their exclusion from the [75/76] hospital, where their effects have often been known to be not only baneful, but even fatal.

7. It is one of the circumstances which softens the lot of the poor, that they are exempt from the solicitudes attendant on the disposal of property. Yet there are exceptions to this observation: and it may be necessary that an hospital patient, on the bed of sickness and death, should be reminded, by some friendly monitor, of the importance of a *last will* and *testament* to his wife, children, or relatives, who, otherwise, might be deprived of his effects, of his expected prize-money, or of some future residuary legacy. This kind office will be best performed by the house surgeon, whose frequent attendance on the sick, diminishes their reserve, and entitles him to their familiar confidence. And he will doubtless regard the performance of it as a duty. For whatever is right to be done, and cannot by another be so well done, has the full force of moral and personal obligation.

8. The physicians and surgeons should not suffer themselves to be restrained, by parsimonious considerations, from prescribing *wine,* and *drugs,* even *of high price,* when required in diseases of extraordinary malignity and danger. The efficacy of every medicine is proportioned to its purity and goodness; and on the degree of these properties, *caeteris paribus,* both the cure of the sick, and the speediness of its accomplishment must depend. But when drugs of inferior quality are employed, it is requisite to administer them in larger doses, and to continue the use of them a longer period of time; circumstances which, probably, more than counterbalance any savings in their original price. If the case, however, were far otherwise, no *economy,* of a *fatal* tendency, *ought to be admitted into institutions,* which, founded on principles of the purest beneficence, in this country, when well conducted, can never want contributions adequate to their liberal support.

9. Hospital affairs ought not to be incautiously revealed.[20]
10. Professional charges are to be made only before a meeting of the faculty.
11. Medical and surgical cases to be distinguished.
12. Principles which authorize the use of new medicines and operations. (See page 71, ante.)[76/77]
13. To advance professional improvement, a friendly and unreserved intercourse should subsist between the gentlemen of the faculty, with a free communication of whatever is extraordinary or interesting in the course of their hospital practice. And an account of every case or operation, which is rare, curious, or instructive, should be drawn up by the physician or surgeon to whose charge it devolves, and entered in a register kept for the purpose, but open only to the physicians and surgeons of the charity.
14. Hospital registers usually contain only a simple report of the number of patients admitted and discharged. By adopting a more comprehensive plan, they might be rendered subservient to medical science, and beneficial to mankind. The following sketch is offered, with deference to the gentlemen of the faculty. Let the register consist of three tables; the first specifying the number of patients admitted, cured, relieved, discharged, or dead; the second the several diseases of the patients, with their events; the third the sexes, ages, and occupations of the patients. The ages should be reduced into classes; and the tables adapted to the four divisions of the year. By such an institution, the increase or decrease of sickness; the attack, progress, and cessation of epidemics; the comparative healthiness of different situations, climates, and seasons; the influence of particular trades and manufactures on health and life; with many other curious circumstances, not more interesting to physicians than to the community, would be ascertained with sufficient precision.[21]
15. By the adoption of the register, recommended in the foregoing article, physicians and surgeons would obtain a clearer insight into the comparative success of their hospital and private practice; and would be incited to a diligent investigation of the causes of such difference. In particular diseases it will be found to subsist in a very remarkable degree: and the discretionary power of the physician or surgeon, in the admission of patients, could not be exerted with more justice or humanity, than in refusing to consign to lingering [77/78] suffering and almost certain death, a numerous class of patients, inadvertently recommended as objects of these charitable institutions. "In judging of diseases with regard to the propriety of their reception into hospitals," says an excellent writer, "the following general circumstances are to be considered."

"Whether they be capable of speedy relief; because, as it is the intention of charity to relieve as great a number as possible, a quick change of objects is to be wished; and also because the inbred disease of hospitals will almost inevitably creep, in some degree, upon one who continues a long time in them, but will rarely attack one whose stay is short.

"Whether they require, in a particular manner, the superintendance of skilful persons, either on account of their acute and dangerous nature, or any singularity or intricacy attending them, or erroneous opinions prevailing among the common people concerning their treatment.

"Whether they be contagious, or subject in a peculiar degree to taint the air, and generate pestilential diseases.

"Whether a fresh and pure air be peculiarly requisite for their cure, and they be remarkably injured by any vitiation of it."[22]

16. But no precautions relative to the reception of patients, who labour under maladies incapable of relief, contagious in their nature, or liable to be aggravated by confinement in an impure atmosphere, can obviate the evils arising from close wards, and the false economy of crowding a number of persons into the least possible space. There are inbred diseases which it is the duty of the physician or surgeon to prevent, as far as lies in his power, by a strict and persevering attention to the whole medical polity of the hospital. This comprehends the discrimination of cases admissible, air, diet, cleanliness, and drugs; each of which articles should be subjected to a rigid scrutiny, at stated periods of time.

17. The establishment of a committee of the gentlemen of the faculty, to be held monthly, would tend to facilitate [78/79] this interesting investigation, and to accomplish the most important objects of it.[23] By the free communication of remarks, various improvements would be suggested; by the regular discussing of them, they would be reduced to a definite and consistent form: and by the authority of united suffrages, they would have full influence over the governors of the charity. The exertions of individuals, however benevolent or judicious, often give rise to jealousy; are opposed by those who have not been consulted; and prove inefficient by wanting the collective energy of numbers.

18. The harmonious intercourse, which has been recommended to the gentlemen of the faculty, will naturally produce frequent consultations, viz., of the physicians on medical cases, of the surgeons on chirurgical cases, and of both united in cases of a compound nature, which falling under the department of each, may admit of elucidation by the reciprocal aid of the two professions.

19. In consultations on medical cases, the junior physician present should deliver his opinion first, and the others in the progressive order of their seniority. The same order should be observed in chirurgical cases; and a majority should be decisive in both: but if the numbers be equal, the decision should rest with the physician or surgeon, under whose care the patient is placed. No decision, however, should restrain the acting practitioner from making such variations in the mode of treatment, as future contingencies may require, or a farther insight into the nature of the disorder may show to be expedient.

20. In consultations on mixed cases, the junior surgeon should deliver his opinion first, and his brethren afterwards in succession, according to progressive seniority. The junior physician present should deliver his opinion after the senior surgeon; and the other physicians in the order above prescribed.

21. In every consultation, the case to be considered should be concisely stated by the physician or surgeon, who [79/80] requests the aid of his brethren. The opinions rela-

tive to it should be delivered with brevity, agreeably to the preceding arrangement, and the decisions collected in the same order. The order of seniority, among the physicians and surgeons, may be regulated by the dates of their respective appointments in the hospital.[24]

22. Due notice should be given of a consultation, and no person admitted to it, except physicians and surgeons of the hospital, and the house-surgeon, without the unanimous consent of the gentlemen present. If an examination of the patient be previously necessary, the particular circumstances of danger or difficulty should be carefully concealed from him, and every just precaution used to guard him from anxiety or alarm.

23. No important operation should be determined upon, without a consultation of the physicians and surgeons, and the acquiescence of a majority of them. Twenty-four hours' notice of the proposed operation should be given, except in dangerous accidents, or when peculiar occurrences may render delay hazardous. The presence of a spectator should not be allowed during an operation, without the express permission of the operator.[25] All extra-official interference in the management of it should be forbidden. A decorous silence ought to be observed. It may be humane and salutary, however, for one of the attending physicians or surgeons to speak occasionally to the patient; to comfort him under his sufferings; and to give him assurance, if consistent with truth, that the operation goes on well, and promises a speedy and successful termination.[26]

As an hospital is the best school for practical surgery, it would be liberal and beneficial to invite, in rotation, two [80/81] surgeons of the town, who do not belong to the institution, to be present at each operation.

24. Hospital consultations ought not to be held on Sundays, except in cases of urgent necessity; and on such occasions an hour should be appointed, which does not interfere with attendance on public worship.[27]

25. It is an established usage, in some hospitals, to have a stated day in the week, for the performance of operations. But this may occasion improper delay, or equally unjustifiable anticipation. When several operations are to take place in succession, one patient should not have his mind agitated by the knowledge of the sufferings of another. The surgeon should change his apron, when besmeared; and the table or instruments should be freed from all marks of blood, and every thing that may excite terror.

26. Dispensaries afford the widest sphere for the treatment of diseases, comprehending, not only such as ordinarily occur, but those which are so infectious, malignant, and fatal, as to be excluded from admission into infirmaries. Happily, also, they neither tend to counteract that spirit of independence, which should be sedulously fostered in the poor, nor to preclude the practical exercise of those relative duties, "the charities of father, son, and brother," which constitute the strongest moral bonds of society. Being institutions less splendid and extensive than hospitals, they are well adapted to towns of moderate size; and might even be established, without difficulty, in populous country districts. Physicians and surgeons, in such situations, have generally great influence: and

it would be truly honourable to exert it in a cause subservient to the interests of medical science, of commerce, and of philanthropy.

The duties which devolve on gentlemen of the faculty, engaged in the conduct of Dispensaries, are so nearly similar to those of hospital physicians and surgeons, as to be comprehended under the same professional and moral rules. But greater authority and greater condescension will be found [81/82] requisite in domestic attendance on the poor; and human nature must be intimately studied, to acquire that full ascendancy over the prejudices, the caprices, and the passions of the sick, and of their relatives, which is essential to medical success.

27. Hospitals, appropriated to particular maladies, are established in different places, and claim both the patronage and the aid of the gentlemen of the faculty. To an asylum for female patients, labouring under syphilis, it is to be lamented that discouragements have been too often and successfully opposed. Yet whoever reflects on the variety of diseases to which the human body is incident, will find that a considerable part of them are derived from immoderate passions, and vicious indulgencies. Sloth, intemperance, and irregular desires are the great sources of those evils, which contract the duration, and embitter the enjoyment of life. But humanity, whilst she bewails the vices of mankind, incites us to alleviate the miseries which flow from them. And it may be proved, that a *lock hospital* is an institution founded on the most benevolent principles, consonant to sound policy, and favourable to reformation and to virtue. It provides relief for a painful and loathsome distemper, which contaminates, in its progress, the innocent as well as the guilty, and extends its baneful influence to future generations. It restores to virtue and to religion, those votaries whom pleasure has seduced, or villainy betrayed; and who now feel, by sad experience, that ruin, misery, and disgrace, are the wages of sin. Over such objects pity sheds the generous tear; austerity softens into forgiveness; and benevolence expands at the united pleas of frailty, penitence, and wretchedness.[28]

A few remarks on the preceding axioms are necessary.

The relief afforded by hospitals, though they are institutions of the most benevolent kind, is procured with difficulty; patients are admitted only 1 day in the week, no matter how dangerous their cases (unless they be accidents), fees [82/83] are often required, the sick are removed from their families, the nurses are strangers. These defects are so manifest that the public have wisely established dispensaries to obviate them. In these establishments medical assistance is obtained with the greatest facility every day; it is afforded to one parent, without removing him from the means of earning support for himself and family, and to the other without withdrawing her from the superintendence of her domestic concerns. Besides, the natural affections, which every philanthropic mind must wish to see cherished, are reciprocally called into exercise, and strengthened, where the parent is the patient, where the wife becomes the nurse, and the children assistants; and medical aid is rendered more efficacious when the mind is relieved from the anxieties necessarily attendant upon a separation from family, and a removal from home. The early application on the first feelings of indisposition prevents the diffusion of contagious diseases; and pestilence, which once stalked forth, spreading terror and desolation around, is now arrested in its progress, or strangled in its birth; and, it is not too much to assert, that the general healthiness of the metropolis, and the less frequent recurrence of contagious disorders, are to be in a great measure attributed to their early suppression in the abodes of poverty, by the activity and vigilance of the medical officers of dispensaries. It is also apparent, that, without the medical assistance thus afforded the poor, the demand on parochial rates would be increased in a very considerable degree, and the medical establishments of every parish would be increased to double their ordinary expenditure.

But it is quite contrary to the objects for which hospitals and dispensaries are founded, to render them subservient to those in affluent circumstances; an abuse which exists in every one of them. This is an imposition on charity, and a direct injury to the profession; yet the medical officers cannot prevent it. It is a fact which cannot be controverted, that a large proportion of the patients admitted into the hospitals (especially of this metropolis), and relieved at dispensaries, are not real objects of charity, and are often the relatives, or personal friends, or servants, of the governors or subscribers; [83/84] and thus the junior members of the profession are seriously injured. This abuse exists in every part of the empire, but to a vast extent in this metropolis. I have often remonstrated with my colleagues, and with governors, on this impropriety; but their reply was ready – "These things are tolerated in every public institution."

It has been a maxim with the faculty, that a practitioner of standing, a senior, should be called over the ordinary attendant. This rule is often violated, and indeed it is not an easy matter to observe it on many occasions. The late eminent Dr. James Gregory, of Edinburgh, has commented with his usual force on this point. He says,

> but mere *standing,* or seniority, superadded to the most complete and regular education in the profession, will neither procure confidence from the public, nor success and employment to any person. We are well accustomed to see many juniors surpass, and most deservedly surpass their seniors, perhaps even their own instructors; and leave them so far behind, that, before half their race is run, they can have no farther hopes of success.
>
> Some individuals soon shew by their talents, and the use which they make of them, that they can profit more by seven years of observation and experience, than others could do in the longest life. And very many soon show that they are incapable of ever improving; from a real natural want of those faculties which would enable them to observe accurately, to compare different observations together, to reason acutely and fairly, and ultimately to draw just and useful practical inferences from what they had observed. Many, not naturally deficient in their intellectual powers, become so from defects or improprieties in their education; especially the want of that general preliminary education, which improves the faculties, while it extends the sphere of knowledge, and directs the attention to proper objects. And many more, who have no such excuse either from natural or accidental defects, never improve, and soon show that they never will, purely by their own fault. They think the knowledge or improvement they had acquired, when they first entered on the exercise of their profession, sufficient for all purposes, or, at least, for their purpose; they find the effort of attention in [84/85] observing, comparing, reading, and thinking, too laborious; and, as they flatter themselves it is unnecessary for them, they soon cease to make it.
>
> Of course, all chance of improvement in them is at an end; they grow older, and yet grow never the better or wiser. On the contrary, as they often become more negligent, they

grow worse in every respect, and really become more ignorant, forasmuch as they acquire no new knowledge, and forget much of what they had formerly learned.

They become a kind of drones, content to do their business in a humdrum workman-like sort of a way; by which they have the best chance of escaping reflections or censure. Their faults are much more frequently sins of omission than of commission. For once that they do any thing positively and immediately pernicious, they miss, from negligence or ignorance, or both, an hundred opportunities of doing good. None but those in the secret have any notion how faithfully many physicians and surgeons go on for thirty or forty years, or longer, if they live longer, employing, even in the commonest diseases, the remedies which they were taught when young, though useless at best, if not pernicious; how faithfully many great and grave writers have transcribed from their predecessors; from generation to generation, the same frivolous, absurd, or dangerous precepts, the same useless or pernicious prescriptions, and the same silly remarks; how tenaciously many practitioners adhere to old recipes, so extravagantly absurd as to contain perhaps fifty or a hundred ingredients, of which probably not more than three or four are of any use; and how manfully they fight against the introduction of other remedies, the most simple, powerful, and safe; which they reprobate, and will not employ, for no other reason but because they are new.

Men of such talents, characters, and habits, whether physicians or surgeons, can neither improve by experience themselves, nor contribute to the instruction of others, and the improvement of their art. They are peculiarly unfit to practise in an hospital, where, on account of the great number and urgency of the cases to be treated, the greatest extent and accuracy of knowledge, the greatest quickness, precision, [85/86] and discrimination in applying it; and, in one word, the greater effort of attention and thought is required. Any deficiencies in them, which in private practice might well have escaped observation and censure, must soon become conspicuous on so public a stage; just like those of a lawyer at the bar; and will not only bring on themselves reproach and contempt, but will in some measure affect the character of the hospital itself. Whatever lessens the confidence of the public in the administration of it, and of the patients who resort to it in the skill of those to whose care their health and lives are entrusted, tends strongly to frustrate the benevolent purpose of the institution, and is, in truth, a very great injury to the public. Some men, naturally of good sense and quick discernment, and active, vigorous minds, who attend accurately to what passes around them, are distinguished even at an early period of life for sagacity, prudence, decision, and quickness in conduct, and a thorough knowledge of the characters of men, and the management of business. They are accordingly respected in the world, and often consulted on nice and difficult occasions by those who are acquainted with them, and who very wisely rely more on the judgment of such men than they would do on their own.

But such men are not the majority of mankind. An infinitely greater number are either so deficient in natural talents, or so culpably negligent in the use they make of them, that they appear to acquire no improvement at all by their experience of men and things. At the age of fifty or sixty, they are a good deal more dull, but not a jot wiser, than they were at twenty-five or thirty. They become as arrant drones in common life as any are in law, or physic, or surgery. No man of sense, who knows them, would ever think of consulting them, or relying on their judgment, in any business whatever, any more than he would think of consulting a lawyer when he was sick, or a physician when he was engaged in a law-suit.

A man of such a character can never deserve respect, or confidence, or employment, even in his own profession: and there are many such in law, in physic, in surgery, and in all the employments of life.[29] [86/87]

It must be unnecessary to enter into serious proofs of the importance of consultations. The mere want of medical assistance, says the distinguished physician whom I have just quoted, is in many cases so bad, as to imply almost certainly very pernicious, if not fatal consequences. In such cases, to withhold it voluntarily would be almost as criminal as to suffer a wretch to perish by withholding food from him. This point being proved, a few words may be said on the utility of numerous consultations. The opinion of Dr. James G. is so excellent upon this topic, that it must be quoted.

> With respect to physicians and surgeons both, and their patients, it is plain that all the good that can be expected from a consultation may be obtained from one of two, or three, or four, at the utmost, at least as well as from one ten times as numerous; and I should think it almost as plain, that much of that good may be prevented, and much positive evil done, by a very numerous consultation.
>
> On this point I presume, without vanity, to know as much as most men. For full half of my life, I have been a professor of physic in the University of Edinburgh, during which time consultations have been a great part of my business, to the number certainly of some thousands. Nineteen times out of twenty, at least, I have been the youngest physician of the consultation; and, of course, when any written directions were to be given to the patient, have had the honour to put them in writing, to the number, I presume, of two or three hundred at least. I can say with confidence, in point of fact, that I never knew any good come of a very numerous consultation; and I doubt much, whether any physician or surgeon of competent experience will give a different account of the result of what he has observed. The conduct of physicians and surgeons, when themselves or any of their families are sick, affords a still better proof and illustration of the same truth, and is indeed supreme and decisive authority with respect to what is useful, or what is useless, or worse than useless, in medical consultations. With us all considerations of economy are out of the question. Bad as we may be thought, we are not such cannibals as to prey on one another. We may all have, for nothing, to ourselves and our families, [87/88] as much assistance in point of physic and surgery as we choose. We feel strongly, that we have not sufficient calmness and firmness to judge and act properly, when the lives of those are at stake in whom we are most tenderly interested: and, as to ourselves when sick, we all know, for it is a long settled point in physic, that every man who doctors himself has a fool for his patient.
>
> For these reasons, we are all accustomed, when ourselves or our families are sick, to ask the assistance, not of all, but of some of our professional brethren. A numerous consultation is a kind of debating society, in which the patient's welfare, which ought to be the only object in view, is nearly forgotten. The illustrations of such consultations by Moliere, Le Sage, Fielding, and many others, were just, though inapplicable at present. In former times the *odium medicum* was as violent as the *odium theologicum,* even matters went so far that the disputants resorted to arms; but there is little danger of modern theorists taking the field in support of their opinions, though they war with words fully as bitterly as their predecessors. – *Op. Cit.*[30]

Section II.   Of Professional Conduct in Private or General Practice

1. The moral rules of conduct, prescribed towards hospital patients, should be fully adopted in private or general practice. Every case, committed to the charge of a physician or surgeon, should be treated with attention, steadiness, and humanity: reasonable indulgence should be granted to the mental imbecility and caprices of the sick: secrecy and delicacy, when required by peculiar

circumstances, should be strictly observed. And the familiar and confidential intercourse, to which the faculty are admitted in their professional visits, should be used with discretion, and with the most scrupulous regard to fidelity and honour.

2. The strictest temperance should be deemed incumbent on the faculty; as the practice both of physic and surgery at all times requires the exercise of a clear and vigorous understanding: and, on emergencies, for which no professional man [88/89] should be unprepared, a steady hand, an acute eye, and an unclouded head, may be essential to the well-being, and even to the life, of a fellow-creature.[31]

3. A physician should not be forward to make gloomy prognostications; because they savour of empiricism, by magnifying the importance of his services in the treatment or cure of the disease. But he should not fail, on proper occasions, to give to the friends of the patient, timely notice of danger, when it really occurs, and even to the patient himself, if absolutely necessary. This office, however, is so peculiarly alarming, when executed by him, that it ought to be declined whenever it can be assigned to any other person of sufficient judgment and delicacy. For the physician should be the minister of hope and comfort to the sick, that by such cordials to the drooping spirit, he may smooth the bed of death; revive expiring life; and counteract the depressing influence of those maladies, which rob the philosopher of fortitude, and the Christian of consolation.

4. Officious interference, in a case under the charge of another, should be carefully avoided. No meddling inquiries should be made concerning the patient; no unnecessary hints given, relative to the nature or treatment of his disorder; nor any selfish conduct pursued, that may directly or indirectly tend to diminish the trust reposed in the physician or surgeon employed. Yet though the character of a professional busy-body, whether from thoughtlessness or craft, is highly reprehensible, there are occasions which not only justify but require a spirited interposition. When artful ignorance grossly imposes on credulity; when neglect puts to hazard an important life; or rashness threatens it with still more imminent danger; a medical neighbour, friend, or relative, apprized of such facts, will justly regard his interference as a duty. But he ought to be careful, that the information on which he acts, is well founded; that his motives are pure and honourable; and that his judgment of the measures pursued is built on experience and practical knowledge, not on speculative or theoretical differences of opinion. The particular circumstances of the case will suggest the most proper mode of conduct. In general, however, a personal and confidential [89/90] application to the gentlemen of the faculty concerned, should be the first step taken, and afterwards, if necessary, the transaction may be communicated to the patient or to his family.

5. When a physician or surgeon is called to a patient, who has been before under the care of another gentleman of the faculty, a consultation with him should be even proposed, though he may have discontinued his visits: his practice, also, should be treated with candour, and justified, so far as probity and truth will permit. For the want of success in the primary treatment of a case is no impeachment of professional skill or knowledge; and it often serves to throw light on the

nature of a disease, and to suggest to the subsequent practitioner more appropriate means of relief.

6. In large and opulent towns, the distinction between the provinces of physic and surgery should be steadily maintained. This distinction is sanctioned both by reason and experience. It is founded on the nature and objects of the two professions; on the education and acquirements requisite for their most beneficial and honourable exercise; and tends to promote the complete cultivation and advancement of each. For the division of skill and labour is no less advantageous in the liberal than in the mechanic arts: and both physic and surgery are so comprehensive, and yet so far from perfection, as separately to give full scope to the industry and genius of their respective professors. Experience has fully evinced the benefits of the discrimination recommended, which is established in every well regulated hospital, and is thus expressly authorized by the faculty themselves, and by those who have the best opportunities of judging of the proper application of the healing art. No physician or surgeon, therefore, should adopt more than one denomination, or assume any rank or privileges different from those of his order.[32]

7. Consultations should be promoted, in difficult or protracted cases, as they give rise to confidence, energy, and more enlarged views in practice. On such occasions no rival- [90/91] ship or jealousy should be indulged: candour, probity, and all due respect should be exercised towards the physician or surgeon first engaged; and as he may be presumed to be best acquainted with the patient and with his family, he should deliver all the medical directions agreed upon, though he may not have precedency in seniority or rank. It should be the province, however, of the senior physician, first to propose the necessary questions to the sick, but without excluding his associate from the privilege of making farther inquiries, to satisfy himself, or to elucidate the case.

8. As circumstances sometimes occur to render a special consultation desirable, when the continued attendance of another physician or surgeon might be objectionable to the patient, the gentleman of the faculty whose assistance is required, in such cases, should pay only two or three visits; and sedulously guard against all future unsolicited interference. For this consultation a double gratuity may reasonably be expected from the patient, as it will be found to require an extraordinary portion both of time and attention.

In medical practice, it is not an unfrequent occurrence, that a physician is hastily summoned, through the anxiety of the family, or the solicitation of friends, to visit a patient, who is under the regular direction of another physician, to whom notice of this call has not been given. Under such circumstances, no change in the treatment of the sick person should be made, till a previous consultation with the stated physician has taken place, unless the lateness of the hour precludes meeting, or the symptoms of the case are too pressing to admit of delay.

9. Theoretical discussions should be avoided in consultations, as occasioning perplexity and loss of time. For there may be much diversity of opinion, concerning speculative points, with perfect agreement in those modes of practice, which are founded not on hypothesis, but on experience and observation.

10. The rules prescribed for hospital consultations, may be adopted in private or general practice.[33] And the seniority of [91/92] a physician may be determined by the period of his public and acknowledged practice as a physician, and that of a surgeon by the period of his practice as a surgeon, in the place where each resides. This arrangement, being clear and obvious, is adapted to remove all grounds of dispute amongst medical gentlemen: and it secures the regular continuance of the order of precedency, established in every town, which might otherwise be liable to troublesome interruptions by new settlers, perhaps not long stationary.[34]

11. A regular academical education furnishes the only presumptive evidence of professional ability, and is so honourable and beneficial, that it gives a just claim to pre-eminence among physicians, in proportion to the degree in which it has been enjoyed and improved: yet, as it is not indispensably necessary to the attainment of knowledge, skill, and experience, they who have really acquired, in a competent measure, such qualifications, without its advantages, should not be fastidiously excluded from the privileges of fellowship. In consultations, especially, as the good of the patient is the sole object in view, and is often dependent on personal confidence, the aid of an intelligent practitioner ought to be received with candour and politeness, and his advice adapted, if agreeable to sound judgment and truth.

12. Punctuality should be observed in the visits of the faculty, when they are to hold consultation together. But as this may not always be practicable, the physician or surgeon, who first arrives at the place of appointment, should wait 5 min for his associate, before his introduction to the patient, that the unnecessary repetition of questions may be avoided: no visits should be made but in concert, or by mutual agreement: no statement or discussion of the case should take place before the patient or his friends, except in the presence of each of the attending gentlemen of the faculty, and by common consent; and no prognostications should be deli- [92/93] vered, which are not the result of previous deliberation and concurrence.

13. Visits to the sick should not be unseasonably repeated; because, when too frequent, they tend to diminish the authority of the physician, to produce instability in his practice, and to give rise to such occasional indulgences, as are subversive of all medical regimen.

    Sir William Temple has asserted, that "an honest physician is excused for leaving his patient, when he finds the disease growing desperate, and can, by his attendance, expect only to receive his fees, without any hopes or appearance of deserving them." But this allegation is not well founded: for the offices of a physician may continue to be highly useful to the patient, and comforting to the relatives around him, even in the last period of a fatal malady, by obviating despair, by alleviating pain, and by soothing mental anguish. To decline attendance, under such circumstances, would be sacrificing, to fanciful delicacy and mistaken liberality, that moral duty which is independent of, and far superior to, all pecuniary appreciation.

14. Whenever a physician or surgeon officiates for another who is sick or absent, during any considerable length of time, he should receive the fees accruing from such additional practice; but if this fraternal act be of short duration, it should

be gratuitously performed, with an observance always of the utmost delicacy towards the interest and character of the professional gentleman, previously connected with the family.

15. Some general rule should be adopted by the faculty, in every town, relative to the pecuniary acknowledgments of their patients; and it should be deemed a point of honour to adhere to this rule, with as much steadiness, as varying circumstances will admit. For it is obvious, that an average fee, as suited to the general rank of patients, must be an inadequate gratuity from the rich, who often require attendance not absolutely necessary; and yet too large to be expected from that class of citizens, who would feel a reluctance in calling for assistance, without making some decent and satisfactory retribution.

But in the consideration of fees, let it ever be remembered, [93/94] that though mean ones from the affluent are both unjust and degrading, yet the characteristic beneficence of the profession is inconsistent with sordid views, and avaricious rapacity. To a young physician, it is of great importance to have clear and definite ideas of the ends of his profession; of the means for their attainment; and of the comparative value and dignity of each. Wealth, rank, and independence, with all the benefits resulting from them, are the primary ends which he holds in view; and they are interesting, wise, and laudable. But knowledge, benevolence, and active virtue, the means to be adopted in their acquisition, are of still higher estimation. And he has the privilege and felicity of practising an art, even more intrinsically excellent in its mediate than its ultimate objects. The former, therefore, have a claim to uniform pre-eminence.

16. All members of the profession, including apothecaries as well as physicians and surgeons, together with their wives and children, should be attended gratuitously by any one or more of the faculty, residing near them, whose assistance may be required. For as solicitude obscures the judgment, and is accompanied with timidity and irresolution, medical men, under the pressure of sickness, either as affecting themselves or their families, are peculiarly dependent upon each other. But visits should not be obtruded officiously, as such unasked civility may give rise to embarrassment, or interfere with that choice, on which confidence depends. Distant members of the faculty, when they request attendance, should be expected to defray the charges of travelling. And if their circumstances be affluent, a pecuniary acknowledgment should not be declined; for no obligation ought to be imposed, which the party would rather compensate than contract.

17. When a physician attends the wife or child of a member of the faculty, or any person very nearly connected with him, he should manifest peculiar attention to his opinions, and tenderness even to his prejudices. For the dear and important interests which the one has at stake supersede every consideration of rank or seniority in the other; since the mind of a husband, a father, or a friend, may receive a deep and lasting wound, if the disease terminate fatally, from [94/95] the adoption of means he could not approve, or the rejection of those he wished to be tried. Under such delicate circumstances, however, a conscientious physician will not lightly sacrifice his judgment; but will urge, with proper confidence,

the measures he deems to be expedient, before he leaves the final decision concerning them to his more responsible coadjutor.

18. Clergymen, who experience the *res angustæ domi*,[35] should be visited gratuitously by the faculty. And this exemption should be an acknowledged general rule, that the feeling of individual obligation may be rendered less oppressive. But such of the clergy as are qualified, either from their stipends or fortunes, to make a reasonable remuneration for medical attendance, are not more privileged than any other order of patients. Military or naval subaltern officers, in narrow circumstances, are also proper subjects of professional liberality.

19. As the first consultation by letter imposes much more trouble and attention than a personal visit, it is reasonable, on such an occasion, to expect a gratuity of double the usual amount. And this has long been the established practice of many respectable physicians. But a subsequent epistolary correspondence, on the further treatment of the same disorder, may justly be regarded in the light of ordinary attendance, and may be compensated, as such, according to the circumstances of the case, or of the patient.

20. Physicians and surgeons are occasionally requested to furnish certificates, justifying the absence of persons who hold situations of honour and trust in the army, the navy, or the civil departments of government. These testimonials, unless under particular circumstances, should be considered as acts due to the public, and therefore not to be compensated by any gratuity. But they should never be given without an accurate and faithful scrutiny into the case; that truth and probity may not be violated, nor the good of the community injured, by the unjust pretences of its servants. The same conduct is to be observed by medical practitioners when they are solicited to furnish apologies for non-attendance on juries; or to state the valetudinary incapacity of persons appointed to execute the business of constables, churchwardens, or overseers of the [95/96] poor. No fear of giving umbrage, no view to present or future emolument, nor any motives of friendship, should incite to a false, or even dubious declaration. For the general weal requires that every individual, who is properly qualified, should deem himself obliged to execute, when legally called upon, the juridical and municipal employments of the body politic. And to be accessory, by untruth or prevarication, to the evasion of this duty, is at once a high misdemeanour against social order, and a breach of moral and professional honour.

21. The use of quack medicines should be discouraged by the faculty, as disgraceful to the profession, injurious to health, and often destructive even of life. Patients, however, under lingering disorders, are sometimes obstinately bent on having recourse to such as they see advertized, or hear recommended, with a boldness and confidence which no intelligent physician dares to adopt with respect to the means that he prescribes. In these cases some indulgence seems to be required to a credulity that is insurmountable. And the patient should neither incur the displeasure of the physician, nor be entirely deserted by him. He may be apprized of the fallacy of his expectations, whilst assured, at the same time, that diligent attention should be paid to the process of the experiment he is so unadvisedly making on himself, and the consequent mischiefs,

if any, obviated as timely as possible. Certain active preparations, the nature, composition, and effects of which are known, ought not to be proscribed as quack medicines.[36]

22. No physician or surgeon should dispense a secret nostrum, whether it be his invention, or exclusive property. For if it be of real efficacy, the concealment of it is inconsistent with beneficence and professional liberality. And, if mystery alone give it value and importance, such craft implies either disgraceful ignorance, or fraudulent avarice.

23. The *Esprit du Corps* is a principle of action founded in human nature, and when duly regulated, is both rational and laudable. Every man who enters into a fraternity engages, by a tacit compact, not only to submit to the laws, but to promote the honour and interest of the association, so far as they are consistent with morality and the general good of [96/97] mankind. A physician therefore, should cautiously guard against whatever may injure the general respectability of his profession; and should avoid all contumelious representations of the faculty at large; all general charges against their selfishness or improbity; and the indulgence of an affected jocularity or scepticism, concerning the efficacy and utility of the healing art.

24. As diversity of opinion and opposition of interest may in the medical, as, in other professions, sometimes occasion controversy, and even contention; whenever such cases unfortunately occur, and cannot be immediately terminated, they should be referred to the arbitration of a sufficient number of physicians or of surgeons, according to the nature of the dispute; or to the two orders collectively, if belonging both to medicine and surgery. But neither the subject-matter of such references, nor the adjudication, should be communicated to the public; as they may be personally injurious to the individuals concerned, and can hardly fail to hurt the general credit of the faculty.

25. A wealthy physician or surgeon should not give advice gratis to the affluent; because it is an injury to his professional brethren. The office of a physician can never be supported but as a lucrative one; and it is defrauding, in some degree, the common funds for its support, when fees are dispensed with which might justly be claimed.

26. It frequently happens, that a physician, in his incidental communications with the patients of other physicians, or with their friends, may have their cases stated to him in so direct a manner as not to admit of his declining to pay attention to them. Under such circumstances, his observations should be delivered with the most delicate propriety and reserve. He should not interfere in the curative plans pursued; and should even recommend a steady adherence to them, if they appear to merit approbation.

27. A physician, when visiting a sick person in the country, may be desired to see a neighbouring patient, who is under the regular direction of another physician, in consequence of some sudden change or aggravation of symptoms. The conduct to be pursued, on such an occasion, is to give advice [97/98] adapted to present circumstances; to interfere no farther than is absolutely necessary with the general plan of treatment; to assume no future direction, unless it be expressly

desired; and, in this case, to request an immediate consultation with the practitioner antecedently employed.

28. At the close of every interesting and important case, especially when it hath terminated fatally, a physician should trace back, in calm reflection, all the steps which he had taken in the treatment of it. This review of the origin, progress, and conclusion of the malady; of the whole curative plan pursued; and of the particular operation of the several remedies employed, as well as of the doses and periods of time in which they were administered, will furnish the most authentic documents on which individual experience can be formed. But it is in a moral view that the practice is here recommended, and it should be performed with the most scrupulous impartiality. Let no self-deception be permitted in the retrospect; and if errors, either of omission or commission, are discovered, it behoves that they should be brought fairly and fully to the mental view. Regrets may follow, but criminality will thus be obviated. For good intentions, and the imperfection of human skill, which cannot anticipate the knowledge that events alone disclose, will sufficiently justify what is past, provided the failure be made conscientiously subservient to future wisdom and rectitude in professional conduct.

29. The opportunities which a physician not unfrequently enjoys, of promoting and strengthening the good resolutions of his patients, suffering under the consequences of vicious conduct, ought never to be neglected. And his counsels, or even remonstrances, will give satisfaction, not disgust, if they be conducted with politeness; and evince a genuine love of virtue, accompanied by a sincere interest in the welfare of the person to whom they are addressed.

30. The observance of the Sabbath is a duty to which medical men are bound, so far as is compatible with the urgency of the cases under their charge. Visits may often be made with sufficient convenience and benefit, either before the hours of going to church, or during the intervals of public worship. And in many chronic ailments, the sick, together with their [98/99] attendants, are qualified to participate in the social offices of religion; and should not be induced to forego this important privilege, by the expectation of a call from their physician or surgeon.

31. A physician who is advancing in years, yet unconscious of any decay in his faculties, may occasionally experience some change in the wonted confidence of his friends. Patients who before trusted solely to his care and skill, may now request that he will join in consultation, perhaps with a younger coadjutor. It behoves him to admit this change without dissatisfaction or fastidiousness, regarding it as no mark of disrespect but as the exercise of a just and reasonable privilege in those by whom he is employed. The junior practitioner may well be supposed to have more ardour than he possesses in the treatment of diseases; to be bolder in the exhibition of new medicines, and disposed to administer old ones in doses of greater efficacy. And this union of enterprize with caution, and of fervour with coolness, may promote the successful management of a difficult and protracted case. Let the medical parties, therefore, be studious to conduct themselves towards each other with candour and impartiality; co-operating by mutual concessions in the benevolent discharge of professional duty.

The commencement of that period of senescence when it becomes incumbent on a physician to decline the offices of his profession it is not easy to ascertain; and the decision on so nice a point must be left to the moral discretion of the individual. For one grown old in the useful and honourable exercise of the healing art may continue to enjoy, and justly to enjoy, the unabated confidence of the public. And whilst exempt, in a considerable degree, from the privations and infirmities of age, he is under indispensable obligations to apply his knowledge and experience in the most efficient way to the benefit of mankind. For the possession of powers is a clear indication of the will of our Creator concerning their practical direction. But, in the ordinary course of nature, the bodily and mental vigour must be expected to decay progressively, though perhaps slowly, after the meridian of life is past. As age advances, therefore, a physician should, from time to [99/100] time, scrutinize impartially the state of his faculties, that he may determine, *bona fide,* the precise degree in which he is qualified to execute the active and multifarious offices of his profession. And whenever he becomes conscious that his memory presents to him with faintness those analogies on which medical reasoning and the treatment of diseases are founded; that diffidence of the measures to be pursued perplexes his judgment; that, from a deficiency in the acuteness of his senses, he finds himself less able to distinguish signs, or to prognosticate events; he should at once resolve, though others perceive not the changes which have taken place, to sacrifice every consideration of fame or fortune, and to retire from the engagements of business. To the surgeon under similar circumstances this rule of conduct is still more necessary. For the energy of the understanding often subsists much longer than the quickness of eye-sight, delicacy of touch, and steadiness of hand, which are essential to the skilful performance of operations. Let both the physician and surgeon never forget, that their professions are public trusts, properly rendered lucrative whilst they fulfil them; but which they are bound, by honour and probity, to relinquish as soon as they find themselves unequal to their adequate and faithful execution.

Section III.   Of the Conduct of Physicians
Towards Apothecaries

1. In the present state of physic, in this country, where the profession is properly divided into three distinct branches, a connection peculiarly intimate subsists between the physician and the apothecary; and various obligations necessarily result from it. On the knowledge, skill, and fidelity of the apothecary depend, in a very considerable degree, the reputation, the success, and usefulness of the physician. As these qualities, therefore, justly claim his attention and encouragement, the possessor of them merits his respect and patronage.

2. The apothecary is, in almost every instance, the precursor of the physician; and being acquainted with the rise and [100/101] progress of the disease, with the hereditary constitution, habits, and disposition of the patient, he may furnish very

important information. It is in general, therefore, expedient, and when health or life are at stake expediency becomes a moral duty, to confer with the apothecary before any decisive plan of treatment is adopted; to hear his account of the malady, of the remedies which have been administered, of the effects produced by them, and of his whole experience concerning the juvantia and lædentia[37] in the case. Nor should the future attendance of the apothecary be superseded by the physician: for if he be a man of honour, judgment, and propriety of behaviour, he will be a most valuable auxiliary through the whole course of the disorder, by his attention to varying symptoms; by the enforcement of medical directions; by obviating misapprehensions in the patient, or his family; by strengthening the authority of the physician; and by being at all times an easy and friendly medium of communication. To subserve these important purposes, the physician should occasionally make his visits in conjunction with the apothecary, and regulate by circumstances the frequency of such interviews: for if they be often repeated, little substantial aid can be expected from the apothecary, because he will have no intelligence to offer which does not fall under the observation of the physician himself; nor any opportunity of executing his peculiar trust, without becoming burthensome to the patient by multiplied calls, and unseasonable assiduity.

3. This amicable intercourse and co-operation of the physician and apothecary, if conducted with the decorum and attention to etiquette, which should always be steadily observed by professional men, will add to the authority of the one, to the respectability of the other, and to the usefulness of both. The patient will find himself the object of watchful and unremitting care, and will experience that he is connected with his physician, not only personally, but by a sedulous representative and coadjutor. The apothecary will regard the free communication of the physician as a privilege and mean of improvement; he will have a deeper interest in the success of the curative plans pursued and his honour and reputation will be directly involved in the purity and excellence of the [101/102] medicines dispensed, and in the skill and care with which they are compounded.

4. The duty and responsibility of the physician, however, are so intimately connected with these points, that no dependence on the probity of the apothecary should prevent the occasional inspection of the drugs, which he prescribes. In London, the law not only authorizes, but enjoins a stated examination of the simple and compound medicines kept in the shops. And the policy that is just and reasonable in the metropolis, must be proportionally so in every provincial town throughout the kingdom. Nor will any respectable apothecary object to this necessary office, when performed with delicacy, and at seasonable times; since his reputation and emolument will be increased by it, probable in the exact ratio, thus ascertained, of professional merit and integrity.

5. A physician called to visit a patient in the country, should not only be minute in his directions, but should communicate to the apothecary the particular view, which he takes of the case; that the indications of cure may be afterwards pursued with precision and steadiness; and that the apothecary may use the discretionary power committed to him, with as little deviation as possible from the general plan prescribed. To so valuable a class of men as the country apothecaries, great attention and respect is due. And as they are the guardians

of health through large districts, no opportunities should be neglected of promoting their improvement, or contributing to their stock of knowledge, either
by the loan of books, the direction of their studies, or by unreserved information on medical subjects. When such occasions present themselves, the maxim
of our judicious poet, is strictly true, "the worst avarice is that of sense." For
practical improvements usually originate in towns, and often remain unknown
or disregarded in situations, where gentlemen of the faculty have little intercourse, and where sufficient authority is wanting to sanction innovation.

6. It has been observed, by a political and moral writer of great authority, that
   "apothecaries' profit["] is become a bye-word, denoting something uncom-
   monly extravagant. This great apparent profit, however, is frequently no more
   than the [102/103] reasonable wages of labour. The skill of an apothecary, is a
   much nicer and more delicate matter than that of any artificer whatever; and the
   trust which is reposed in him is of much greater importance. He is the physician
   of the poor in all cases, and of the rich when the distress or danger is not very
   great. His reward, therefore, ought to be suitable to his skill and his trust, and it
   arises generally from the price at which he sells his drugs. But the whole drugs
   which the best employed apothecary, in a large market town, will sell in a year,
   may not, perhaps, cost him above 30 or 40 lb. Though he should sell them,
   therefore, for 300 or 400, or a 1,000% profit, this may frequently be no more
   than the reasonable wages of his labour charged, in the only way in which he
   can charge them, upon the price of his drugs.

7. Physicians are sometimes requested to visit the patients of the apothecary, in his
   absence. Compliance, in such cases, should always be refused, when likely to
   interfere with the consultation of the medical man usually employed by the sick
   person, or his family. It would be for the interest and honour of the faculty to
   have this practice altogether interdicted. Physicians are the only proper substi-
   tutes for physicians; surgeons for surgeons; and apothecaries for apothecaries.

8. When the aid of a physician is required, the apothecary to the family is
   frequently called upon to recommend one. It will then behove him to learn fully
   whether the patient or his friends have any preference or partiality; and this he
   ought to consult, if it lead not to an improper choice. For the maxim of Celsus
   is strictly applicable on such an occasion; Ubi par scientia, melior est amicus
   medicus quam extraneus.[38] But if the parties concerned be entirely indifferent,
   the apothecary is bound to decide according to his best judgment, with a
   conscientious and exclusive regard to the good of the person, for whom he is
   commissioned to act. It is not even sufficient that he selects the person on
   whom, in sickness, he reposes his own trust; for in this case, friendship justly
   gives preponderancy; because it may be supposed to excite a degree of zeal and
   attention, which might overbalance superior science or abilities. Without any
   regard to any personal, family, or [103/104] professional connections, he should
   recommend the physician whom he conscientiously believes, all circumstances
   considered, to be best qualified to accomplish the recovery of the patient.

9. In the county of Norfolk, and in the city of London, benevolent institutions have
   been lately formed, for providing funds to relieve the widows and children of
   apothecaries, and occasionally also members of the profession, who become

indigent. Such schemes merit the sanction and encouragement of every liberal physician and surgeon. And were they thus extended, their usefulness would be greatly increased, and their permanency almost with certainty secured. Medical subscribers, from every part of Great Britain, should be admitted, if they offer satisfactory testimonials of their qualifications. One comprehensive establishment seems to be more eligible than many on a smaller scale. For it would be conducted with superior dignity, regularity, and efficiency; with fewer obstacles from inter-est, prejudice, or rivalship; with consider-able saving in the aggregate of time, trouble, and expence; with more accuracy in the calculations, relative to its funds; and, consequently, with the utmost practicable extension of its dividends.[39]

Dr. Percival recommends the formation of district dispensaries, and, in an excellent pamphlet on the Farming of Parishes by surgeons, a practicable plan, now acted on in Warwickshire, for superseding that wretched mockery of medical aid, called "Parish Doctoring," by District Infirmaries, is suggested by Mr. H. L. Smith.

Mr. Phelan, of Clonmell, proposes to establish district infirmaries and dispensa-ries in Ireland, in the proportion of one to every 40,000 inhabitants, and that each ought to be attended by a physician, surgeon, and resident apothecary, as in the English hospitals and dispensaries. (On the Medical Charities of Ireland, 1835.)

The Commissioners of the New Poor Law Act of 1834, have divided several parishes into unions, and appoint the lowest medical bidders, however young and inexperienced, as medical officers. The average pay allowed is half-a-crown a head; but in cases of cholera, or other epidemics, the pay is only allowed for a certain number of sick, and the rest are to be attended for nothing! This Act must be amended speedily, as it is most injurious to the sick poor. [104/105]

Notes

1. *Editors' note:* See Footnote 1, Chapter 3.

2. *Author's note:* It will generally be found that there is disease in the brain or spinal marrow (cerebro-spinal system), or in some of the organs in the abdomen; as the stomach, liver, bowels, spleen, kidneys, womb, &c.

3. *Editors' note:* "About everything in general and another thing in particular."

4. *Author's note:* This is rarely the case at present; it, however, sometimes happens.

5. *Editors' note:* "If you wish to ignore new inventions, if you wish to ignore their elegance of novelty, then do not criticize with utterances of defilement and envy. It is permissible for you to extol, but not to [literal translation] put shoes on the feet of others." (The sense here appears to be: send others packing or into exile.)

6. *Editors' note:* "What number of years I have lived are the more mature years? Actions create (respected) old age; the actions and not the years are what you should enumerate."

7. *Editors' note:* The literal translation of this phrase is somewhat misleading. The general sense is as Ryan relates, that those who rely solely on experience, without appropriate scientific education, cannot really make sense of what they observe and hence remain ignorant.

8. *Editors' note:* "Many inconveniences encompass a man of advanced years" (Horace, *De Arte Poetica* 3.60.153).

9. *Editors' note:* "Old age, arriving late, debilitates the virility of the soul and alters vigor" (Virgil, *Aeneid*, 9.610–11).

10. *Editors' note:* Giorgio Baglivi (1668–1706) experimented in musculo-skeletal physiology; Marie-François-Xavier Bichat (1771–1802) was a prominent anatomist and surgeon. By including an example whose career came after Gregory's death, Ryan shows here that he has freely interpolated his own comments among those taken directly from Gregory.

11. *Editors' note:* Latin version omitted; Ryan's own translation follows.

12. *Author's note:* The obligations of this oath will be fully explained after the insertion of the whole of the British Moral Statutes.

13. *Editors' note:* Latin version omitted; Ryan's own translation follows.

14. *Editors' note:* Latin version omitted; Ryan's own translation follows.

15. *Editors' note:* "Thus the medical arts, undaunted, discern the workings of remedies, that sometimes bring forth health and other times are pernicious."

16. *Editors' note:* This portion of the Hippocratic Oath is commonly rendered as, "And whatsoever I shall see or hear in the course of my profession, as well as outside my profession in my intercourse with men, if it be what should not be published abroad, I will never divulge, holding such things to be holy secrets."

17. *Editors' note:* "You should proceed with the utmost modesty in treating women; and if you must touch or examine the breast or genitals, ensure trust by speaking smoothly and professionally, with a calm spirit, either by custom or formula. The worst rumors that could be attributed to you by corrupt gossip, understandably, are that you have a slippery hand, an impure spirit, and violate chastity. The very names of certain diseases that principally afflict girls and women, might be dangerous to them if you reveal their disgrace. Secretly, as if confided into your heart, hide any knowledge of sexual transgressions (*literally:* the wanderings of human genitalia). Be cautious in entering the homes of widows, of youthful virgins, of wanton wives, and of attractive servant women. I have witnessed foolish women and girls practically captive to insane lust, searching for a cure to their passions; and I have seen mothers disgraced by the deeds of virgins incautiously entrusted to the supervision of male tutors."

18. *Author's note:* Analysis of Medical Evidence.

19. *Editors' note:* Ryan's "sections" correspond to chapters in the original work by Percival (1803).

20. *Editors' note:* Paragraphs 9–12 of this section are shortened from Percival's original and the general topic only is given.

21. *Author's note:* There are no such case-books or registries in the London hospitals, even in 1835, and such as are kept are locked up, and students are seldom allowed to see them.

22. *Footnote in Percival (1803), copied here by Ryan:* See Dr. Aikin's Thoughts on Hospitals, p. 21. *Added by Ryan:* This suggestion is now universally adopted throughout the United Kingdom.

23. *Author's note:* This is now done by the medical committees of hospitals and dispensaries.

24. *Author's note:* Seniority, in my opinion, ought to be determined by the date of admission into the Colleges or Apothecaries' societies. Suppose Sir Astley Cooper [a prominent London surgeon of Ryan's day] were to go and reside in some provincial city, he could not be considered junior to some young provincial surgeon.

25. *Author's note:* It is usual in London, that all members of the profession, who choose, and all medical students, may be present.

26. *Author's note:* The barbarity of some old hospital surgeons during operations, even on children, is disgraceful to such individuals, and to a civilized country. *Editors' note:* Both Percival and Ryan practiced in the days before anesthesia was introduced.

27. *Author's note:* It is most improper to interfere with the religion of the sick, by circulating tracts, bribing patients, giving them flannel, &c. The sick ought to have free will in the selection of clergymen of their own.

28. *Footnote in Percival (1803), copied here by Ryan:* See two Reports, intended to promote the establishment of a Lock Hospital in Manchester, in the year 1774, inserted in the author's Essays Medical, Philosophical, and Experimental. Vol. II. p. 263. 4th edition. *Editors' note:* "Locked Hospitals" referred to asylums for the insane. Ryan here omits a further paragraph of 27 and 28–31, all of which relate to insane asylums.

29. *Editors' note:* Ryan here fails to cite the source for this long quotation from James Gregory, even though he later references another quotation "op. cit." as if he had already given the source.

30. *Editors' note:* Ryan here resumes quoting nearly verbatim from Percival.

31. *Editors' note:* Ryan here omits a brief anecdote about Philip of Macedon that is included in Percival's original text.

32. *Author's note:* This rule is right, though seldom observed. In London all the surgeons act as physicians, and also throughout the British dominions.

33. *Footnote in Percival (1803), copied here by Ryan:* See rules xix. xx. xxi. pp. 80, 81.

34. *Author's note:* Seniority is, in general, determined by the date of admission into the respective colleges. An individual, who is not a physician or surgeon according to law, may be the oldest resident, but does not belong to the profession.

35. *Editors' note*: narrow circumstances at home; poverty.

36. *Editors' note:* From this point on, Ryan, without explanation, omits the Roman numerals used by Percival to order his ethical pronouncements. We have reinserted the numerals according to Percival's 1803 edition.

37. *Editors' note:* benefits and harms.

38. *Editors' note:* "Where equal in knowledge, a friendly doctor is better than a stranger." We could not locate this passage in the works of Celsus.

39. *Editors' note:* This concludes the verbatim quotation from Percival.

Chapter V
American Medical Ethics

Professional Reputation

The following description of professional reputation was published by Dr. Godman, Professor of Anatomy and Physiology in Rutger's Medical College in 1829, and is so graphic and admirable, that I am induced to copy it. This essay was read, by appointment, before the Philadelphia Medical Society, Feb. 8, 1826, and is extracted from a volume, entitled, "Addresses delivered on Various Occasions. By John D. Godman, M. D., &c.," and politely sent to me. It affords me much pleasure to place American Medical Ethics before British readers.

Our profession is coeval with the distresses and sufferings of the human race, and its respectability is as universal as the benefits it is capable of conferring, when rightly administered – those engaged in the discharge of its duties having always been tacitly considered by their fellow-men, as beings peculiarly set apart from the rest of mankind, and worthy of an estimation, not conceded to persons employed in merely secular affairs.[1] The real excellence and usefulness of our art – when worthily practised – has always tended to increase the confidence and admiration of the public; and, if medicine have not attained a degree of perfection and immunity from censure, equal to its venerable age and importance to society, this results from circumstances, which, however they may have injured, are entirely extrinsic to the profession.

[105/106] Yet our useful and excellent science presents a great number of obstacles and difficulties to her votaries, which are only to be surmounted by well-directed and most persevering efforts. A mistake made in the outset, may exert its prejudicial influence on the whole of your subsequent course; therefore it is desirable, that your principles and views should be both early and correctly formed.

The members of our profession are subjected to many temptations from ambition, which are scarcely to be resisted. Few, perhaps none of us, are willing to look upon our art as a mere mode of obtaining subsistence, whatever be our situations. We hope to gain a reputation, or fame, by the exercise or improvement of it; and this is the unseen, but ever operative cause, which urges us forward in our variously deviating careers. This desire of fame – this hungering after the approbation of the wise and good of our species – this wish to be singled out and placed above the great mass of our fellow-creatures – is a perfectly natural feeling, and of kin to immortality. To this cause we are indebted for the noblest exertions of human genius; it was this feeling which incited all the great of former days to the actions which still live on the page of history; – and the same breath will continue to enkindle from their ashes, fires which shall warm, cherish, and enlighten, universal society.

There are two kinds of fame, between which it is necessary for you to know the distinctions. The first, and only excellent, is that which tempts the wise and good man to become *great;* whose influence is not only felt during the existence of the possessor, but leaves behind it a holy light, undimmed and undiminished by the lapse of ages. This fame is built upon the solid basis of usefulness, genuine worth, and high desert. Its growth is not rapid, but its maturity is perfect; at first, it is the applause of those who are emphatically called "the few;" it is not gained until many privations and toils have been endured; yet, like the ascending sun, it surely attains a meridian altitude, and disperses by the potency of its irradiations, all clouds which would obscure or intercept its brightness.

The other kind of fame, is "base, common, and popular." [106/107] It is never the result of great intellectual exertion – often it is produced by accident, and it frequently is awarded to great vice. At first, it may *appear* bright and dazzling; but this light is the phosphorescent gleam hovering over putrefying substances, compared with the intense, steady, and sun-like ray of that first mentioned. This second fame, is the clamorous plaudit of the deceived or ignorant crowd; it is sustained solely by the breath of the vulgar herd, and would sink for ever in a purer atmosphere.

The fame that you should desire to win, is that which rewards the exertions of generous and virtuous minds. But, you should not only feel the proper emulation – you must be aware of the best mode of attaining your object. Let the intellectual capacity be what it may, or the impulse of ambition never so strong, much time may be wasted in ill-directed and desultory efforts, without the proper training and preparation; even giant strength may be rendered worse than useless, for want of skill to direct its exertions.

A first requisite to your success, is a good education, concerning the best mode of gaining which, wise men have differed in opinion. As the great object is to enlarge the mind, stock it with images, and train it to habits of investigation and sound reasoning, "a classical education" may be stated, as of the various modes, one of the best adapted for the discipline and development of the intellectual powers[.] Of this education we consider the study of those languages, whence not only our technical phrases, but our mother tongue itself are derived, as a most essential and vitally important part.

In speaking thus, we are conscious of advancing an opinion directly opposed to notions which, of late, are becoming very general and fashionable. It is easier, however, to declaim against the ancient languages, than to learn and employ them; as to the indolent, it is far more agreeable to demolish a noble edifice, than to erect even a comfortable shed.

The correctness of the opinion we have advanced, is not supported by assertion merely: it will bear close examination – and equally resist the subtilties of sophistry, or the ruder shocks of ignorance. Other preparatory branches of educa- [107/108] tion have their specific value by the aid of mathematics, the mind is sobered, sharpened, subtilized – taught to abstract itself and become concentrated on a point: to reach out and grasp the almost inconceivable combinations of numbers, or the ineffable extensions of space. But it is with man that physicians have to do – in all his varieties – his excellence, his errors, and his sufferings; it is with the hidden springs of the passions and emotions of our race that we wish to become acquainted; it is with the defaced, not destroyed, image of the Creator, that we are to be continually engaged. We cannot comprehend man better, than by understanding the manner in which he communicates his sensations and wishes to those around him; learning from the context of his thoughts and modes of expression, the nature of the mind whence they spring; and having gained thus much, become better able to make ourselves and our profession more useful and acceptable.

We can neither acquire nor impart knowledge, without the use of words. These, however imperfect, are the signs of our ideas; hence, he who is acquiring a language – if taught aright – is, at the same time, accumulating a vast store of objects for the future exercise of his intellect, and is also forming habits of reflection and discrimination rarely to be attained in any other way.

Independent of other advantages, the language of Judea, Greece, and Rome, are particularly worthy of regard, as containing the most sublime exertions of genius – the most valuable body of truth – and, moreover, as being the fountains, whence the now widely flowing streams of knowledge gushed forth to animate and adorn the world, after the prolonged and dreary periods of its cheerless gloom. In the tongue first mentioned, we see language in its ancient and original form, venerable alike for its simplicity and force. By it are we instructed of the origin of our race, and the commencement of human society. In the Greek, we see language refined to the highest degree, and are furnished, through it, with models in almost every exercise of human intellect. It is not only the tongue by which the invaluable observations of the primitive father of our science are preserved; but we have also [108/109] delivered to us, in the same language, the words of Him "who spoke as never man spake." In the Latin tongue, we have an inexhaustible storehouse of intellectual gratification; it is, moreover, the true language of science – the ideas attached to the words being fixed, and freed from the mutations to which a living language must always be subject – it is the key to a great number of living dialects, forms a large part of the body and substance of our own tongue, and constitutes, along with the Greek, almost the whole mass of the language consecrated to the use of our profession. Hence, those who enter upon the study of medicine without having learned either of these languages, necessarily meet with numerous difficulties, which the instructed have not to encounter; and, even with the most assiduous attention, a large amount of their professional reading must remain unintelligible.

Opposers of classical education object, that the time necessary for the acquisition of learning, might be more profitably employed, as if the student were not learning to think and judge correctly; at the same time, filling his mind with ideas, and becoming well versed in the history and characters of men. Let it be remembered, that any or all of these languages may be studied while the memory is vigorously retentive and the judgment unformed. The exercise afforded by such studies, develops every faculty of the mind; the memory is replete with words, and if the studies be correctly pursued, the mind becomes acquainted with the things to which they relate; the habit of patiently investigating, and understanding the philosophy of modes of expression, teaches proper care, and gives us greater skill in our own language; and the attention is awakened to its true value and meaning, which otherwise might be neglected from habit.[2] [109/110]

You may inquire how the acquisition of such knowledge is to assist *you* in becoming distinguished in your profession. The answer is easy; nothing is more essential to the success of a physician, than a facility of communicating his own sentiments, as well as of understanding the sentiments of these who consult him. He must approach persons of every rank in society, and commune with every variety of intellect. Possessing a well-grounded knowledge of language – fully acquainted with the true value and nature of his own, which is derived from various other tongues – he is always prepared; whether by speech or writing, he addresses those with whom he is concerned successfully, because he is sure of making his wishes or opinions plain and intelligible to all. Classic learning has another influence, not less powerful or beneficial, on the human mind. The books, which should be studied, continually present the most excellent sentiments and morals, conveyed in the most refined style; and the superiority of such refinement over coarseness and vulgarity, will imperceptibly lead the student to an habitual imitation of them. The virtues of the good and the wise, and the examples of the truly great, will invite to a similarity of behaviour; while the conduct of an elegant scholar, will be a perpetual recommendation to the intelligence and acquirements of the physician.

Every commender of the learning we have spoken of, exposes himself to the charge of being prejudiced, or having too much veneration for mere antiquity. Instead of attempting to disprove such an unfounded accusation, let us employ the words of a celebrated author on the same subject. "I have," says he, "a great reverence for posterity; nor do I think lightly of the learned men of the present day. There are many, I know, who adorn our age, who would have ornamented any period; yet among the whole number, I have not known one who did not cultivate and honour ancient learning, whose wisdom was not similar to that of the ancients, or who did not admire and observe their precepts; from which, in proportion as you depart, you wander from nature and truth."[3] [110/111]

Many of the younger members of the society now present, are, ere long, to receive the honours of the profession, as a testimonial of their diligence and faithfulness as students, and will then be preparing to take their stand among the guardians of the public health. The boisterous sea of the world is attended with comparatively few dangers, to those who have not trifled away their time, and who set forth under the guidance of correct principles. Though you know not in what haven you may ultimately cast anchor, the possession of a sound moral and professional education, will insure your safety through all the turbulence you may be exposed to, during your voyage.

The greatest evil to be guarded against, when you commence your efforts for professional distinction, is impatience and instability of purpose. It will be wrong in you to anticipate that business can, in any situation, immediately follow your annunciation of being ready to receive it, more especially, where you are to meet with competition from a member [111/112] of the profession already established. The first half-year of a young physician's residence in a strange place, is the most trying part of his probation; for, should he mistake fewness of calls for neglect of his merits, or suppose that he will never be employed, because he is not immediately preferred to another, he is in danger of becoming unsettled, restless, neglectful of his books, society, and acquaintances, thus sacrificing the very means of his eventual success. If, however, we recollect how much people are prejudiced by education and habit, we shall find no fault with them for not employing a stranger on his first arrival, neither should we suppose their prejudices to be immovably fixed. A proper degree of patience, and an improvement of those opportunities that are every where presented of winning confidence, will, in no long time, yield us that opening, which is the great requisite to future profit and eminence. Wherever we attempt to establish ourselves – except under very extraordinary circumstances – some time must be passed in acquiring the confidence of those around us, by the recommendations of our friends, by our own deportment, and the event of such cases as may be incidentally thrown into our hands. However unimportant any such particulars of conduct may individually appear, they are of great moment collectively viewed; as every circumstance in the appearance, conversation, and character, of a newly-arrived physician, is of deep interest to those among whom he wishes to remain.

It will be unjustifiable to trust your success, in the slightest degree, to accident, because accident has occasionally given currency to men neither remarkable for education, talent, nor judgment. Accident *has*, at times, given a man of the highest merit an introduction to extensive business; but we must never forget, that accident cannot sustain our reputation, nor minister to our continued success. Our greatest care must be to acquire reputation by a diligent cultivation of our talents; though we should never neglect to improve any accidental success, in all honourable ways, to forward our professional views. If a character for skill and discernment be acquired suddenly, we must not attempt to increase it, by endeavouring to extend this reputation to the utmost stretch of possibility, [112/113] but by displaying new instances of talent and intellectual strength, thus substantially augment our capital of fame. The fortuitous elevation of men destitute of true greatness of character, is almost universally followed by reverses, as sudden and severe as this elevation was great. This is frequently exemplified in the fate of those who have a great air-built reputation, and much verbal fame; who, instead of modestly refusing a part of the honours proffered them, and exerting themselves *to prove* that they are worthily praised, receive the whole as no more than their right, and leave their admirers in a short time to discover, that their extraordinary pretensions are unfounded, and that their reputation is nothing better than the gratuitous offering of ill-judging and partial friends.

Next to an acquaintance with the principles[4] of the profession, and correct moral feeling, the young physician should most rely on the exercise of what may be summed up by the word *manners*. This embraces his general intercourse with society; in which he should display that habitual ease and cheerfulness, which results from correct habits of thought and action, and that kind attention to the wishes, prejudices, and necessities of those he associates with, which shows that he possesses the most generous and elevated feelings. To be accessible and attentive, without familiarity or cringing – to be mild, gentle, and forbearing, without sinking into tame submissiveness – to be ever ready to act when called on, without being officious or intrusive – and to do full justice to all those with whom he is professionally concerned, will insure a physician a degree of public respect, that may at length [113/114] amount almost to idolatry; filling every bosom with kindness towards him, and every mouth with his praise.

The sagacious Lord Bacon has given a rule for increasing our knowledge, and insuring conversation with all sorts of persons, which is one of the best that could be devised, and one more positively conducive to popularity can scarcely be imagined. This is, to learn something from all persons, whatever is their occupation, when we chance to be with them; this is always to be accomplished by inciting them to speak of what they know. As every man is better acquainted with his own business than we can possibly be, by inducing him to converse on the subject, we not only gain some valuable ideas, but we win his regard by manifesting an interest in what so peculiarly interests him. By adopting this method of Lord Bacon's, you need never suffer from tedium in the company of unlettered men, nor need you in the slightest degree to descend from your place, while judiciously exciting their remarks. This is not merely applicable to your intercourse with persons of inferior standing in society; the most learned, refined, and accomplished men, are equally pleased to find, that their pursuits, avocations, and interests, are interesting to you. You may frequently induce such persons to display before you, a stock of knowledge which otherwise would have been withheld, and you will part mutually satisfied, instead of being in ignorance of, or prejudiced against each other.

This rule may be observed with the utmost sincerity, and without the slightest approach to the meanness of flattery. The information thus to be acquired, will, in general, be far more easy of attainment, as well as more valuable, than can be gathered from many books; and you will, at the same time, be forming a more profound acquaintance with human nature, and also gain friends. Kindness, uniformly produces kindness; confidence inspires confidence. If we examine ourselves, we shall find that we are as excitable in this way as others, for we never deliver our thoughts with more force and feeling, than when we reply to interrogations on subjects in which our minds are most deeply engaged.

In all your intercourse with society, as well as in all your [114/115] thoughts and actions, cultivate an habitual tenderness of regard for TRUTH. By this, I would not pretend to warn you against the disgrace of falsehood; but, that you should guard against a habit, which is almost as common as the human family is numerous, of suffering apparently harmless exaggerations to escape. Truth and falsehood, like light and darkness, are opposite extremes – the one is as excellent as the other is base. But there are a great many aberrations from truth, which the world does not consider to be absolutely false; as there are many deviations from honesty, which, by a similar laxity, are not considered as positively dishonourable? If my wishes could influence, you should begin your career with resolving to adhere to the full purity of truth, and the perfect honourablenesss of honesty; so that when the day of your success arrives, you may look back on the means by which it was attained, without breathing a sigh of regret, or suffering one blush of shame.[5]

Our profession has long been subjected to the charge of "envy, hatred, malice, and all uncharitableness," among its members; and unfortunately, too much of the charge is well-founded. We cannot at present enter into an investigation of the causes by which this state of things has been produced, although it does not affect the profession to the degree which persons commonly suppose.[6] To lessen this evil, and avoid meriting such an accusation, make it a rule never to [115/116] speak of a professional rival, unless you can speak to his advantage; if he have merit, allow him the whole of it, and give your sentiments of his talents with the unaffected earnestness of truth. Do not imagine that your acknowledgment of his merits, will hide his defects, or obscure your own good qualities. Grant that he adopts a contrary course, speaks ill of you, or throws out insinuations intended to be prejudicial to your interests; – then is your triumph complete. Think you that men will not contrast his mean and soulless conduct, with your manly and honest candour? Think you that he will not more deeply condemn himself, by attempting to misrepresent you – that society will not visit his ungenerous conduct on his own head, while the profession silently spurn him from their confidence?

Should you be eminently successful after others have failed, avoid pushing your triumph so far as to wound the feelings and outrage the pride of your less fortunate competitors. Your success is sufficient for you, and by judicious deportment, you may compel a man to respect, if he does not esteem, who might otherwise cherish against you, a spirit as stern as hate – as inexorable as the grave. If, after such success as we have alluded to, you hear of disparaging suggestions made against you, by one you have set aside or over-shadowed, you are neither obliged to know, nor resent it[7]; you would owe it to the dignity of your own character, to recollect that some allowances are to be made for mortified feeling, as well as that no malicious insinuations can stand against the daily repetition of actions, which prove you exempt from a grovelling and miserably irritable disposition.

That you will not attain the professional elevation you desire, without struggling against hosts of difficulties, and encountering every degree of opposition, is most certain. "It may be, that the iron grasp of poverty, for a considerable time, will impede your progress and enfeeble your efforts. Against rivalry and opposition, your armour of principles and determined perseverance will afford every security, and [116/117] poverty itself may be made to minister to your success, by urging you to the display of your noblest powers.[8] Look at the men of talents who now lead the van of our profession, and are considered as its ornaments. Who are they? Men born to fortune and reared in the lap of luxury? No. Men who have been elevated by protection and patronage? – who have been favoured by circumstances, or raised by accident? No. They are, most frequently, those who have emerged from poverty, if not obscurity. Many of them have been nursed in sorrow, and baptized with tears; – they have protected and patronized themselves, until the great and powerful have become proud to rank as their friends; they have *made* the very circumstances, which superficial observers suppose to have been the *causes* of their elevation. It is the triumph of talent, of genius, to rise in proportion to the magnitude of difficulties; to trample the opposition of malignant mediocrity into the dust; and gaining its merited elevation, to raise the profession it has chosen to a corresponding degree of eminence.

Since the commencement of the present session of the society, some of our young friends, who entered on the career with hopes as warm, and eyes as bright as ours, have been called away to the "narrow house," and their spirit-stirring bustle of youthful expectation, has been exchanged for the solemn quiet of the tomb.[9] While we sympathize with such as mourn over youth snatched away in its blooming – and warm hopes chilled by the icy hand of death; while we sorrow over the mental anguish of those, whose far distant parents were not permitted to minister to their last earthly wants, or receive their dying sighs – let us not forget to be thankful, that we are still spared to usefulness and virtue. Yet a little [117/118] while, and the mighty ocean of oblivion will whelm us in its fathomless depths, sweeping away every trace of our existence. This is not matter of regret. All nature tends to one common point – disintegration and change of form; – the cloud-capped and tempest-braving mountains, towering in seemingly indestructible grandeur, are hourly yielding their atoms to the earth and the air. Virtue alone survives all change; the immortal mind bids defiance to the destructibility of matter.

Build then your monuments, imperishably, on the love of mankind, by sincerely devoting yourselves to the cause of humanity – to the honour of your profession and country – to the faithful service of your friends – to the humble worship of God: thus, the necessary evils of life will pass over you unheeded; and the inevitable shaft of death, while it stills for ever all mortal disquiets, shall be unable to disturb the serene and ecstatic composure of your intellectual being. [118/119]

Notes

1. *Editors' note:* Omitted here is a footnote copied from Godwin's original text, appending a quotation in Spanish from Don Juan de Zabaleta.
2. *Footnote copied from Godman's original:* In urging the importance and necessity of classical learning to those destined for the profession of medicine, it is no part of my intention to state, that the manner in which the languages are most generally taught is either the best or even the true one. It is but too common to ascribe the faults of teachers to the thing taught; to prejudge and condemn an unexplored region, because the ways leading thereto chance to be foul.
3. *Editors' note:* Omitted here is a footnote copied from Godman's original, giving this quotation in Latin, and crediting it to "And. Dacier." This may be André Dacier, who published a French translation of Hippocrates' works in 1697. Following this quotation Godman appended a further quotation in English, credited to "Harris's *Hermes,*" extolling the study of ancient languages. This would appear to be a reference to James Harris, *Hermes, or a Philosophical Inquiry Concerning Universal Grammar*, London, 1771.
4. *Footnote copied from Godwin's original:* "I would not be understood, in what I have here said, or may have said elsewhere, to undervalue *experiment,* whose importance and utility I freely acknowledge in the many curious nostrums and choice receipts with which it has enriched the arts of life. Nay, I go further – I hold all justifiable practice in every kind of subject, to be founded in *experience,* which is no more than the results of many repeated experiments. But I

must add, withal, that the man who acts from experience alone, though he acts ever so well, is but an empiric or quack; and that, not only in medicine; but in every other subject." – Harris's *Hermes*, 352.

5. *Editors' note:* Here is omitted a footnote copied from Godman's original containing a quotation in Latin from Marcus Antoninus.

6. *Editor's note:* Here is omitted a footnote copied from Godman's original containing a quotation in Spanish from la Zabaleta.

7. *Editors' note:* Here is omitted a footnote copied from Godwin's original, containing a quotation in Latin from Marcus Antoninus.

8. *Editors' note:* Here is omitted a footnote copied from Godwin's original, containing a Latin quotation attributed to "Reinwardt," whom we have been unable to identify further.

9. *Footnote copied from Godwin's original:* Several had then recently died of small-pox.

Chapter VI
Medical Education

Degrees – Diplomas – Medical Appointments – Success –
Reputation – Eminence – Moral and Physical Medicine – Art
of Prescribing

It is now universally admitted by all our medical institutions connected with education, that those intended for the study of the healing art should receive a good general and classical education. The professors of all our universities and medical schools, the examiners of all our colleges and apothecaries' societies, delivered their evidence in proof of this point, before Mr. Warburton's Parliamentary Committee on Medical Education and Practice, in the summer of 1834. All the witnesses proved that a knowledge of the Greek, Latin, English, French, German, and Italian languages; of mathematics, logic, moral philosophy, natural history and philosophy; in fine, that the course of general education required by the universities for degrees in arts, is indispensably necessary to those intended for the medical profession. This extensive course of preliminary instruction has long been required by the universities of all candidates for admission into the learned professions; and is now exacted by most of the colleges of surgeons, and societies of apothecaries, of those intended for the practice of medicine.

No class of men stand in so much need of extensive erudition and knowledge. They may attend all ranks of society from the lowest to the highest, and ought to be exceedingly well-informed in general as well as in medical literature.

A greater character for learning and science cannot exist than that which constitutes an accomplished physician; the most extensive study, and the most comprehensive mind, are therefore requisite. The youth intended for medicine should have all his senses acute, and in the utmost degree of perfec- [119/120] tion; because it is from the perceptibility of the senses alone, that the human mind is stored with all those sublime ideas which shine so conspicuously in the future life of a medical practitioner. When the senses are obtuse, dulness is the consequence; and a dull student never made a brilliant physician. The medical student should, therefore, have the clearest powers of perception, so as to receive accurate impressions, and possess tenacity in their retention.

He ought to have a good memory and great reflection. The intellectual faculties should be copiously enriched by indefatigable industry, and unceasing study; with the most extensive ideas of sensation and reflection; with memory, invention, and genius – so as to be capable of arranging a variety of ideas in strict logical order. When all demonstrative facts relating to our art are acquired, reflection, acute reasoning, and profound judgment, will decide the manner in which they are to be applied in the treatment of diseases. What a vast field of knowledge is comprehended in the medical sciences! How many days and years of labour and industry are required! What sedulous diligence is absolutely necessary! We must, therefore, pursue with an ambitious zeal the various branches of the medical and collateral sciences; we should study all, as all united form, with experience, the greatest medical characters.

We set out in pursuit of professional distinction when the buoyancy of youth, and the vigour of imagination lift us over every impediment, and break down every barrier. Hope tints the distance with the most glowing and flattering colours, and the mind revels in delightful anticipations of pleasure, fortune, and renown. A moderate experience in the cold realities of life proves that we have been dreaming, and teaches that if these good things are ever to be realized it is only when years of patient endurance have elapsed, and after the fires of youth have been well nigh expended in the service of our fellow-creatures. Accident may sometimes realize the expectations of youth, but the most universal rule is, that wealth and fame from professional exertion is the slow, though sure reward of long labour and persevering industry. This circumstance is of the greatest advantage to society, and to [120/121] our profession; but those who have yielded too much to the dominion of hope and fancy, are frequently so much affected by discovering the truth, as to suffer an entire revulsion of feeling, and sink from the most brilliant flights of imagination to the lowest depths of despair. This despondence is permitted sometimes to prey on the mind until it produces neglect of business or harsh misanthropy; and the unfortunate sufferer is continually tortured with notions of the ingratitude of mankind, the neglect of merit, the low state of professional character; while he is letting slip the best opportunities to convince himself of the contrary by efficiently performing those duties his profession enjoins, and society requires. Be then prepared to discover, that the world yields neither wealth nor distinction except as the price of industry and great deservings. Stop not to consider whether men are ungrateful or merit is neglected; but perform the actions that create a claim to their gratitude; declare your merits by the faithful discharge of your duties: and then you will find such complaint impossible.

If such were not to be the result, policy would dictate the propriety of concealing our mortification. The voice of repining and discontent is ever painful and offensive to others; and the same persons who warmly sympathize with a noble spirit struggling against misfortune, and, though broken-hearted, looking calmly on the approach of inevitable fate, despise the creature who is continually vexing their ears with fruitless and peevish complaints, or venting selfish ejaculations against the characters of those who have lived beneath a brighter sky, or been wafted along by more propitious gales.

Of this you may feel perfectly assured, that really meritorious conduct cannot go altogether unrewarded; neither can the fire of true genius be entirely smothered. The time must come when perseverance in the conscientious discharge of high duties will secure the remuneration and respect it is entitled to; the mind that has been wrought up by the study of proper objects, and is sustained by a determined enthusiasm, to effect great purposes, may for a time be weighed down by poverty or misfortune; but, like the giant of ancient fable, its struggles will convulse the superincumbent mass, [121/122] and must eventually shake off every hindrance to perfect success.

If, in offering these considerations to you on the present occasion, we appear diverging too far from the beaten track, we trust you will pardon the zeal that urges the laying before you, what reason and. experience induce us to hope may be to your advantage. Being exclusively devoted to the service of those who are engaged in the study of medicine, we

may be allowed in some degree to identify our feelings with theirs, and be anxious to spare them suffering, not less than to aid in insuring their success. Whatever defect there may be in manner, there is none in feeling; nor is there the slightest departure from fact in stating
For you, ye studious, I strive,
For you, I tame my youth to philosophic cares, And grow
still paler o'er the midnight lamp.[1]

We may, however, practise any one department suited to our taste or inclination; but we should study and know the whole. One practitioner prefers surgery, another obstetricy, a third aural, a fourth ophthalmic, a fifth dental surgery; others devote themselves to diseases of the brain and mind; some to maladies of the chest; more to those of the heart; many to disorders of the digestive system – stomach, intestines, liver, &c.; more to genito-urinary diseases; others to local affections – lithotrity, gout, &c. But every physician and surgeon should be conversant with the nature and treatment of every species of human disease. It is perfectly impossible to separate diseases into medical, surgical, and obstetrical; as any one disease, no matter what may be its situation, can and does derange and disorder all the organs and functions in the human body. Away, then, with the absurd and untenable distinctions of medicine, surgery, and obstetricy; the profession as a science, is one, and indivisible.

It is highly advantageous to the cultivator of the healing art, to visit different universities and schools, and to learn the principles and practice of medicine under the most celebrated [122/123] professors. It is for this reason, that students who commence in London, repair to Edinburgh, Dublin, Paris, Vienna, or Pavia. When a man of superior merit is to be found in any part of the world, his fame has great influence on strangers, and remote countries send him disciples and admirers. Few are ignorant of the great afflux of students from many nations to the lectures of Boerhaave, Morgagni, Monro, Hunter, Frank, Cullen, Scarpa, Abernethy, Cooper, &c., &c. It has often been a great advantage to have studied under such masters; and, even in our own time, under an Abernethy, a Cooper, a Dupuytren, a Bell, a Colles, &c. Lucrative and important appointments have been given to the pupils of these eminent individuals. There is, however, a more independent spirit abroad in our day; and few are ready to swear by the word of a master. Moreover, the opinions of all eminent practitioners are generally published, and may be procured with the utmost facility. This incalculable improvement was effected by the medical periodical press, and is now general throughout the world. The periodical press publishes the lectures and opinions of the most celebrated and eminent practitioners of different countries, and presents them on such low terms that every student and practitioner may possess them. It saves a vast expense which would otherwise be incurred, in the purchase of numerous original works; and affords the greatest facility of learning the opinions of celebrated physicians and surgeons of all countries.

The medical periodicals also form a complete encyclopaedia of the healing art, and prevent the necessity of procuring many expensive works. It is for this reason that every scientific practitioner, and every industrious medical student, possess them. A good library is, however, indispensable to medical practitioners. It ought to contain the best systematic, monographic, and elementary works, of this and foreign countries.

A well-selected museum is also a source of much instruction and information. It is likewise essential to associate with literary and scientific individuals, and to join the medical societies and academies. Such are a few of the requisites for [123/124] obtaining a proper knowledge of the principles and practice of medicine.

Medical Degrees, Diplomas, and Medical Appointments

It is now universally admitted, that the present curricula of medical education and practice in the British dominions, require the interference of the legislature for their improvement. This position is amply attested, by the appointment of a parliamentary committee in 1834, to examine into the actual state of medical education and practice, before which all the representatives of the universities, colleges of physicians and surgeons, and societies of apothecaries, in this kingdom, were examined; and all admitted that there was the greatest need of improvement. It has been long complained of, and with reason, that there is too much facility in obtaining degrees and diplomas, and that the period of study was much too short. A great number of professors attested this fact before the Parliamentary Committee. The period of 2.5 years, required for acquiring medical knowledge by the Royal College of Surgeons, in London, and also by the Apothecaries' Company, is much too short; and the examinations are by no means probationary, nor the diplomas a sufficient guarantee to the public, of the possessors' competency as medical or surgical practitioners. Some of the universities are equally liable to the same charge of neglect and dereliction of duty. The title of doctor is too often obtained by incompetent individuals, who necessarily dishonour it. Some of the foreign universities absolutely sell their degrees, and it is now of daily occurrence to pay a few pounds into a certain bank, and receive a doctor's degree by post, from some of the German universities. This was the custom at some of the Scottish universities, until a recent period, (1826). It cannot be denied, that too many illiterate persons are still admitted into the profession in the different schools of this country, and such indiscriminate admission has considerably lowered the character of the faculty. This observation particularly applies to the College of Surgeons and Apothecaries' Company of London. These bodies require the most superficial preliminary and medical education, [124/125] and often allow the most illiterate persons to obtain their testimonials. No experienced physician or surgeon will declare; that 1 year's attendance on the medical or surgical practice of an hospital or dispensary is sufficient for acquiring a practical knowledge of medicine and surgery? Who is it that will admit, that the science of medicine and surgery can be properly studied in two or three winter sessions of six months each? The time has nearly arrived when the good of the public and profession will enforce a uniform system of education, and one sufficiently comprehensive. Such will most probably be the result of the intended Medical Reform Act, which will be proposed by Mr. Warburton in the next session of parliament. It is now generally known, that almost all medical appointments in England, Scotland, and Ireland, are filled by jobbing and bribery. The majority of the physicians and surgeons selected, do not possess any, or sufficient, or superior merit; and obtain their situations by intrigue, family influence, or purchase.

Such is the case as regards our court, hospital, army, navy; and all other public medical appointments. Men of talent and genius can never succeed in obtaining places in this kingdom; while political and religious feeling, with intrigue and money, throw them into the shade. In France, a concours or medical jury, decide who is best qualified, among several candidates, for professorships and other medical situations. This excellent custom excites emulation, and rewards superior merit with a proper recompense. There is no means so favourable to the progress of science, and nothing so likely to foster that ambition which is so laudable in medical practitioners. What efforts are impossible to a young physician or surgeon, which inspire the hope of an honourable and profitable appointment, and the desire of conquering able rivals by extent and variety of knowledge? We find, accordingly, that some of the French professors had successively contested for appointments, so often as six different times, before they succeeded in attaining one of the objects of their ambition. It has been urged against concours, or public examination, that a timid individual might become so confused as to acquit himself so badly in his answers, that one of inferior talent or acquirement might be [125/126] elected. It has certainly happened, that the public voice has not always approved of the decision of the jury. Nevertheless the decision of a medical jury after the public disputation, examination, and competition of rival candidates, would be infinitely superior to the election of professors and medical officers by private interest, and non-professional governors, which is the usual mode in this country. It is also notorious, that many individuals have often purchased their situations, though totally incompetent to discharge the duties belonging to them. It has also been long remarked, that the medical situations, filled by the government and public societies, are disposed of through favour and interest, and seldom on the grounds of merit or talent. Thus it was the custom, until within a few years, to place the fellows of the London College of Physicians in hospitals, dispensaries, and other public appointments, though, in general, far inferior to rival candidates.

Again, the surgeons appointed to all lucrative situations, as hospitals, &c., are mostly the relatives or apprentices of former hospital surgeons, who transmit appointments as ancestral property. Moreover, we find the colleges, excluding all who practise obstetricy or pharmacy from their council and examining boards. We observe, too, that the apprentices of the Dublin College of Surgeons, preclude surgeons of all other colleges, from infirmaries, and all situations of fame and emolument in Ireland. The result of all this monopoly is a carelessness and indifference on the part of the bulk of the faculty to the proper cultivation of science; they reasonably declare we have not the least chance of appointment under the present system, and there is no stimulus to exertion and study. The consequence is, that fewer improvements have been recently made in medical science in this kingdom, than in any other in the world. What have we done in introducing new and powerful medicines? Our neighbours have introduced all of them. What discoveries have we made, except those of Sir Charles Bell? The stethoscope and lithotrity do not belong to us. It is true, that we can boast of Sir Charles Bell's splendid discovery on the functions of the nerves, and of the application of transfusion, by Dr. Blundell, to the preservation of life after hemorrhage; but with these exceptions our [126/127] improvements in physiology, pathology, and therapeutics, bear no comparison to those of vicinal nations. It is also true, that, Sir Astley Cooper tied the aorta, but *cui Bono ?*

Such are a few of the evil results of monopoly and abuse in the medical profession in Great Britain and Ireland. Talent and merit are unrewarded. But better times approach, a complete reformation in education, and the practice of medicine is at hand.

Entry of a Junior Medical Practitioner into the World

A young physician or surgeon who has passed some of the best years of his life in the lecture rooms, hospitals, and libraries, expects, on receiving his degree or diploma, to acquire fame and fortune on commencing his professional career. He is well acquainted with the principles of his art, and closely follows them. He employs all that he has heard and read in the investigation of the nature and treatment of disease. He is a scrupulous observer of the rules of art, and fears to break them in their application; he examines with great care, and gives his opinion with diffidence or fear. He is cautious and timid in the use of remedies. The better informed he is, the less presumptuous or rash he will be in practice. He is in general diffident in himself, hesitates or is too anxious about his patient, and is for some years before he acquires that confident assurance, which is the result of experience. He generally supposes that he has mistaken the disease, and has not done enough for its removal. This is the ordinary feeling of the best informed medical practitioners in the commencement of their career; but experience will, in time, remove it. The complication or intensity of disease may be such that no remedial means can preserve life. The contrast between well-informed and ignorant practitioners is very remarkable.

It has been long observed that the most uninformed in every science and art are, in general, the most confident and self-sufficient. This is also the case with pretenders to medical knowledge. No case is too dangerous for them; they can cure all diseases, though they kill a vast number of patients. They attribute all their failures to the intensity of disease, and not to their own ignorance. As they have had no pre-[127/128] liminary or medical education, which is so necessary to prepare the mind for the investigation and estimation of morbid phenomena, they treat diseases by names, and not according to their real nature. Men of unlettered ignorance usually pride themselves, on imagined practical experience, or on what Cullen aptly termed "false facts," and blindly pursue the same erroneous system without a particle of sound judgment. In fact, nothing appears beyond their ability, or too difficult for them to attempt. They can cure all diseases – a power only possessed by empirics; they ascribe their success, in cases in which nature effected a cure, to their active interference, but they never suppose that their failures are in any degree the result of their incapacity. They do not know that experience, without a proper scientific education, is useless.

It cannot be denied, that unless a medical practitioner knows the anatomy and physiology of the human body, or, in other words, the phenomena of the animal economy, his observation and experience are based on ignorance, and must be erroneous. Such is the advantage of experience, unaccompanied by proper medical information. On the other hand, the well educated junior or senior practitioner, who has had the most extensive experience and information, is always diffident in his own powers, and doubtful of the result of the disease. He never professes to cure every patient.

But I must return to the career of the young physician or surgeon. On receiving his diploma, he is most anxious for fame, and thinks the world ought to be instantaneously aware of his qualification to practise. He, however, speedily discovers that the world esteems him as a student, and that he is not a more important individual than he was, the day or month before he received his qualification. He speedily discovers that patients will not consult him. Some suppose him inexperienced, and too youthful; others have their own medical advisers; and all will prefer practitioners of standing to our young Aesculapius. He now thinks that his youth is a great or invincible obstacle to his success, and he sighs for the time in which the public will reward him for all the labour and expense he has incurred in the study of his art. Nevertheless, every other member of the profession had the same difficulties [128/129] to contend with, and only surmounted them by time, attention, skill, and unexceptionable morality. He does not imagine that a physician, extensively engaged in practice, has any cause to complain. He does not know that such an individual is deprived of all pleasure, recreation, or amusement, and regrets the period of his studies, when he enjoyed liberty. Thus, man is never content with his condition.

The first success or failure of a physician depends on the opinion entertained by the public of his talent. How embarrassing and difficult is the debut of a physician in the world! How careful he must be in gaining the confidence of his first patients, in investigating the symptoms of their diseases, and in employing remedies! He must recollect, that others of his faculty have been consulted on former occasions, and that their manners and conduct will be compared with his own. When the disease is dangerous, and he effects a cure, he will greatly benefit his reputation. His religion and zeal compel him to attend the poor and miserable, as well as the comfortable and affluent; and he may gain reputation by both. He will often succeed after his seniors have failed; because they in general belong to the old school, and he to the new. He should not expose their errors, unless glaring and dangerous to the life of the patient. He is always ready to afford relief to suffering humanity, with or without remuneration. In the same manner, a surgeon should never refuse to perform a doubtful operation when it is required. The patient ought to have every chance. Either physician or surgeon can explain to the relations the danger of any case.

The essential requisites of a physician are, in addition to the possession of real virtue, knowledge of the world, and polished manners, an earnest desire for fame, a great love of study, a good knowledge of charlatanism, and what is vulgarly termed humbug, a good share of small talk, and an audacity which nothing can disconcert. The critical reader will inquire, why a medical practitioner should be acquainted with charlatanism and humbug? The answer is, because it is the nature of mankind to esteem quackery, as it professes to effect cures of incurable diseases. All classes of society encourage and patronize empiricism; but an immense crowd [129/130] of fools constitute the public. In all countries, the most absurd quackeries have been patronized; but this is easily explained, when we dispassionately observe the varied degrees of intelligence, and the love of life and of the marvellous which influences all ranks of society.

It must be confessed, though a stain on the dignity of the medical profession, but equally remarkable in every class of society, that some Esculapians have acquired success, fame, and opulence, by fostering the follies and prejudices of the world; and by acting as arrant humbugs. Such men are, however, despised by their brethren, who, in general, are actuated by the love of humanity and reputation. The former are despised by the honourable cultivators of medicine, notwithstanding their titles, station, and influence.

The medical profession has great reason to complain of the follies and injustice of the public. The mass of fools forming the public, with few exceptions, will ascribe professional success to chance or to nature; and, if the patient dies, some will not hesitate to declare that the doctor killed him. They will unhesitatingly proclaim the incompetency of the medical practitioner, unmindful of the well-known unfortunate truism, that many diseases and many constitutions are utterly incurable.

M. Lorry, a celebrated member of the profession has happily exposed the folly of the public in this respect. He says—"A physician fond of study has spent many years in the schools, hospitals, and anatomical rooms; he has passed the best years of his life in the infected air of hospitals: the pallor of his countenance, and the meagreness of his figure, attest the multiplicity of his vigils; and what remuneration compensates for these labours?"

Here the ignorant man of the world declaims against the stability of one of the noblest of human sciences, and unblushingly confounds medicine and empiricism. Thus there are many, whose lives have been saved by medical practitioners, who forget it, and even denounce the medical faculty.

If the public do not derive the usual advantages from medical practitioners, they should blame themselves for encouraging illiterate pretenders to medical knowledge, nom- [130/131] inal doctors, and empirics. Such impostors are very numerous in all countries; and if the public do not take the trouble of distinguishing the learned from the ignorant, the fault is their own. If the nobility and gentry prefer the St. John Longs and such basket makers, the Hahnemanns, and other empirics; the woodmen of Molière, to well-educated and eminent medical practitioners, they must not decry the benefits of the healing art, and depreciate the Bells, Coopers, Dupuytrens, &c., &c.

In general, young medical practitioners are most attentive to their patients, and render them the greatest services. So do all duly-educated members of the medical profession, who are influenced by the code of ethics laid down for their guidance. There are a few, however, among the eminent in all countries who are a disgrace to the profession. They take advantage of public credulity, which in all ages has rendered the worthy part of mankind dupes to the artifices of the knavish; who, unrestrained by principle, are ever eager to profit by the unsuspicious disposition of generous minds. Among the various kinds of imposture practised on society, quackery has always been the most successful, in consequence of the extreme respect paid to the professors of the healing art. These have now and then availed themselves in acquiring reputation by the foibles of the public.

Reputation of a physician. When the desire of reputation is inspired by the love of fame, it is allowable, because it tends to the good of society. It has been often observed, that those physicians who have not a laudable ambition to rise to eminence by their talents and honourable conduct, seldom or ever become renowned. They remain every-day characters, and never obtain the highest places in the profession. When a thirst for gold is the only object of professional reputation, it leads to meanness and disreputable behaviour. We see empirics, illiterate and professional, amassing great wealth at the expense of every virtue which adorns the true medical character.

The public is the cause of this, as it awards reputation by caprice, and this is the reason that charlatans share, in common with educated physicians, reputation and renown. They [131/132] announce, with unblushing effrontery their marvellous powers, and all ranks of society, both high and low, are ready to believe them. Kings, nobles, and commons, patronize the empiric, no matter how preposterous his pretensions. Let him be an animal magnetizer, a homeopathist, an extractor of all diseases with a stimulating liniment, or metallic tractors, it is all the same. Witness the patrons of a Long and a Hahnemann, of a Graham and his celestial beds, from which a noble race of offspring was to follow; witness a scion of nobility attesting, three years since, that he saw a quack extract quicksilver from a lady's brain! Look to the noble patrons of Hahnemann, who believe the most salutiferous doctrine, that the millionth part of a grain of magnesia or of sugar could not only cure the inhabitants of a city, but an empire, however extensive. What difficulties, on the contrary, has the modest man to surmount before he can obtain the smiles of public favour.

There, is, however, in medicine, different kinds of reputation which lead to fame and fortune. One gives himself up to scientific pursuits, another distinguishes himself by writing or teaching, a third is renowned as an excellent physician, a fourth as an operative surgeon or obstetrician, &c.

It rarely happens that one individual possesses these different claims to celebrity. The healing art is too abstruse and extensive for the greatest genius to comprehend it in all its parts. One is superior in medicine, another in surgery, and another in obstetricy, chemistry, botany, &c.

When a physician acquires reputation by merit, he should never attempt to increase it by resorting to the artifices of charlatanism, which invariably lower him in the estimation of his brethren, and, through them, of the public. He should avoid all species of puffing, whether having nurses or relations to trumpet his abilities, or superficial works to circulate gratuitously among his connexions, and their friends. If his fame be founded on talent and merit, it rapidly increases without the artificial helps – it enlarges as it progresses – *vires acquirit eundo.*

If the cacoethes scribendi should seize one, let him write something rare, new, or – nothing: *scribe rara aut nova, vel nihil.* The style and composition ought to be correct and [132/133] elegant. Citations may be numerous and acknowledged, though some writers too often give the whole of other men's works and opinions as their own. These are generally detected and exposed by reviewers and critics. It is no excuse to the reader that a work is written or printed in a hurry, and full of blemishes; the public would have waited for it with patience, and seldom care if it had never been published. An author, should deliver his opinions with modesty, and expose what is true, dubious, or erroneous. He must rarely condemn others, or censure the living or dead with undue severity. He must praise sparingly, and vituperate more sparingly. There are some writers and critics who praise no one, and vituperate every one. They find fault with every thing – style, composition, opinions, arguments, observations, and conclusions. They hold, that every thing is to be found in the works of the ancients, and that there is nothing new under the sun. These critics raise a host of enemies against them, and seldom have any friends. They should adopt the ethical axiom: – "it is not right always to praise or dispraise"—"Nec laudare semper, nec semper vituperare decet. Lauda parce, vitupera parcius. Lauda recentes, lauda veteres, vel utrosque carpe, si licet. Stet sua cuique reverentia, suus honos. Non viventes auctores enormiter laudes, ut vicissim lauderis. Stet verbis, stet titulis, stet sua encomiis mensura. Unico libello scriptores omnes, omnes amicos non alliga. Nec æmulorum, nec mortuorum laudes dissimules, nec excedas."[2]

When criticism is unjustly severe, it often becomes pointless, and only serves the circulation of a work.

There is no work free from some objection, and there is a vast number of modern publications deserving of severe criticism. Thus one man writes a pamphlet or book, for the express purpose of making himself known by advertisements in the public journals; another does not state a fact that was not well known for centuries before his time; another gets himself puffed in a sly paragraph of some newspaper; and in these, and a thousand other ways, do medical practitioners attempt to captivate the public. These schemes, *ad captandum vulgus,*[3] are derogatory to the dignity of medicine, though [133/134] they are now unblushingly laid by some of our highest physicians and surgeons. We daily observe their advertisements by the side of those of empirics, and their paragraphic puffs are most abundant.

Few of the many medical works now published in this country are successful, and few pass through future editions. Many of them are still-born from the press; others speedily die. This failure of success has influenced publishers so much, that they rarely purchase a copyright of a medical work, and will take no part in the risk incurred in bringing out books of this description. It was lately stated by one of the most extensive medical publishers in London, that few monographs had a circulation of five hundred copies. Does this want of circulation arise from the mediocrity of books, or from the distaste for medical literature? I think from the former. But a second edition is called for; a new title-page, preface, or chapter or two are added; the public is gratefully thanked for its patronage; and this species of charlatanism is invariably discovered and censured.

Besides these modes of gaining fame, we find physicians and surgeons recommending apothecaries, and vice versa. The female portion of their own and other families become puffers of their skill, and often succeed in increasing their fame. A lady of title has been known to make the reputation and fortune of her medical attendant.

People of fashion are led by notoriety and imitation of each other, and thus the physicians to royalty are indispensable to the nobility. No other will do.

In other instances, a reputation and a name are made by many petty artifices – by gossiping, a talent for flattery, the influence of medical and other patrons, and, lastly, by religious partisanship. These means will often raise a medical man in public estimation, though he does not possess a ray of talent; while the man of genius, of erudition, and of practical knowledge may, without them, remain all his life in "illustrious obscurity." He sees numbers of his inferiors who, by their address, policy, and worldly wisdom, rapidly pass him by, and leave him in the shade.

Nevertheless, what a galaxy of men of genius and knowledge [134/135] have risen to eminence and fame. It is unnecessary to multiply examples; but be it observed, that no physician arose to greater eminence since Hippocrates than Boerhaave. A letter was addressed to him from China with this superscription – "To the Great Boerhaave, in Europe." It reached him.

To make oneself known. It has been long observed, that great talents are not the surest or speediest way to acquire much medical reputation. It is, in general, a matter of the greatest difficulty, for a young physician, or surgeon, to make himself known in a large city, where there is a prodigious number of doctors of one sort or another. What difficulty, what address, and what labour are necessary, before a junior can obtain a prominent place among the multitude; or raise the edifice of his reputation! Years must elapse, in general, before he has accomplished his object. In fact, as Dr. Baillie truly observed: "he cannot earn bread until he has not teeth to eat it."

Let us examine the means which are to be employed to succeed in practice, and those that ought to be avoided. The credulity and folly of the public are the causes which impede medical practitioners in their career. The public never distinguish between the educated and uneducated professors of the healing art. Those who make the greatest promises – and they are the most ignorant – are surest of favour and success; while those who ought to be patronized, are entirely neglected. Such is the general result, but there are some few exceptions.

It sometimes happens, that distinguished or celebrated individuals have patronized a young medical practitioner, more on account of vanity, than a love of science. They are not always particular in their selection, and often prefer the ignorant to the learned; indeed, it generally happens, that genius is neglected, while ignorance finds the most powerful protectors.[4] The natural protectors of a young physician, or surgeon, [135/136] are his masters, or those practitioners who have acquired eminence and celebrity. The estimation in which they are held, easily allows the junior to clear the foundation of his reputation. Every medical practitioner who wishes to succeed, must be determined to accomplish his object, and to surmount a host of difficulties. He ought to have a good knowledge of the world, and accommodate himself to the follies and foibles of those with whom he comes in contact. He must possess policy, politeness, tact, and, as I have already observed, some knowledge of charlatanism, and perhaps, "humbug." Such are the means of acquiring great celebrity and fortune. Indeed, it rarely happens that talent – the enemy of artifice – leads to celebrity.

I cannot agree, however, with M. Monfalcon (Dict. Des. Sciences Med.) in his advice, which is similar to that of Lord Chesterfield to his son, as regards society in general: "that all young medical practitioners who wish to gain a reputation rapidly, should bespeak the good opinion of the fair sex – be assiduous, complaisant, gallant to them, but nothing more. They should study every thing to please, and flatter feminine vanity. The first thing necessary, is a perpetual adulation; the second, an absolute devotedness. There are some whose love of study and self-denial, unfit them for paying proper attention to the fair sex, and they can never become doctors a la mode. When their vanity is flattered by a fashionable doctor, they proclaim him in all places as a charming man and a learned physician. He takes care to sympathize with them on account of their nervousness from imaginary disorders; he advises amusements, change of residence, travelling, &c. and often assists in rendering the husband the very humble slave of the will of the wife. Some medical practitioners have enjoyed the faculty of addressing small-talk most agreeably to the other sex. It would be invidious to mention names, though it might be easily done. But there is no reason why grave, sad, and discreet doctors should not possess amiability as well as all other individuals."

Medicine is a science by no means incompatible with the usages of the world; it does not exempt those who practise it from being distinguished for that politeness, amenity, and those graces which charm society. [136/137] It has long been admitted, that the art of pleasing and the art of healing have intimate relations to each other.

If a medical practitioner does not possess politeness, and those graces which distinguish a gentleman, he will be very much impeded in his career. Manners and personal appearance are every thing with the majority of the world. No man ought to be more polite or possess a greater knowledge of the world than a medical practitioner. He should avoid eccentricity, or peculiarity of manners.

There are some who are so passionately fond of study, that they cannot live without their books, and they withdraw themselves from society to pursue their learned researches. This constant study gives them an embarrassed air and a timid demeanour in company, and is very injurious to their success in the practice of their profession. No professional man ought to neglect every legitimate means which is calculated to increase his reputation, or to obtain public approbation. He must renounce all amusements and pleasures, as society calls for him every instant; he cannot enjoy repose without interruption, and his sleep only continues so long as others do not disturb it; but the incessant toil of his profession requires relaxation to preserve his health. He must occasionally frequent places of public amusement, though he is advised by his seniors, to leave balls, fetes, routes, &c., to the unthinking portion of society. A medical practitioner, who frequents such amusements too much, is usually considered by the public, as having little to do in his vocation. It is true, that more serious matters generally require his attention; *urgentiora ilium vocent negotia*. The world is so severe, that it will sometimes censure a physician or surgeon who cultivates those arts and sciences which have no immediate connexion with his profession. It also requires him to avoid all other pursuits except his vocation. This is often a great hardship, as when a medical practitioner can scarcely support his family by his calling, and might acquire wealth by following other business. It is for this reason that many members of the profession are now obliged to engage in various commercial and other speculations.

"Scientific physicians," says M. Monfalcon, "sacrifice [137/138] your health and fortune to become learned and clever; sit over your books, become pale in the hospitals, meditate night and day on the most difficult points of your art, study like Boerhaave, fourteen hours a day for sixty years, renounce all the pleasures of life, and the charms of society, observe a complete self-denial, and if you disdain to make yourself known, you will be forgotten, neglected, and unappreciated; and you will never approach the fame of an empiric."

Charlatanism will always be successful, as it governs man by the first of all his interests, the love of life and the fear of death. Its origin commenced with man, and every one is in turn a doctor.

> Fingunt se medicos quivis idiota sacerdos,
> Judæus, monachus, histrio, rasor, anus,
> Miles, mercator, cerdo, nutrix et orator,
> Vult medicus hodie quivis habere manus.[5]

But the surest means of success are profound knowledge of science and of the world. Some obtain fame and reputation by the first, others by the second, and some by possessing both. Another means of obtaining success is forming a matrimonial engagement.

Marriage of a medical practitioner. There are many reasons to prove the necessity of marriage for a medical practitioner; it gives him more importance and maturity; it leads the world to overlook his youth and age, and it gains the confidence of families and mothers, who, unless he had entered into this state of life, would often decline his services. Hoffman was of opinion that a medical practitioner ought not to marry too early, unless he did so very advantageously, because a wife and establishment would occupy a large portion of the time, which ought to be devoted to study. This last objection is futile, as we find Racine was married, and was alternately pleased with his children and his books; Haller was assisted in his literary pursuits by his wife, and was one of the most voluminous authors of his time.

It is, however, the general opinion at present, that a medical practitioner ought to marry as soon as he can do so prudently and advantageously. This was the subject of a [138/139] thesis, sustained by Dr. Treyling at the University of Ingolstad, in which he discussed it in the form of two questions: (1) Ought a physician marry? (2) What sort of a wife is suited to him? This writer declaimed against marriage, cited all the arguments against it, and contended that the public prefer unmarried physicians! He passed in review all the disagreements of married persons, and dwelt more especially on the danger of female infidelity.

Accidit et hoc viro præsertim medico, quod si juvenculam sibi junxerit, hancque formosam, habeat, quod metuat, illud Epicteti dicentis: qui formosam duxerit, habebit communem. Cum enim medicus densa praxi obrutus, nec domûs nec uxoris custos esse valeat, quid? Si hæc interim hospitalis sit, et Dianam æmulata cornifica metamorphosi maritum cervina superbum corona in Actæonem transformat, hæredesque ipsi aperat, non nisi adamitico cum ipso sanguine conjunctos? Ita ut non semel saltem tacitè secum murmurare querelus debeat: haud ego mi uxorem duxi, tulit alter amorem. Sic vos non vobis.[6]

Such was the opinion of Epictetus. It is not, however, true, that he who marries a beautiful woman, enjoys her in common with others, nor does it happen that the husband always wears horns, or has a spurious heir. Handsome women are objects of general admiration, but when their minds are properly cultivated, they spurn the slightest equivocal compliment, or impertinent remark. It must be admitted, however, that if, on the other hand, there is a want of high moral feeling, the wife of a medical man, actively engaged in practice, has ample opportunity and time to degrade herself and dishonour her husband. Nevertheless, there are no more indiscretions in the families of medical practitioners than in all others; and, therefore, marriage is as suited to them as others; and, indeed, much more so, for the reasons already assigned. It is a state prescribed for all men, for "it is not good that man should be alone."

In fine, a medical practitioner seldom succeeds as an obstetrician, or is consulted in diseases of women until he is married, or considerably advanced in life. The virtuous of the other sex have a great objection to very young practitioners during parturition; and even the poor, who are compelled to [139/140] obtain medical attendance from public charities, very frequently decline it, when young medical students are ordered to attend them. "Marriage is honourable in all," and particularly so to the members of the medical profession.

Means of obtaining the confidence of the sick. It is useless for a medical practitioner to assume a grave exterior, and to possess the most profound scientific knowledge, unless he obtain confidence of the sick. Without this the greatest talent loses much of its power; and with it, every thing is possible to mediocrity. The principal means of obtaining this confidence, is the art of persuasion; which is a gift not always possessed by genius. The medical practitioner ought to be a man of the world, as well as minister of health. Some will prefer a grave, others a cheerful physician; but whatever may be the talents and acquirements of a medical practitioner, he cannot preserve the confidence of his patients, except by success; and a few unfavourable events may injure or ruin a reputatation, however well established. The public is, in general, disposed to attribute the impotence of medicine to its practitioners.

Another cause of loss of confidence, is when a practitioner is unable to explain the nature of disease, or fails to remove or alleviate it. The sick, or their friends, will not be satisfied without further advice; and if recovery happens, the whole merit is awarded to the new attendant. Thus it is manifest, that "knowledge is power," and often triumphs over all other means of securing the confidence of the sick.

Rewards of medical practitioners. – If the office of physician, or surgeon, exposes every day to the disdain of ignorance, to the forgetfulness of ingratitude, and to the outrages of calumny; if reputation entirely depends on the caprices of the multitude; if to practise, requires a medical man to renounce all pleasure, and even his liberty – he finds, however, in the exercise of his profession, many rewards for his sacrifices. The esteem of a small number of learned and sensible men, and the firm conviction that he has done his duty towards the sick, console him against the jealousies of his rivals, and the unreasonableness of the world. A quiet conscience is the [140/141] recompense of a medical practitioner, who honourably discharges his noble functions. He is delighted with the good he has done, and with the conviction that he cannot do too much. The poor implore his aid; he is the harbinger of hope and consolation in the abodes of misery and distress; and is blessed by the afflicted whom he relieves.

When, under Divine Providence, he rescues a patient from the jaws of death, and conducts a convalescence after a dangerous disease or operation, his success is an ample reward for his cares and anxieties. He that is saved becomes his friend for life, and lauds his skill and judgment.

It has long been remarked, that physicians and surgeons, in general, with few exceptions, seldom acquire a large fortune; nevertheless, the fruit of their labours is not subjected to such sudden vicissitudes as affect commercial men, and precipitate them from extreme opulence to extreme poverty. They are placed in tranquility and a happy mediocrity, which, of all states of life, is that which is most accordant with happiness. Respected by society – esteemed by men of learning – desired by the rich and poor – the cultivators of medicine to the remainder of life are venerated, loved, and honoured. Many of the profession have, however, acquired immense wealth. Sir Astley Cooper was in the receipt of £20,000 a year; and it is said, that one year his income, by his profession alone, amounted to £25,000. Few deserved more success, but fewer, as the world now exists, if any, will ever acquire so much.

Relations of medical practitioners towards each other. The etiquette of the profession requires that its members should live on friendly terms with each other, a reciprocal indulgence ought to make them excuse the errors they commit, and especially those of junior practitioners. They ought to be always on good terms. It would be shameful and scandalous in one member to expose the errors, or depreciate the merits of another; for no one would like this to be done to himself. Calumny and detraction are base vices, which ought not to characterize any medical practitioner. No one ought to injure another "by look, word, or suspicious silence," or under any pretext whatever interfere with the patients of [141/142] another. It is an axiom, that we should refuse attendance on patients under the care of another practitioner, unless in co-operation with him, or until his visits are discontinued, or when there is danger of death. Nevertheless, this vice is too common with necessitous practitioners. They are often jealous at the success of others, and depreciate, or calumniate their rivals. There is no envy so great as that of medical practitioners: "non est invidia supra medicorum invidiam." We rarely observe true friendship existing among them: envy and interest oppose it. But the real philosophical physicians despise such views, and entertain the strongest friendship for each other. A little reflection will convince any rational practitioner, that no man can be treated badly without resenting such conduct; and that never was there a wiser maxim than, "do unto others as ye would they should do unto you."

Some cannot speak without wounding the reputation of others. Many are bitter in their remarks, prodigal in their criticisms and sarcasms, imperious and cutting in their decisions; and these make a host of enemies for themselves. The wounds inflicted on self-love never heal; and it shows little common sense or discretion, to lose a friend for a remark. This however is often done; and some few of our most scientific practitioners have suffered severely by it. They have lost much consultation practice, and very justly, on account of such unprofessional behaviour. Thus, some engaged in consultations act uncourteously towards the first attendant; wound his vanity and self-love; lower, or endeavour to lower him in the opinion of his patients, his friends, or attendants. Others arrive before the hour appointed for the consultation, or several hours afterwards; and then act in the same disgraceful manner. A few encourage their trumpeters to sound their praises to the relations, acquaintances, or menials of the sick; or bribe nurses, with the view of depreciating the practitioner in attendance. These, and numerous other acts of immorality, are disgraceful to all who are guilty of them. They are, however, of too frequent occurrence in all large cities, and among unprincipled practitioners in all [142/143] situations. But no respectable or honourable physician or surgeon would be guilty of resorting to them.

Medicine of the mind. One of our celebrated poets has asked,

Canst thou minister to a mind diseased,
Or pluck from memory its rooted sorrow?

The answer may be given in the affirmative in many cases. There are cases in which we minister to diseased minds with perfect success. Indeed, the art of reading the human heart is indispensable to the medical practitioner; and it is often all that he can do. Medicines do not always cure disease; sage advice or discourse, which convinces the reason; and the proof of friendship, which touches the heart, are often the most powerful means of preserving health and life. He who knows the characters of the passions, can best moderate them, diminish their influence, and often prevent their fatal results. Those who have studied the human mind, can mostly discover the cause of the emotions of patients, the value of their answers, and the confidence that ought to be placed in them. Without this quality, we cannot satisfy or govern hypochondriacs, or remove their fears and apprehensions. Every one knows the influence of the mind over the body in health and disease. It was a profound knowledge of the human heart which led Hippocrates to discover the love of Perdicas for Phylla; and Galen, that of the Roman lady for Pylades. The importance of the moral influences in therapeutics is so great, that the ancients considered the mind, with philosophy and rhetoric, as medicinal remedies; and that their influence on the body was most advantageous. It cannot, indeed, be denied, that moral medicine is in many cases more powerful than physical. Who does not know, that the declaration of an eminent physician or surgeon, "that recovery is certain," will do more good than the whole materia medica ? We often see this in the practice of obstetricy – the assurance of safety and speedy relief has the most astonishing effect in expediting parturition.

It was on this account that philosophy and medicine were cultivated by the ancient physicians, and that modern universities require an excellent preliminary education of those [143/144] intended to study medicine. It is really a melancholy reflection to any one of education and acquirements, to think, that nine-tenths of the profession in this section of the British empire are admitted as medical practitioners if they can translate the first chapters of Gregory's "*Conspectus Medicinæ Theoreticæ*," and those of "*Celsus de Medicina.*" And such is the classical ignorance of those entering the profession, that divers translations of the above are published, and even some that are absolutely interlineal! No knowledge of English composition, of logic, of rhetoric, of natural or moral philosophy, of mathematics, and the other studies included in humanity are required; and every one who can translate the above is admitted by the great manufactory for doctor-making, the Apothecaries' Society in London. What notion of moral medicine have individuals thus educated? Can persons thus educated be influenced by that consummate knowledge of the human mind so essential in the practice of medicine? What will foreigners say to the preliminary education which extends to a power of translating the first and third books of "Celsus de Medicina," and the first nine chapters of "Gregory's Conspectus de Medicina?" Surely, no one can consider this a sufficient proof of classical or general knowledge. I grieve to indite these remarks, but truth obliges me.

It is different, however, in the universities. In these institutions the preliminary education is as extensive, and indeed more so, than that required in other countries. They adopt the adage

Adde quod ingenuas didicisse fideliter artes,
Emollit mores, nec sinit esse feros.[7]

Physical medicine. We now approach the usual occupation of medical practitioners. We bring them to the bed-side, to investigate the nature and treatment of disorders and diseases. All general and medical studies – all our former remarks, are subservient to this most important ministry.

The physician or surgeon bases his diagnosis, on the history of his complaint given by the patient. He considers the age, sex, temperament, habit, occupation, residence, and constitution of the patient, and the season; and he next inquires, [144/145] what is the most troublesome or painful sensation? The patient replies, that he has pain in some particular part, which at once directs the attention of the inquirer to the seat of his malady. It is in the head, chest, abdomen, or extremities. A momentary glance at the countenance, chest, or abdomen, will enable a scientific practitioner to determine the seat of the disease. He can also conclude in a few minutes, whether the malady is a disorder of function, an unpleasant sensation, pain, spasm, &c.; or a disease, a change of structure, congestion, inflammation, or the usual consequences of the last.

In either case, disorder or disease of one organ will derange the whole, and may cause painful sensations in all, and a multitude of symptoms. In every possible case, there is disorder of function, change of structure, or a combination of both.

In disorders of function, the nervous power, or innervation, is anormal, [*sic*] or deranged; and the internal and external use of sedatives is indicated.

In disease, the circulatory system is affected; there is congestion, inflammation, suppuration, ulceration, cicatrization gangrene, &c. Disorder or disease may affect all parts of the body, in consequence of the universal distribution of nerves and blood-vessels. It is most important to determine the existence of the one or the other, as the treatment must differ. In disorder of function, without change of structure, there will be some unnatural sensation, which may cause pains in all parts of the body, and be removed by the internal and external employment of sedatives. In disease, general and local bleeding, low diet, counter-irritation, purgatives, and diaphoretics, will be necessary. In complicated cases, both classes of remedies must be employed.

It is not understood by society, that a learned and scientific practitioner examines the head, chest, and abdomen, in a minute or two; and patients suppose that they must detail every sensation they have felt for months or years. All complaints are either disorders or diseases, or both combined; and the latter are acute or chronic. The simple question, what is the matter? – where have you pain? – what distresses you most? – will enable the inquirer to discover the seat of the malady. He then determines whether it is disorder or dis- [145/146] ease, and prescribes the appropriate remedies. The fundamental principles of medicine are now happily fixed, or it would be impossible to arrive at a correct notion of disease.

Many patients cannot express their ideas correctly, and jumble the most opposite symptoms together; they enter into long digressions, and confound the most different objects. They complain of all diseases. The question, where have you most pain? – or, where did it commence? will lead the practitioner to the seat of the complaint. In my hospital and dispensary practice, I inquire, before a large number of students, which of all your complaints is worst? and what was the first symptom? I also adopt this plan of interrogation in private practice. It affords a light which guides me through great darkness; and by it I distinguish the circumstances that preceded the attack, as well as the nature of the complaint itself. I have fully described the method of examining patients, and distinguishing diseases, in my edition of "*Hooper's Physician's Vade-mecum*," and "*Lectures on the Physical Education of Children*, 1835," already quoted.

There are some patients who exaggerate their sufferings, and the expression of pain is not always real. Others put insidious questions to their medical adviser, in order to learn his opinion of their condition, as hypochondriacs, nervous, and parturient women. The former suppose that they labour under one or many fatal diseases; the last are anxious to learn, are they in a safe condition, and how soon they will be well, or delivered.

We also observe many who simulate a variety of diseases with great success. But the pathological symptoms which belong to the functions, however modified by the influence of the brain, are the signs on which we can place absolute confidence. Physical signs will never deceive us, and we should always judge by them. The statements of patients, or of medical practitioners, cannot always be relied on, while physical evidence is invariably correct. I might mention numerous instances in which patients and practitioners would have deceived me, had I not solely depended on the physical phenomena.

In examining patients, we should inflect the voice, and [146/147] avoid putting our questions too abruptly, lest we confuse and embarrass them. There is an excellent chapter in the work of M. Double on Semeiotics, as to the method of examining patients, and further information may be derived from the production of M. M. Merat, Creuveilher, and Martinet (see Professor Quain's translation). M. Double divides the subject into two distinct parts – the knowledge of the history preceding the disease, and the circumstances which appertain to the disease itself. The physician ought to investigate all circumstances relative to surrounding objects, season, temperature, and medical topography where he practises. He then determines the age, sex, profession, occupation, or mode of life, of the patient; the passions, habits, general health, or ordinary state of functions, state of health before the invasion of the disease, the hereditary diseases of the father, mother, and family; the antecedent diseases from the period of infancy: and, lastly, the general effects of medicines upon the constitution.

The practitioner should sit by the bed-side, so as to see the patient's countenance, and then examine the external appearance of the body, attitude, movements, coloration of the skin, &c. He then compares the existing and natural states of the functions, and proceeds to examine the external senses, respiration, circulation – including the state of the heart and pulse, digestion, secretions, excretions, generation, sensibility, irritability, voice, voluntary motions, sleep, intellectual faculties, and temperature of the body. Every part of the human body is thus examined and passed in review. We also know all that belongs to the symptoms, mutations, and sympathies – all the circumstances that can modify the signs and prognosis of diseases. The nature and treatment of the complaint are then easily determined.

It is right to visit the patient at the paroxysmal periods, or when he is worse; and also at other hours.

It is worthy of remark, that individuals who suffer from the effects of libertinage, are often ashamed, and disposed to mislead their medical advisers. A large experience warrants me in stating, that the abuse of the generative function, is a most common cause of indigestion, hypochondriasis, neuralgic, or [147/148] painful, and nervous complaints. It requires great delicacy and caution in alluding to this function at all times, but especially when youths are our patients. The excitation of the generative organs at puberty, and often during the whole time of celibacy, leads to the adoption of natural or other means for its subdual. There are, of course, exceptions to this general proposition. Those who abuse or over-exert this function, are affected with severe indigestion, flatulence, lowness of spirits, fear of death, dread of impotence, and innumerable unpleasant sensations. Such cases are of frequent occurrence from the age of puberty to the decline of life, and sometimes even in old age. The celebrated Tissot has given numerous illustrations in his work on Onanism, and a vast number has fallen under my own observations. I have slightly alluded to such cases in my work on population, marriage, impotence, and sterility.[8]

I am convinced, by multiplied experience, that the genital function is, in general, too much excited, and consequently debilitated. The profession, in general, overlook this fact, and very improperly fall into the vulgar error, of ridiculing those who have abused this function, and rendered themselves impotent or sterile. Such persons hesitate to consult many, and those who have not offspring, are too often derided by their acquaintances. They avoid consulting respectable medical practitioners – for the reason above stated – and apply to unprincipled empirics. They are, as a matter of course, duped and deceived by these wretches, and often seriously injured, or their health destroyed. Several have applied to me under such circumstances. I feel convinced that the generative function is one of the most powerful in the body, and influences mankind in general, more than any other. Every male, high or low, makes it the theme of conversation, and the most modest and virtuous women are greatly annoyed at infecundity.

In delivering an opinion to a patient, the physician, or medical practitioner, ought, in general, to predict a favourable [148/149] result when circumstances permit; avoid alarming him, and hope for the best. He may, however, express his real opinion to the relatives, so as to give them an opportunity of consulting others, if they think proper. In most cases, the prognosis must be guarded, as so much depends upon the mutations of disease, punctuality of attendants in following directions, on the seasons, situation, and compliance of the sick. Medical practitioners have often been most unjustly ridiculed by the world, on account of the ambiguity of their opinions as to the issue of acute diseases; but no man can be positive as to the prognosis. He may state, he is most happy to say, that at this moment there is no danger: but diseases undergo numerous changes, and that it is utterly impossible to predict, with certainty, the result.

This is the fact and truth, in the majority of human infirmities. The diagnosis or distinction of diseases, is also a matter of grave importance. If the disease is mistaken, censure will assuredly follow. Thus, suppose it is stated, that a young female of irreproachable character is affected with gonorrhœa, or is pregnant, when there is really no such disease or condition; the practitioner would be most severely and justly censured. It is a common error to assert, that female children from the age of six to fourteen years have been violated, and infected with gonorrhœa, when their condition depends on physical diseases. I have fully described such cases in another part of this work; and they are noticed by Underwood, Hamilton, Dewees, Sir Astley Cooper, Jewel, and numerous French authors.

In fine, I have to observe, that I have already described the moral qualities which are necessary to a medical practitioner. I may, however, remark here; that he ought to possess sensibility, humanity, amenity, amiability, and compassion. He must make the greatest allowances for the conduct of those in pain and suffering. Thus, a medical practitioner must display all the finer feelings of human nature, and all the moral excellencies which distinguish the true cultivators of his art, while he is affording aid at the painful process employed by nature at the nativity of his species. Women are naturally timid at this period, and require all the sympathy [149/150] and kindness that can be shown towards them. I have fully dilated on the duties of the profession under such circumstances, in my Manual of Gynæcology, or Obstetricy.

I may state, in this place, that medical practitioners ought to be most patient, humane, benevolent; and never resent any offensive remarks that may be applied to them by the sick. Such remarks are invariably apologised for by patients.

Lastly, we ought to be cool and undisturbed in the worst cases. Perturbation of mind betrays weakness and ignorance. If agitation and indecision characterize us, patients will lose all confidence in us.

One who knows his profession well, will be never agitated. He will employ the resources of his art; and if these fail, he will remember, that all medicine is from God, and that it may not be accordant with his divine will, that recovery should happen in every case of disease.

Rules for prescribing. The art of prescribing medicines demands especial attention, and requires to be well understood. Diseases are cured with remedies, and not with words. There are certain rules for prescribing, which ought to be followed as closely as possible, and ought not to be deviated from, except in certain circumstances. The precepts of Gaubius are simple, and I have added some remarks in my translation of the New Practical Formulary of Hospitals, 1835.[9]

1. The practitioner, whom prudence ought always to guide, should never prescribe a medicine without being able to give satisfactory reasons for so doing. The first question he ought to put to himself should be, is it or is it not necessary to administer medicines in the present case?

When it is thought that the powers of nature are sufficient to effect a cure; that the disease is absolutely incurable, or that the cure of the disease would produce a greater disease, it is generally thought proper to abstain from prescribing medicines, as well to prevent injuring the patient as uselessly tormenting him. Medici plus interdum quiete, quam movendo et agendo, proficient.[10] Nevertheless, as in some cases it would be inhuman to abandon the patient, and in others impolitic to shew the imperfection of our art, under such circumstances those substances which, if they are not really useful, are not [150/151] injurious, ought to be administered. This precaution should principally be attended to in the treatment of women and young girls.

2. When on the contrary it is necessary for the practitioner to prescribe, he ought, in the first place, to determine what he should do, and the medicines he ought to employ, &c.; questions, the solution of which he will find in his therapeutic knowledge. He should always recollect, that his end is the cure of his patient, as promptly and in as agreeable a manner as possible.

3. He ought always to make choice of the most efficacious remedies, and those best calculated to the attainment of his end; and this not only with respect to their nature, but likewise to the forms under which he prescribes them. Those medicines ought always to be employed whose actions are the most certain, and which are not likely to cause any bad effects. In certain hopeless cases, a medical man may try extreme remedies, but always with reserve, and be careful to announce to the relatives, the uncertainty of their results.

4. Medicines which are but imperfectly known ought never to be used, if the same effects can be obtained by the employment of those substances which have been

sanctioned by usage; and if it is necessary to prescribe new remedies, it should be done with the greatest prudence.

5. The use of substances which, by being kept, easily become changed, and which age has rendered inert or prejudicial, ought to be carefully avoided. In acting in this manner, a practitioner runs no risk of not obtaining the desired effects, of uselessly fatiguing the patient, and even producing serious accidents. It is for this reason that a practitioner, if he does not compound his own medicines, should have his prescriptions compounded in those houses where the sale is rapid, by which means he would stand every chance of having them well prepared, and their ingredients fresh.[11]

6. All things being equal in other respects, he should prefer indigenous medicines to those of foreign countries; they [151/152] are more easily known, and less likely to be adulterated.

7. The use of medicines of a low price should be preferred, provided they are as efficacious as those of a higher. Nevertheless, when a practitioner has to treat a rich patient, who thinks, as it sometimes happens, that medicines are only efficacious inasmuch as they are costly, he ought, in a certain degree, to comply with this ridiculous prejudice; because, as we have before said, the influence of the imagination is not to be despised.[12]

8. There are likewise cases in which, on account of prejudices, or individual repugnances, the practitioner is obliged to disguise, in different manners, the substances he prescribes. At one time it is their name he is obliged to change; at another, their taste and odour he will find necessary to mask by proper mixtures. But he must always be careful to be perfectly intelligible to the pharmacopolist, and not alter the therapeutical properties of the medicines he orders.

9. It is advisable, as much as possible, not to make use of those medicines whose odour, taste, &c. are very disagreeable; or, at least, to use them in small quantities, and disguised as we have already mentioned. And here it may be proper to remark, that this plan should be strictly followed in treating women and children, and in every case.

10. Before prescribing any medicine it is indispensable to find out, by every possible means, the idiosyncrasy of the patient. For it sometimes happens that a medicine, in other respects judiciously chosen, may, on account of certain individual dispositions, impossible to be foreseen, become useless or even prejudicial. For example, castor oil, one of the mildest purgatives, and one most commonly used, even for children, acted as a poison on all the individuals of a whole family which one of us was called in to attend. Gaubius relates an example of a man upon whom a small dose of the powder of crabs' eyes produced all the symptoms of poisoning by arsenic. Instances of this kind are too common to require enumeration. [152/153].

11. It sometimes happens that the patient is strongly prejudiced in favour of certain medicines, either because he has seen them administered with success in cases which he considers similar to his own, or for some other cause. If the practitioner thinks that the use of the medicine desired will not be injurious, he ought to comply with the wish of the patient; in other cases, without positively refusing, he should endeavour to make his patient comprehend that there would be danger

in complying with his request, and he should endeavour to gain time until his patient has changed his mind, or that his state will allow of the administration of the medicine desired.[13]

12. Temperaments, which modify in so powerful a degree the progress and character of diseases, likewise merit a particular attention with regard to the therapeutical means employed. In strong and robust individuals, endowed with a sanguine temperament, the sanguineous evacuations, diluents, in a word, the antiphlogistic treatment is much oftener employed than in persons of a weak, and irritable constitution, and of a lymphatic and nervous temperament; in the last-mentioned cases, tonics and antispasmodics are more frequently administered. It will nevertheless be conceived, that this is far from being an invariable rule.[14]

13. Attention ought also to be paid to the effects of habit; and it should be remembered, that organs frequently submitted for any length of time to the influence of a medicinal substance, become so accustomed to it as to be insensible to its effects. This is forcibly exemplified, amongst other examples, in the enormous quantities of opium which certain individuals can take, without experiencing any immediate accidents, as is seen in numerous instances in the east. Therefore, when it is necessary to apply the same substance for a long time, the dose should be gradually increased for it to make any impression on the organs. It is especially in medicines which [153/154] act upon the nervous system that this phenomenon is remarkable. There are, on the contrary, some medicines, whose action is slow and gradual, which require some time for them to develop themselves, and their effects are not manifest until after they have been administered for a long time. Their effects are much less weakened by use than those whose actions are more prompt. Nevertheless, after some time, quantities may, without any danger, be administered, which, in the first instance, would have been followed by alarming symptoms.

14. In prescribing a medicine, the consideration of the circumstances which may tend to favour or modify its action should not be neglected. Thus, in administering a sudorific, the patient should be placed in a warm situation; because, if he is exposed to cold, diaphoresis will not be produced.

15. Before introducing a medicinal substance into the intestinal canal, the practitioner ought to examine attentively the pathological and physiological state of this organ, the nature and extent of the diseases of which it is the seat, &c.; for a medicine which would be inoffensive, and even salutary, if the stomach were in health, may become fatal if this organ is diseased; it is necessary in this case to associate the medicine with others which weaken its local action, or even abstain entirely from its use.[15]

I shall now proceed to examine the modes by which medicinal substances are made to act on the economy.

1. When medicines have only a decided action upon those organs with which they are put in contact, they ought, as far as is practicable, to be applied to the diseased part, at least, where it is not desirable to obtain general effects by revulsion; and in that case it is always a healthy part that ought to be acted on.

2. When the influence of a medicine can be propagated through continuity of organs, the nearest parts to those affected are to be acted on, in order that their effects may be as marked as possible; because the influence of these [154/155] substances is as much less strong, as the parts to which they are applied are distant from those of which a change of their actual state is required.

3. Those medicines which act by sympathy are generally introduced into the stomach; because this organ has the most direct sympathetic connections with all other important organs.

4. When medicinal substances act through the absorption of their molecules, they are generally administered through the medium of the stomach. But they may sometimes be introduced into the economy, by putting them in contact with some other part of the mucous surface of the alimentary canal; it is on this account that enemata are sometimes administered, and frictions made on the gums, &c.

5. Formerly, advantage was taken of the absorbent power of the mucous membrane which covers the aerial passages, to cause the same result; and the patient was made to respire the vapour of those substances, under the influence of which it was desirable to bring him.

6. In fine, there are cases in which medicines are caused to penetrate into the economy by applying them to the skin. But as the presence of the epidermis is a powerful obstacle to the absorption of medicinal molecules, their action would be very slow, and even almost useless, if a simple application alone, was pursued. To obviate this inconvenience, it is necessary to make them penetrate the pores of the epidermis, by means of frictions more or less violent, or by raising this membraneous layer to a certain extent, and thus putting them in immediate contact with the surface of the dermis or skin. The first of these methods, that of friction, has been for a long time known, and is called iatraleptic. M. Christien, of Montpellier, has much extolled it, and put it in practice with success in a great many cases. The second is named, by M. Lembert, the methode endermique, who, conjointly with M. Bailly, has made numerous experiments, at the hospitals of La Pitié and Cochin, in applying upon a blistered surface, different medicinal substances, capable of acting by absorption. The result of. these experiments – repeated since by a great number of French and foreign practitioners – leaves no doubt of the efficacy of this mode of applying medicines, which [155/156] appears to be very advantageous when their irritating action is dreaded upon the mucous membrane of the gastro-intestinal canal, or when it is wished to prevent the alteration which the digestive faculties may produce in them. It would be, nevertheless, advisable to employ, in this manner, those medicines only which are susceptible of acting effectively in very small doses, such as morphia, strychnine, &c.

The knowledge of the doses in which medicines are administered, is called Posology; but it is impossible to determine them.

The doses in which medicines are administered, differ according to their nature and their degree of activity. It would be difficult to establish fixed rules in this respect, as experience alone must be our guide.[16] We shall only observe, that the doses of the same medicine ought to vary according to the effects that are wished to be produced, and according to the age, sex, and temperament, &c. of the patient.

The effects of a medicine frequently differ, according to the quantity in which it is administered. It is thus that the greater part of astringent, tonic, and exciting substances have only a local action, when given in small doses; whilst, on the contrary, in large doses, they extend their influence over the whole of the economy. Opium, taken in small quantities, is a very energetic sedative; in larger doses, it becomes excitant, and produces cerebral congestion when the dose administered is too large. Digitalis, in large doses, acts directly on the intestinal canal, as is proved by the vomiting and alvine evacuations which follow its administration. In smaller doses, on the contrary, its local effects are no longer observable, but are replaced by general phenomena, such as quickening the action of the heart, and augmenting the secretions, especially the urine. Some antimonial preparations are, according to the doses in which they are administered, alternately emetic, purgative, diaphoretic.

The doses of medicines ought always to be proportioned to [156/157] the age and strength of the patient. It is worthy of observation, that, in general, the weaker a patient is, and the more under the adult age, the more characteristic are the effects of a determined quantity of a medicine. It therefore follows, that to obtain similar effects on an adult and an infant, very different doses must be employed.

The following table, drawn up by Gaubius, may serve as a guide to young practitioners in the administration of active substances at the different epochs of life;

For an adult where the dose is 1 dr	
Under 1 year	1–15th to 1–12th
2	1–8th
3	1–6th
4	1–4th
7	1–3rd
14	1–half
20	2–3rds
From 20 to 60	1

although it should always be remembered, that this rule allows of deviation according to circumstances.[17]
Above this age, the inverse gradation must be followed.

The constitution of women is, in general, less strong than that of men; it will therefore be seen, from what has been already said, that the doses administered to them must be less; but it would be impossible to state in what exact proportion.

The doses of medicines ought likewise to be modified according to the temperament and idiosyncrasy of the patient; for it will readily be conceived that a very irritable person, endowed with what is called a nervous temperament, could not, without inconvenience, bear the dose of certain medicines, of excitants, for example, which could be given with advantage, to one of a lymphatic constitution.

It is, therefore, highly important to adapt the doses to different constitutions. There are certain individual dispositions, at first unknown in their nature, the whole of which form idiosyncrasy, and which prevent the [157/158] same substances – given in the same doses, and under the same circumstances, from acting in the same manner, and with the same energy in all individuals. It is in this way, that a small quantity of opium will produce in some persons all the symptoms of narcotism, whilst in others it would act insensibly. Half an ounce of a neutral salt of any kind, in some cases, produces abundant evacuations, and even superpurgation; whilst, in other cases, two ounces of the same substance would have scarcely any effect.

In fine, the effects of medicines being modified by habit, as I have before observed, it is of importance to have regard to this consideration, as often as it is necessary to continue, for any length of time, the use of a medicinal substance, or when we wish to administer, in large doses, certain very energetic preparations. Some of our popular pharmacological writers, have given the preceding as their own, thinking that the old standard medical works are seldom seen: These authors are to be pitied, and ought to remember, "palmam qui meruit, ferat."[18]

Following the precepts I have laid down, I think that every one will be enabled easily to modify the doses, according to the exigence of the case, and the observations he must have made individually in this respect.

Medicines are either simple or compound. Those are called simple which can be administered in such manner as nature offers them, or which are formed of one substance, of which the intimate nature may, in other respects, be more or less complex – such as ether, the acetate of morphia, &c. The second, on the contrary, are the result of a combination of several simple medicines.

Simple medicines in general ought to be preferred to compound and, when recourse is had to the latter, simplicity should be sought after as much as possible. The following maxim ought always to be present in a practitioner's mind: *Superflua nunquam non nocent*,[19] and those substances only should be united, whose reciprocal action and influence on the animal economy are well understood.

Medicinal substances are mixed together, or *compounded*, for the attainment of divers ends: [158/159]

First. To augment the action of the principal medicine which is intended to be exhibited. This may be attained:

(A) in mixing different preparations of the same substance. When all the active principles of a medicine are not soluble in the same liquid, and when it cannot be administered in substance, recourse should be had to this kind of combination. It is in this manner that infusions and decoctions are made more active by the addition of a small portion of the tincture or extract of the same plant.

(B) In combining medicines of the same species, that is to say, those which taken separately produce the same effects, but with less energy than when combined. This augmentation of activity is only evident in a certain number of medicines. According to the observations of Valisniéri, twelve drachms of cassia produce a purgative effect almost equivalent to that of four ounces of manna. But if eight drachms of cassia and four of manna are united, the effects obtained are much more marked, and even may be said to be double. The mixture of diffusible aromatic substances is equally susceptible of modifying the action of each individually.

(C) In uniting a medicine with a substance of a different nature, which exercises no action on it, but which renders the economy in general, or the stomach, or any other organ, more sensible to its influence. It is much easier to prove this than to explain it; therefore I shall content myself with giving a few examples. The mixture of ipecacuanha and jalap render the purgative effects of the latter much more energetic. The action of certain purgatives is increased by the addition of a bitter. Cullen remarks, that, in mixing a bitter substance with an infusion of senna, the same effects would be obtained in administering a small dose of this purgative as in employing a large dose of it alone. The influence that opium has over mercury is likewise very remarkable. It appears in some cases, that after the general effects of mercury have completely ceased, they reappear under the influence of opium.

Secondly. To diminish, or to correct in some degree, the too irritating effects of a medicine. This indication is fulfilled:

(A) by mixing a medicine with another which augments or diminishes its solubility. It is by this means that [159/160] the addition of a small quantity of an alkali diminishes the tendency of certain drastics to produce colic; and in mixing camboge with an insoluble substance, nausea is prevented, by rendering the solution more difficult.

(B) By its mixture with a substance susceptible of preserving the stomach, or the economy in general, from deleterious effects. There is a great number of substances which, when they irritate the intestinal canal too violently, cannot be absorbed, and are expelled without producing the desired effects. Squills and antimonial preparations, for example, do not act as diuretics or diaphoretics when they cause purging and alvine dejections. In such cases, it is necessary to know how to combine substances capable of remedying this local action, and of correcting such effects. Opium frequently fulfils this indication; at other times aromatic stimulants are used, or mucilaginous and emollient substances, which envelope, in some degree, the active ones, and thus diminish the local action which is dreaded.

Thirdly. To obtain, at the same time, the effects of two or more medicines:

(A) In employing substances which, though they act differently, produce frequently the same result when combined. To augment the secretion of urine, for example, medicines, whose modes of action on the economy are entirely different, are combined, such as calomel and squills. The former acts, as most mercurial preparations do, as an active absorbent; whilst the latter acts principally on the urinary organs.

(B) In combining substances of which the action is entirely different, and which are designed to fulfil several indications at the same time. It is with this view that purgatives are frequently united with antispasmodics, narcotics, tonics, mercurials, &c. The use of tonics often occasions constipation, and it is necessary to combine a purgative medicine to counterbalance this effect. In the treatment of ascites, and chronic dropsies in general, there are cases in which the practitioner finds it necessary to support the strength of his patient, while at the same time he causes abundant evacuations. This is effected by uniting tonics and excitants with drastic purgatives. [160/161]

Fourthly. To obtain effects which, if taken separately, would not result:

(A) In uniting, medicines whose actions are essentially different, and which, by their combination, produce other effects than those they would have produced singly, without acting chemically on each other. This effect appears to me inexplicable; but examples of the kind are too numerous for a doubt to be entertained on the subject. We see that opium and ipecacuanha, administered together, produce neither the narcotic effects of the one, nor the emetic effects of the other, but act as a powerful diaphoretic.

(B) In combining substances which act chemically on each other, and which give rise to new compounds, or which render the active principles of one of them null. In making, for example, acetic acid act upon ammonia, a new product is formed, the action of which is very different from that of the two other bodies taken separately. In the anti-emetic potion of Riviere, citric acid is mixed with the carbonate of soda. This latter is decomposed by the citric acid, and disengages the carbonic acid which it contains.

(C) In mixing substances which augment or diminish the solubility of the principles which contain the medicinal properties. This indication may be fulfilled by the aid of substances which act either chemically or mechanically. Thus the tartaric acid, or cream of tartar, becomes more soluble, and, consequently, more active, by the addition of the acid of borax.

Fifthly. In fine, to give them a form more agreeable or efficacious. Substances mixed with medicines, with a view either to render their taste or odour less disagreeable to the patient, or to prevent a too prompt decomposition, or in order to facilitate their action, vary according to the nature of the medicines employed, their degree of solubility, the end proposed, and, to a certain point, the caprice of the patient. Nevertheless, a choice ought to be made of such as would not annul the efficacy of the principal medicines. We shall have occasion to revert to this subject hereafter.

Such are the different objects that are had in view, when a mixture of several simple medicines is made to form a compound. According to the effects that these different sub- [161/162] stances are wished to produce, they are called by the following names the *base*, the *adjuvant* or *auxiliary*, the *corrective*, and the *excipient or intermediate*. The base is the principal medicine; the adjuvants are those added to facilitate and accelerate its action; the correctives are destined to reduce the too energetic action of the base; the excipients serve as a vehicle in which it may be taken; and the intermediate, a kind of excipient, is intended to render it miscible in water.

It is often useless to employ at the same time the whole of these elements in the formation of a compound medicine. Many substances want no adjuvant to facilitate their action, and others are administered very well without any corrective, or even without any vehicle. It also frequently happens, that the same substance fulfils at the same time several of these indications. For example, the adjuvant may serve both as a corrective and a vehicle. These last considerations are so much the more important, as simplicity is one of the most essential conditions in the composition of medicines.

Pharmaceutical preparations are divided into two great classes.

First, *Officinal preparations*, that is to say, those whose composition is laid down in the pharmacopoeias, and which are generally those kept in shops.

Second, *Magistral preparations* are those whose composition is indicated by the practitioner, and which the apothecary prepares from the formula given.

A *formula, or* pharmaceutical prescription, is the indication of the names and doses of substances which enter into the composition of a magistral preparation, to which is generally added instructions for its administration.

First. Clearness and conciseness are two essential conditions in writing prescriptions.

Second. They ought to be written in a legible hand, and in Latin, or perhaps in the vulgar language.[20]

Third. At the commencement of the first line, the sign Rx, or R, which is an abbreviation of the Latin word *recipe*, should be placed. [162/163].

Fourth. Each substance should be indicated by its scientific or pharmaceutical name, according as the one or the other is more generally known, and less liable to be mistaken for any other. The names of medicines ought always to be placed under each other, taking care to put but one in the same line.

Fifth. The order of arranging them is of little importance; nevertheless it would be well to place the most active ingredients first.

Sixth. The quantity of the dose ought always to follow the name of the medicine, and be placed in the same line, leaving a small interval between them. The following signs are those which have been established by use:[21]

The quantity of each of these weights is generally indicated by Roman cyphers. When the same dose of several different substances is used, they are united by a brace, and the letters ana, or āā are placed before the designation of the common quantity of all.

Seventh. The prescription should he ended by indicating the mode of the preparation of the medicine, and the manner of its administration. When the preparation presents nothing particular, it is merely necessary to write the letters F.S.A. *(fiat secundum artem)*. In other cases, the mode of preparation, should be indicated as briefly as possible; then it is to be dated and signed with the initials of a physician, and the name of a surgeon.

Before examining the different pharmaceutical preparations, and the forms under which they are administered, I think it necessary to call the reader's attention to the errors that may be committed in compounding magistral preparations; errors which arise from three principal sources, namely:

First. The association of substances which do not combine, or do not form compositions of a proper consistence. Many [163/164] substances, insoluble in water, cannot be administered in a liquid form, without the aid of an *intermediate*, such as a mucilaginous or albuminous substance, which keeps their molecules in suspension. If the *intermediate* is neglected, the formula will not accomplish the desired effects. This would be the case in ordering camphor and the balsam of capaiba in pills, without adding a proper *intermediate;* because these two substances, mixed together, would form a syrupy consistence, and it would be impossible to make them into pills, unless a small quantity of the coagulated yolk of an egg was added.

Second. The association of substances which mutually decompose each other, by which means their action is changed, or entirely destroyed. Every time that two salts are mixed in solution, which, by an exchange of their bases or acids, may be formed into one soluble and one insoluble salt, or into two insoluble salts, a decomposition necessarily takes place.

Third. The method indicated for the preparation of medicines is insufficient to attain the end proposed, or is of a nature to change or destroy the action of some of the substances employed. Certain medicines are only soluble in alcohol, ether, or oil; and others are only soluble with the aid of caloric; and some lose their active principles by ebullition. It is therefore of the highest importance not to order in an infusion of cold water a substance which is only soluble in warm, and not to order a decoction of medicines which ebullition alters, and which lose their virtue by this process, &c.

The forms under which medicines are administered vary according to the nature of the substances, and the use that is wished to be made of them. These forms are solid, soft, liquid, and gaseous, and most of the pharmaceutical preparations have a special destination; some are always employed externally, and others internally, whilst there is a certain number which serves at the same time both for external and internal uses.

In the preceding remarks, I have offered but a very brief outline of those principles which guide scientific practitioners in prescribing medicines; as I consider this information of great importance to students and junior practitioners. I could not treat of them more fully in a work of this kind; but refer the reader to the excellent Pharmacologia of Dr. Paris, in [164/165] which he will find the most ample account of this branch of medical science in our language.

It has been truly observed, that disease terminates in health, death, or another disease. I may observe, that, when the period of convalescence arrives, it must be managed with judgment and skill. Relapses are more dangerous than the original disease, as the constitution is generally too much enfeebled when they occur. The greatest attention must therefore be paid to diet, regimen, and other hygienic precepts, and curative means. When death approaches, the medical practitioner has certain duties to perform which are indispensable, and have been already described.

Duty of Physicians to the Dying. If a physician have judgment, and all the other requisites essential to his character, if he take a true interest in the health of the sick, he will afford consolation to the dying. The respect due to humanity, and to the dying, imposes the obligation of consoling those departing from life, while their senses remain; and in reminding them of a better world, in which suffering and sorrow will be no more. We ought not to quit a patient in danger, unless with a calm and serene air. When the fatal hour is about to strike, the physician is bound to recommend the consolations of religion. The sublime doctrines of Christianity calm agitation, and smooth the avenue of death. Imprudent zeal, on the contrary, has often hastened the fatal event.

It is also right to inform the patient, his near relatives, or the clergyman who attends him, of the necessity of arranging his affairs, so as to secure his family their rights, and prevent litigation. This is a delicate affair; but it must not be forgotten by the medical attendant. Let him also take care that his patient, "is of sound mind, memory, and recollection, when disposing of his property." Instances have fallen under my own observation, in which testators did not really know what they were signing; and I am convinced that such wills would be annulled, if contested.

Notes

1. *Author's note:* Godman's Addresses, 1824. *Editors' note:* Ryan here takes a long quotation from another of Godwin's addresses, "Monitions to Students of Medicine," dated November 1824; John D. Godman, *Addresses Delivered on Various Public Occasions* (Philadelphia: Carey, Lea & Carey, 1829):16–18.

2. *Editors' note:* "It is not right to praise or criticize. Praise sparingly; criticize more sparingly. If permissible, praise new things and old. Carp on both sides. Your respect remains, as does your honor. Do not praise a living author with enormity in order that you in turn will be praised. Your measure depends on words, titles, and praise. Do not praise or implicate all writers with a single booklet, neither because of jealousy, nor lack of praise or excesses."

3. *Editors' note:* for luring the public.

4. *Author's note:* Gresset has well remarked on this custom: "Des protégés si bas, des protecteurs si bêtes."

5. *Editors' note:* "Any amateur imagined themselves doctors. A priest, a Jew, a king, an actor, a barber, a hag, soldier, merchant, laborer, wet nurse, and an orator. Anyone at all today wished to be a surgeon."

6. *Editors' note:* "And it came to pass that especially for a medical man, being married to a young and beautiful girl was a cause for anxiety. Epictetus described this: he whose wife is beautiful, has her in common with others. When a physician is deeply occupied by his practice, how will he be a successful guardian of either his home or his wife? If meanwhile she is hospitable, and by emulating Diana transforms her arrogant husband like Actaenon into a stag wearing a crown of horns, and by herself produces heirs, not of the husband's blood? Just as at least not once silently with to be responsible for murmuring complaint. By no means I commanded my wife, another bears love. Thus you all not by you all." The latter part of the passage is difficult to translate due apparently to the author's garbled version of nineteenth century medical Latin. The allusion is to the myth of the hunter Actaenon coming upon Diana naked, and the enraged goddess turning him into a stag who is then consumed by his own hounds. Here, though, the "crown of horns" alludes metaphorically to the cuckolded husband. Epictetus wrote in Greek rather than Latin and a corresponding passage could not be located among his works.

7. *Editors' note:* "Moreover, because the noble and more trusted arts are the most extensive, they should not be undertaken except by those with more refined manners" (somewhat freely translated).

8. *Author's note:* See also "De l'Onanisme et des autres Abus Veneriens, consideres clans leur rapports avec la Sante. Par M. Deslandes." 1835.

9. *Editors' note:* Hieronymus David Gaubius, 1705(?)–1780.

10. *Editors' note*: "Sometimes quiet and rest are as effective as the movements and actions of the physician."

11. *Author's note*: Some writers think it high treason to act on this rule. They shall be nameless, though they deserve exposure and censure.

12. *Author's note*: Nothing will do with some of the asinine part of society, but expensive medicines.

13. *Author's note*: Every patient, according to our English notions of liberty, is allowed to play the fool as much as he pleases. He may swallow two or two hundred of Morrison's pills should he think proper, and poison himself.

14. *Author's note*: Some physicians bleed every patient excessively. Every disorder or disease with them is inflammation. They have gone deservedly to the tomb of all the capulets [*sic*].

15. *Author's note*: The routine practice is to prescribe strong aperients or purgatives in all cases. This is a grievous error.

16. *Author's note*: All tables of doses are absurd; they merely mean, that the smallest quantity is prescribed. Ten times the amount is often given, according to the violence of the disease, or power of the constitution.

17. *Author's note*: The safest plan is to prescribe moderate and repeated doses, until the desired effect is produced.

18. *Editors' note*: "Who deserves honors, has earned them."

19. *Editors' note*: "Superfluous ingredients sometimes harm"

20. *Author's note:* I do not assent to this doctrine, because patients would be horrified on being ordered the poisons, mercury, &c.

21. *Editors' note:* We omit a table in which Ryan indicates the approved symbols for writing pounds, ounces, drachms, scruples, grains, and pints, as many of the typographic symbols are no longer in use. Ryan also indicates the conversion factors among these units of measure.

Selections from Ryan's Obstetrical Writings

A Professional Duties of the Obstetrician, or Accoucheur

1 From Manual on Midwifery, *1828 Edition*

[237] The first rule to be observed by the practitioner of obstetric medicine is, to obey the summons to give his personal attendance to the parturient female, as soon as possible. This duty admits of no compromise, for it is of the first importance to ascertain the presenting part, before the disruption of the membranes; because, if preternatural, it can be more easily rectified before the completion of the first stage; for in such a case, the practice is to rupture the membranes, when the os uteri is sufficiently dilated, to seize the feet, and deliver the infant by the operation of turning. If the membranes have burst, the uterus is in close contact with the body of the infant; and if the pains be violent, as it mostly happens, the operation of turning cannot be performed; as the womb would be lacerated, which is a most fatal occurrence. Again, an early examination is most necessary, as flooding may destroy the patient, or the feet of the infant may be expelled, the body retained in the vagina, and such pressure made on the navel string as to impede the circulation of the blood between the mother and child, and kill the latter.

The obstetrician, on his arrival, should ascertain all delicate inquiries from the nurse or other female attendant; and not in the sick-chamber. He should learn the duration of labor, the state of the bowels, and the history of the case, and impress the necessity of admittance to the patient as soon as possible. After admission into the sick chamber, he should approach the patient with his countenance contemplative, cheerful, nor grave and melancholy; his look pleasing, mixed with mildness and humanity, and expressive of a sincere desire of affording her alleviation. He ought to be polite and attentive, never proud, insolent, haughty or grave, which would alarm the patient, [237/238] who is generally timid and dejected. There should be nothing indecorous in his aspect or conduct, nothing rough or uncouth. He should display those good manners and politeness that characterize the well educated medical practitioner and the polished gentleman. Sympathy engages the confidence and affection of the patient; she feels the approach of the practitioner, who displays it, like that of an angel; while that of an unfeeling man, as of an executioner.

Of all diseases incident to humanity, those of the female are most deserving of sympathy and attention. The weakness and peculiar delicacy of the female constitution, call forth our greatest tenderness and compassion, and on no occasion so powerfully as in the agonies of child-bed, when she is stretched upon the rack on which she is laid by nature. On entering the parturient chamber, the practitioner should approach the patient with all the mildness and serenity of manners he may possess, and assure her, that from the history he has heard from the nurse, he is perfectly confident of her safety. After a few pains have occurred, he should pass his hand over the bed or body clothes on the abdomen, in order to ascertain whether there be pregnancy; for it has often happened that females and their medical attendants have expected delivery for days, when there was no pregnancy. Drs. Hamilton and Blundell record such cases, and I can bear testimony to the same fact. I have known a medical man remain four successive days and nights in attendance on a patient, who was not pregnant, so confident was he of delivery; yet the pains gradually ceased, and 8 years afterwards she expelled an immense quantity of hydatids from the uterus. After having ascertained the fact of pregnancy, we should next proceed to discover all the essential characters of labor, by an examination through the vagina, called by the French "the touch." Before we proceed to examine, the proposal ought to be made through the nurse, its impor- [238/239] tance strongly dwelt on, the length of time the labor has continued, the urgency of the pains, and above all, the necessity of ascertaining whether the labor be natural. Some women of high delicate feelings will not permit an examination, until the labor becomes severe; but then we need little persuasion or reasoning, to induce them to comply. A medical friend informed me, that a patient of his on entering her chamber, covered her face with her hands and exclaimed, "You shall not touch me; I'll lose my life first." He withdrew, but in a short time, when the pains became severe, he was anxiously recalled. In such cases, an intelligent nurse will be able to report whether the labor were natural or not. Other women are greatly terrified at the approach of the medical attendant, if never attended previously by an accoucheur; they are alarmed, lest they should experience violence and harsh treatment, which are now unknown, and those foolish fears are carefully fomented, rather than appeased, by the nurse, whose peculiar province she considers invaded. This is mostly the case in first labors, for women once attended by a medical man, who can never be guilty of rudeness, violence, or harsh treatment, without a gross dereliction of duty and breach of established obstetric rules, will never submit to the ignorance and temerity of female attendance afterwards. Surely every educated female must be convinced, that an illiterate and ignorant nurse knows nothing of the mode in which nature accomplishes the wonderful mechanism of labor. How then can she render assistance, in cases of difficulty and danger? What idea can such a one form of the almost innumerable difficulties that may impede labor? Hence the extensive employment of medical practitioners in every civilized nation in modern times. All doubts and fears, which a parturient female entertains must be obviated by reason and good sense, and the examination by the vagina instituted as soon as may be convenient. The patient is to lay on her left side, a [239/240] coverlet being thrown over her, and her hips as near the edge of the bed as possible, her knees drawn up towards the abdomen, and the bosom bent downwards and forwards, towards her knees. The nurse and another female attendant should be in the apartment, the light is to be excluded in some measure, and the curtains drawn close. The index and middle finger of the right hand, the nails being pared closely, are to be lubricated with some oleaginous substance, which is generally prepared by the attendants, as lard, butter, pomatum, olive oil, and passed under the right thigh, during the next pain from the perineum into the vagina, when they are to be directed downwards and backwards, as the orifice of the womb in the first stage of labor is low down, towards the sacrum. The orifice of the womb will be found dilated or not; if it admit the point of the finger, it is to be considered dilating, and will be pushed down during the pains. If it be dilated as large as the surface of a shilling, the head or other presenting part of the infant may be guessed at; and if the pains be those of labor, the vagina will be more or less lubricated with the increased mucous of parturition. Then the hand should be withdrawn, when the pain ceases, and wiped under the bed-clothes with a napkin, previously in readiness by the nurse.

This introduction of the fingers is to be accomplished as speedily and gently as possible, and the greatest delicacy is to be observed. The examination gives no pain, and hence removes the dread which many women, either from some misconception or former harsh treatment, entertain of this operation. A vast deal of useless suffering and fatigue will be saved the patient by an early examination; for if the labor be preternatural, it can be rectified readily before the rupture of the membranes. If labor be about to commence, the practitioner is generally consulted, and his superintendance requested, concerning the arrangements of the sick chamber, and of the bed especially. The nurse usually attends to the adjustment of the bed, but often [240/241] enquires, especially of young men, how the bed is to be arranged, in order to discover whether they have had much practical experience. In the better ranks, the mattress is covered with a skin of red or brown leather, and over this a folded sheet or blanket is applied, to absorb the moisture. A sheet is pinned over all, to keep them in their places. Others recommend a folded blanket alone, having turned up the pallet and blankets towards the head of the bed, and this is most prevalent among the middle and lower classes. A coverlet is to be thrown over the patient, or more covering, if the season require it. The woman is to put on a night wrapper, or other loose dress; but when the head of the child has descended low down into the cavity of the pelvis, she must undress and remain in bed. The chemise may be folded up on the hips, and its place supplied by a loose petticoat. The bedclothes ought to be light and comfortable, the chamber to be kept cool; no fire in summer, although a small fire tends to ventilate the chamber; and no more persons should be admitted into the apartment than are absolutely necessary. One female friend, to whose kind and sympathizing ear the poor sufferer may communicate her sorrows and anxieties, and the nurse, are quite sufficient for every useful purpose. The relation of all bad cases and frightful stories should be avoided, especially during a first confinement; as they do infinite mischief, by depressing the mind of the suffering patient, who naturally becomes alarmed, lest her case might be equally unfavorable; she loses confidence in her own powers, and retards, or perhaps entirely impedes, the process of labor. All nurses, and some medical men, are guilty of this imprudence, in narrating such stories. Dr. Dewees observes, "the poor suffering woman is entitled to all the consolation, a well grounded assurance of a happy termination of her case can afford; yet she must not be betrayed by false promises, as to a speedy issue; for it requires great experience to be able to state with certainty, [241/242] when any case will terminate. Her mind should be kept as free from anxiety as the nature of her situation will permit; therefore no conversation should be indulged in, which might for an instant excite her apprehensions. Conversation should be cheerful, and free from idle discussions of danger from similar situations, and should be as void of levity or want of feeling and sympathy, as of gloominess. Levity ill suits the situation of a woman in labor, and moroseness or ill humour is quite brutal, when the poor sufferer has a reasonable claim to pity and compassion."[1] The patient's mind should be diverted by a well chosen and general conversation, always combined with a confident assurance of her safety. The practitioner can afford no assistance during the dilatation of the womb in natural labor, and therefore he ought to withdraw from the parturient chamber, as his presence is a restraint; and his absence will allow the evacuation of the bowels and bladder, and also abridge the period of his watching. He may even visit other patients, as the first stage of labor may occupy several hours. The woman may walk about, sit, lay in bed, or sleep, during the dilatation of the womb; and she may have any light food, such as tea, toast, coffee, water gruel, sago, arrow root, tapioca, broths, &c., but no cordials, unless she be really debilitated, which is not the case once in a thousand instances. The exhibition of ardent liquors, although a popular custom, is highly improper; for no woman ever dies of weakness, during natural labor. The middle and lower classes are greatly prejudiced in favour of their use, or rather the abuse, of ardent and fermented liquors during labor; and hence the frequency of inflammations, fevers, and deaths amongst them. There is no medicine necessary during the first stage of labor, except the bowels be confined, when they are to be regulated by some castor oil, or a clyster. If the pains be trifling and very inert, for twelve to twenty hours, a dose of tincture of opium will be of advantage. Every accoucheur should [242/243] carry about him some tincture of opium, some ergot of rye, a catheter, a tracheal pipe, and a lancet. The opium is invaluable in allaying irritability, produced by false or spasmodic pains, and the other articles are equally valuable under certain circumstances. During the dilatation of the uterus, or first stage of labor, it is an established rule, that the practitioner need not examine by the vagina, more than two or three times. Frequent examinations, or attempts to dilate the vagina or uterus prematurely, a common practice with nurses, are highly injurious, by inducing irritation, inflammation, and swelling of these parts, which not only oppose invincible obstacles to delivery, but also lay the foundation of many of the fatal fevers and inflammations so common after parturition. The late Dr. Osborne, of this metropolis, used to remark, "that the practitioner should sit quietly and observe nature." The less manual interference in natural labor the better. Nature is the best obstetrician – "a meddlesome midwifery is bad." How many thousand women are delivered annually in the face of the globe, without any assistance![2]

2   From Manual of Midwifery, *1841 Edition*

[226] No obstetric instrument or manual operation ought ever to be performed until the necessity is explained to the relatives of the parturient woman or to herself, when she possesses a strong mind; so that an opportunity may be afforded of calling in another medical practitioner, if convenient, should it be desired by the woman, her relatives, friends, or acquaintances. Every medical practitioner, whether old or young, ought to observe this precept.

The woman or her friends have an undoubted right to have the opinions or assistance of as many medical practitioners as they please.

It is scarcely necessary to state, that the medical attendant may be considered too old or too young – or that there may not be sufficient confidence reposed in him. [226/227]

It is always to be remembered, that women in labour have, in general, the greatest dread of manual or instrumental operations; and that religion and humanity command every thing to be done to appease their fears and alarms, as these are always more or less prejudicial to their condition, as well as to restoration to health.

It may be necessary, as was well observed by Denman, to explain the object of using the forceps, or other blunt instrument, and to prove that no injury can be inflicted upon the woman or infant, when the instrument is employed with caution and judgment.

> In some cases of great apprehension, I have also showed them, upon one of my knees, all I intended to do with the forceps. – *Denman's Aphorisms*[3]

I have long followed this excellent advice, and observed to women, that the instrument was blunt, and not cutting, that it could pass where the hand could not, and that its branches might be considered as artificial hands, and would not injure the woman or infant, and that the latter might or would be born alive.

This explanation almost invariably removes all fear and apprehension.

There are, however, some nervous, hysterical, irritable, and impatient women, who, impelled by their fears, impatience, or sufferings, implore us to deliver them with instruments, long before there is a real necessity.

Such individuals may be encouraged, by fixing some remote period at which it is most likely they may be delivered, as six to twelve hours, and unless they are, and when twenty-four hours of actual and regular labour have elapsed, that then instruments will be employed.

They should also be told, that most women are delivered within twenty-four hours, without instruments; that others remain in labour a day or two, and are safely delivered without artificial interference; and that the good old remedies, time and patience, ought to have a fair trial – and that nature is the oldest and safest obstetric practitioner.

The forceps, or any other obstetric instrument, ought never to be used clandestinely, or without the knowledge of the woman, or of her friends.

There is always great caution required in the use of obstetric instruments, as great fears are entertained by most women, and great blame attached to the obstetrician, unless the woman and infant do well ultimately, and unless he possess a high reputation.

It is always necessary to impress upon the woman the imperious necessity of remaining quiet during the use of obstetric instruments, for when she is restless, there is more or less danger of injuring her, however experienced the obstetrician may be, unless extremely cautious.

When she moves suddenly, or changes her position, though the operator removes his hand off the instrument, it may contuse her slightly, and give her pain.

It is therefore manifest, that instruments ought not, as a general rule, to be applied without her knowledge, and not until having impressed her with the indispensable necessity of her remaining quiet and motionless.

Every woman endowed with common sense, will obey the advice of her medical attendant, when she places proper confidence in him.

B Conflicts Between Preserving the Lives of the Infant and Mother: from *Manual of Midwifery*, 1841 Edition

[241] This operation [induction of premature labour] has been performed, when the length of the superior aperture of the pelvis, from before backwards, the antero-posterior, or sacro-pubic diameter, measures only from two to three inches instead of five, as in the ordinary sized woman. The object in such cases is laudable, and not criminal; namely, to save the life of the infant, and the operation totally differs from the induction of criminal abortion, which is never attempted by the medical faculty, who justly consider it foeticide or homicide, but cases occasionally occur, in which women are so deformed, that although their infants may be extracted alive at seven months and a half, they must be destroyed either by the efforts of nature in strong labour at the full time, which cannot expel them, or by art; and it has frequently happened, that as many as seven and more infants have been thus successively destroyed. A most important question here arises, as to the morality of such women having children at all, and if they have, whether the former ought not be saved by the induction of premature labour. When they cannot be saved by this means, it was the opinion of the late revered and virtuous Dr. Denman, and most others of his contemporaries and successors, that however dangerous the caesarean operation, which was almost mortal in his time, it should be performed upon a woman who had so many infants successively destroyed, when in a future labour. The induction of premature labour, is, however, preferred at present.

I may here observe that there are not theologians, juris-consults, or physicians, even of these enlightened times, who will venture to determine the propriety of separating husbands and wives, or of their using any means to prevent the great object of matrimony. But is infant after infant to be destroyed, either by omission or commission? The result is the same – the death of human beings. In a moral point of view, it matters not whether an infant is starved, exposed to cold, strangled, or so treated or neglected, that its life is destroyed. A woman is assured by competent medical authority that she can never have a living infant at the full time, though she may have seven or eight infants in succession, every one of whom will be destroyed by the efforts of nature or by art, unless saved, when possible, by the induction of premature labour. [241/242]

I have given this subject very mature consideration, and my deliberate conviction is, that husbands and wives are, in general, equally culpable: sometimes the one, and sometimes the other, but mostly the former; in causing such repeated destruction of human life. I have repeatedly known instances of both, in which either party was perfectly indifferent on the subject; but there are many exceptions in which most persons deeply deplore the want of offspring. But however extreme the cases under consideration may be, the medical faculty will not interfere with the dictates of nature, nor violate the laws of religion. What a contrast exists between them and the modern anti-populationists, who insanely propose to limit families, or, in plain English, prevent procreation among healthful individuals, on the fallacious grounds of want of sufficient nutriment, although there is not a single civilized country on the face of the globe at present, to which this immoral, erroneous, and inhuman doctrine can apply. I have fully exposed and denounced this infamous doctrine in another work, "*The Philosophy of Marriage*," 1839, 3rd edition, and its baneful consequences in another volume, "*Prostitution in London, Paris, New York, &c.*", already quoted. I need scarcely observe, that if the medical faculty were to recommend such a doctrine, even in the extreme instances which gave rise to these remarks, they would, in all probability, do much more harm than good to the interests of society at large. They, therefore, always follow nature's laws and dictates, and assist her as much as possible.

[245] Embryotomy is required when the pelvis is so small by conformation or diseased contraction, that a full-grown infant cannot be extracted through the natural passage, either with the hand alone, or with any blunt instrument. The opinion in Continental Europe is, that the infant ought to be dead or destroyed by the pressure of the womb during its contractions or labour pains, before perforation of the skull is attempted, and in this opinion I most fully concur; for I boldly deny that there is any text in either the Old or New Testament which justifies the destruction of the infant in the womb, under any circumstances whatever. But, nevertheless, the obstetric rule in this country is, that the head of a living infant should be perforated with a sharp instrument, its brain pierced, lacerated, and evacuated, the skull reduced in size (see *Craniotomy*), to save the life of the mother; and lamentable cases have happened in which, after the partial destruction of the upper part of the skull and brain, the mutilated infant was brought into the world alive, to perish in a few minutes. Most persons ignorant of medical knowledge may doubt this statement, but it is still positively correct, as every well-informed medical practitioner must admit.

I must here observe, that in all cases requiring craniotomy, the pressure of the womb in forcing the infant's head against the contracted passage of the mother, will very speedily and inevitably destroy life, by causing congestion of the brain, or apoplexy, and in a preponderating majority of instances long before the life of the mother is, or can be in danger of destruction; and it is therefore manifest that a living infant should not be invariably destroyed in general, under the false and unjustifiable impression of saving the life of the mother, which I repeat, is seldom endangered for hours or days; and when it really is, in extreme cases, the destruction of the infant will seldom, if ever, preserve it.

It is well known to obstetricians, that in all difficult labours in which the infant cannot easily pass into the world, it is sooner or later, almost if not universally, destroyed by the pressure of the womb long before the life of the woman is in danger. Every experienced obstetrician must assent to the truth of this statement, as he must have observed in practice, or read of cases in which the infant was destroyed in a few hours, or has not been felt to move for one or more days, without the powers of the woman having been exhausted, or her life endangered; so that, as a general rule, he may wait the destruction of the infant by uterine action in most cases, and ought not to effect it, if ever, by cutting instruments. Such is the rule I have adopted for nearly twenty years; and I have never lost a woman on whose infant I performed craniotomy after death, for I never did while it was living, nor never shall, because I conscientiously consider it unwarrantable, unscriptural, and unjustifiable to act otherwise, for the reasons above assigned, and for others I shall adduce in the account of the caesarean operation hereafter.

Neither can I agree with the French and other European obstetricians, that the caesarean or other incisive operations already mentioned, ought to be performed on a woman to save the life of her infant, for the purpose of baptizing it, as it would be a barbarous proceeding, [245/246] in my opinion, to subject the mother to dangerous, indeed almost fatal operations, for the extraction perhaps of a dying or dead infant, nor do I believe that there is any scriptural authority to warrant such a proceeding, or that alludes to intra-uterine baptism. But when this is considered necessary, it can be readily accomplished in most cases by passing the tube attached to the modern double injecting, or enema syringe into the womb, and in contact with some part of the infant, and the water be applied freely. There are also many cases in which the water might be applied with the finger, when the infant is low in the pelvis.

I cannot agree to the opinion, that father, mother, or obstetrician, is ever justified in warranting or effecting the destruction of a living infant in the womb under any circumstance, even before or after intra-uterine baptism, and consequently I differ from the position of Professor Hatin of Paris, a most able obstetric author, "that embryotomy ought to be performed when the mother refuses every other operation." I maintain she has no right, or any one else, to destroy the life of the infant in the womb. (See Caesarian operation hereafter).

I also differ from Dr. Blundell and many other eminent British obstetricians on this subject.

This is not the place for theological or religious discussion, but I cannot help observing that the commandment, "Thou shalt not kill," cannot be broken without incurring great culpability. I am well aware that all religionists act differently from the principles of the Bible in modern times, but whether they are justified remains to be determined. It is a point of great importance in all cases requiring cutting operations, either on the infant or mother, to determine whether the former is alive or dead, which can be done, in most cases, by means of modern inventions, *as the* stethoscope and metroscope, (see *Signs and Detection of Pregnancy*, p. 159), which will enable the obstetrician to arrive at correct conclusions in general.

[254] It is now an established obstetric rule in all civilized countries, that no woman should be allowed to die undelivered, and that in all cases of parturition, however difficult, attempts may be made to save the lives of mother and infant, or to save the life of the one, even by destroying that of the other. However laudable may be the first part of this proposition, the last part of it, as to the destruction of the life of either woman or infant, is clearly contrary to the divine precept, "thou shalt not kill;" and I fearlessly maintain, that no case can occur in civilized society, not even homicide, in which any human being is justified in destroying the life of a fellow-being in any rank of society, more especially on account of the disqualification of such being or beings for the performance of a natural function, either parturition, respiration, digestion, &c., under any circumstances whatever.

In rude ages, the husband was suffered to murder his children, but at the present enlightened period, I am unacquainted with any scriptural, civil, statute, or other law, in any civilized country, which allows either husband or wife, or father or mother, to destroy the life of either or of their offspring, whether before or after birth. There is no law, which warrants kings, queens, legislators, judges of any court, criminal, civil, naval, military, ecclesiastical, &c. &c., or physicians, surgeons, or any class of society, to destroy, either directly or indirectly, by whatever means, the life of a fellow-subject, or any member of the human family – not, I repeat, even for the horrible crime of homicide. This is the general opinion of many of the ablest British and other statesmen, judges, theologians, jurisconsults, physicians and medical practitioners, as well as of the leading portion of our free, most powerful, and unequalled public press, and of all enlightened individuals, at the approach of the middle of the nineteenth century, AD 1840. Nevertheless, I am grieved to admit, that there are as yet some eminent members of the medical profession in this and other civilized countries, who maintain that the life of the infant in the womb may be sacrificed to preserve that of the mother, although they have forgotten to specify, in an accurate manner, for the best of all reasons, because they could not, the class of cases which would warrant their inhuman and unjustifiable conclusions. I feel convinced that no well-informed member of the medical profession could attempt such a classification, or precisely describe the [254/255] particular cases, amidst the human family on the face of the globe, in which the destruction of the infant would or could save the life of the mother, or vice versa. It is for this reason that the French, German, and most European and American obstetricians prefer symphyseotomy and gastro-hysterotomy to craniotomy, although preferred to the former operations by a majority of British medical practitioners. But so great is the difficulty in determining cases requiring any of these operations, that a consultation must always be held before resorting to any of them.

It is supposed, very erroneously, by some persons, that the husband possesses the prerogative of deciding upon the preservation of the life or death of his wife or infant, or the father of the life of his victim of seduction or concubine, and infant. I think, however, that were husband or father, to cause the death of either, he would be found guilty of murder, or infanticide, or manslaughter, according to the laws of this country. (See Author's *Medical Jurisprudence*, 1836, *Articles – Abortion and Infanticide*.[)]

In my opinion, a medical practitioner is equally amenable to the laws, and is never justified in consulting a husband or father concerning the preservation of the life of either wife or mother, or of the infant in the womb, as every person endowed with common sense will admit. No one but an educated and experienced medical practitioner can be a competent judge of the nature of any difficult case in midwifery, and no husband or father or any other person has the slightest right, in my opinion, to offer any suggestion as to the treatment, or as to saving of the life either of the woman or her infant – a question which must be entirely decided by medical opinion, and one on which even the most eminent of the faculty are still very much divided.

It was properly referred by the medical faculty of France to the doctors in theology at the Sorbonne, in Paris, A.D. 1648, who decided as follows, and I think correctly, according to the Bible: – "Nous sous signés docteurs en théologie de la faculté de Paris, sommes d'avis, que si l'on ne peut tirer l'enfant sans le tuer, l'on ne peut sans péché mortel le tirer; et qu'en ce cas la, il faut tenir à la maxime de St. Ambroise, – 'Si alteri subveniri non potest, nisi alter lædatur, commodius est neutrum juvare.'" – Délibré à Paris le 24 Avril 1648. "We the undersigned doctors in theology of the faculty of Paris, are of opinion, that if the infant cannot be extracted without killing it, it cannot be extracted without committing a mortal sin, and that in such a case, it would be best to hold the maxim of St. Ambrose, – 'If one cannot be assisted without seriously injuring the other (by wound, blow, or otherwise, *see lædo*), it is best not to assist either.'" – Delivered at Paris, April 4, 1648. This is still the doctrine of the Roman Catholic Church. Another great objection to embryotomy was, that the infant could not enter Heaven without baptism. But it was contended by Thomas Aquinas, that the infant could not be baptized in the womb, for, according to Scripture, it should first be born, that is, it should be *natus* before it could be *renatus*, reborn by baptism. This difficulty was, however, over-ruled by the Sorbonne doctors in 1773, who declared that baptism was valid if the water touched any part of the infant's coverings. They decided, "*dum modo infans sit vivus, et arte seu industria medicorum possit aqua* [255/256] *ad ejus corpus immediate pervenire.*", "Whilst the infant is alive, and by the skill of medical practitioners, the water is brought in contact with its body." They might have added, or in contact with the membrane which covers it, and lines the womb, which is now considered part of the infant. It was formerly undecided whether the outer surface of the membrane, which encloses the water and infant, belongs to the womb, as it is closely attached to it, or to the foetus.

The decision of the Sorbonne doctors removed one objection to embryotomy, but admitted the former.

In this country, obstetricians are generally in favour of embryotomy, whether the infant be alive or not, as the more valuable life of the mother, they contend, ought to be preserved. It is said, "the tree should be preferred to the fruit." But the French, German, and American obstetricians are in the favour of the caesarean operation.

[260] It is also a strong objection [to embryotomy] that we have no certain and no positive signs indicative of the death of the infant, except those afforded by the stetho-

scope and metroscope, see p. 159; and hence the European and American authors hold that the sacrifice of the infant is murder; and that in cases of extreme deformity, its mutilation will not save the mother. On the whole, the most respectable and eminent of the foreign obstetric authors are unanimous in preferring gastrohysterotomy to embryotomy. They contend, that if the former operation were performed at an early period, the woman's life would not be so much endangered as by the latter.

Professor Lizars, the eminent surgeon of Edinburgh, has frequently performed gastrotomy with success, and the wound in the abdomen healed by the first intention. Hull informs us, in a note in his valuable [260/261] translation of Baudelocque's work, that of two hundred and thirty one women, operated on by gastrohysterotomy in this and foreign countries, one hundred and thirty-nine recovered; and the recent reports of the German practitioners are still more favourable.

Denman observed, that in cases where the infant should be invariably destroyed, a question ought to arise, whether a woman who was warned of this, again becoming pregnant, ought to be relieved by embryotomy. He, as well as Burns, Hull, and Dewees, are advocates for gastrohysterotomy, while a few others despise the "silly theological discussions, concerning the question of saving the life of the mother or infant," and agree as to the destruction of the latter. Henry VIII. was asked this question before the birth of his son Edward; and with that barbarity and cruelty, for which he was so remarkably distinguished, he exclaimed in a rage, "Save the infant, for it is easier to get wives than children." So he found it. The operation was performed on the mother, to whom it proved fatal. The same question was put to Bonaparte, by Dubois, before the birth of his son, and he answered the terrified accoucheur, "treat the Empress as you would a shopkeeper's wife, in the Rue St. Martin; but if one life must be lost, by all means save the mother."– *O'Meara's St. Helena.* Both were saved.

I have already argued in this article, that no private individual either high or low, has any right or power to advise a medical practitioner; nor has the latter any right whatever to consult a husband, or father, or mother, emperor, empress, king, queen, or any other individual, as to the destruction, or as to the destroying the life of a human being under any circumstances whatever. All the human race are equal in this respect, and there is no exception. The world would have gone as well if neither Henry VIII nor Napoleon never existed, so far as religion, morality, and the rights of humanity were or are concerned. They were mere mortals, like the rest of their species, but somewhat differently situated, and they had, however, no more right to sacrifice the life of a human being, in a social point of view, than the humblest of their subjects. (See pp. 254, 255.)

It will, however, appear, in a succeeding part of this article, that prolicide is unhesitatingly advised by one of the most eminent British obstetricians now living, in women who require the Caesarian operation, and also the removal of a portion of the Fallopian tube and other equally indefensible operations, to induce sterility; on the immorality and illegality of which I shall comment hereafter. (See also pp. 241–243). These recommendations are in unison with the principles of modern new lights, who would prevent generation, or destroy its products, contrary to the antiquated dictates of nature and her Divine Master. Such are the new moral world gentry; some march of intellect philosophers, and many others, who are desirous of destroying the effects of licentiousness and the proofs of infamy, – even at the horrible alternative – the destruction of human life. I have exposed the utter fallacy of these infamous doctrines, in my works on Marriage, Prostitution, and Medical Jurisprudence, already quoted; and likewise in the strictures in the concluding part of this section. But I am proud to record it, that an overwhelming majority – indeed there is only one solitary exception amongst the enlightened and learned members of the faculty, [261/262] who do not agree with me in opinion. They never have, never can sanction the destruction of the foetus in the womb, under any pretence whatsoever, not even to prevent the necessity of the Caesarian operation. They will not do evil that good may follow. They will not commit prolicide, which is justly considered homicide according to the divine, civil, and criminal laws of this and all other civilized countries; notwithstanding the falsely imputed infidelity laid to their charge, or the large bribes which they are so frequently offered.

It is with much pain that I indite these strictures, but when I find a doctrine lightly and flippantly proposed, as will appear hereafter, by an individual whose opinion has great influence upon the rising members of the profession, both junior practitioners and students, in this kingdom and elsewhere, although diametrically opposed to that of the medical faculty in all civilized nations at present, and too well calculated to lead to the commission of the most atrocious crimes, the deaths of most women and their offspring, who may be subjected to fruitless attempts to destroy the embryo in the womb, which generally kill both woman and offspring, (see all modern works on Medical Jurisprudence), the laws of nature, humanity, religion, medicine, and of civilized society, compel me to expose and denounce its inhuman and baneful influence. I neither mean nor intend any personal or professional offence, but medical science and practice is a republic in which every member is entitled to his own opinion, and so also is civilized society. I shall also, in candour and truth, add, that there is not a member of the profession in any country, of whom I entertain a higher opinion as a practical obstetrician, than of my opponent, yet I cannot agree with him on the point under consideration.[4] I shall adduce many additional cogent reasons hereafter, and shall therefore dismiss the subject for the present.

But reverting to the Caesarian operation, I have to observe, that there is not as yet, so far as I know, a correct history of the antiquity or fatality of gastrohysterotomy, in any of our ancient or modern works on obstetricy; a knowledge of which is indispensably necessary to form an accurate opinion of the danger of this most formidable and fatal operation. I trust I shall therefore be excused for attempting a summary of its history.

[264] *Indication*. – The operation [Cæsarian section, or gastrohysterotomy] is necessary when the sacro-pubic diameter of the brim is reduced to an inch and a half, and even when this strait is not so much contracted, when we cannot succeed in extracting the infant by any other mode of proceeding. [264/265]

When the antero-posterior diameter of the brim of the pelvis measures from two inches and a half to two inches and three-quarters, the French propose the forceps, version, or section of the pubic bones, as already stated. But in such cases, the attempts made with the hand or forceps, often contuse or lacerate the vagina or uterus. The vulva, vagina, and all parts contained in the pelvis may be inflamed, swollen, indurated, or gangrenous; the womb may be entirely detached from the vagina, and the woman so prostrate or debilitated, that she may die on the slightest attempt being made to save her life. In such a case are we to attempt to relieve her, or leave her in despair? I think with most modern obstetricians, that efforts should be made to save her life; because it is utterly impossible to determine, in most cases, the exact state of the organs in the pelvis, or whether they are so diseased as to destroy life in a short time. But while ever the slightest hope remains, we are, I feel convinced, bound to operate; and M. Velpeau did so, in such a case, in 1833, contrary to the opinion of many, but with the approbation of Maygrier, Moulin, Halma-Grand, and Bientot. The woman died soon after the operation; but he still persists in believing that conscience dictates similar conduct under similar circumstances. In extreme cases, the first incision may destroy life, as was the result of a remarkable example related to the Medical Society of London, a few years since. But if all signs were fatal, and death near at hand, I should not operate; as an incision would most probably only extinguish life, when the patient is moribund. I shall, however, immediately show that women supposed to be dead for two hours were re-animated by the operation; and in one case in which version was performed, the life of the infant was saved; the mother, though apparently dead, recovered, and was alive four years afterwards.

The operation is also indicated when the pelvis is contracted in the manner already stated, and the woman more than seven months pregnant; also in those cases in which the cavity, or inferior aperture of the pelvis is so contracted as not to admit of delivery by any other means, as if there were tumours in the cavity which it is impossible to remove; and likewise when the bodies of twins are united together, and cannot be extracted on account of their size. (*vide ante* p. 223.)

The operation should also be performed soon after the death of the woman in other cases, as the infant has been saved at the lapse of twelve, twenty-four, and forty-eight hours after the mother had expired.

The Princess of Schwartzenberg died at Paris of a burn, and next day the infant was found alive. Gardien relates a similar case, in which the operation was not performed until forty-eight hours after death, and the infant was living. Cangiamila states in his *Embryologia Sacra*, that twenty-one infants were saved in this way in four years; and that the operation was performed twenty times at Syracuse in eighteen months.[5] Numa Pompilius enacted a law, which still exists, in a work entitled *Legregia Diget*. lib. xx., which commanded the physicians to open the bodies of pregnant women after death with the intention of preserving the citizens of the state, 600 AC. The same law prevailed in Venice in 1608 and 1721, which punished medical practitioners severely unless they used the same caution in operating on the dead as [265/266] on living women. In 1749, the King of Sicily punished medical attendants with death who omitted the operation on women soon after they expired, (see *Author's Manual of Medical Jurisprudence, 1836.*)

Ebel states, that an infant was born after the interment of the mother, whose body was exhumed in consequence of a judicial inquiry. (*Burns's Midwifery, 1837.* Several witnesses attested, that a woman who died at seven o'clock a.m., and appeared so in the evening, was found to have given birth, next day, to an infant. (*J. Univ. Med. des Sci. Med. Tom.* 7, p. 149). Sarrois states, that a living infant was extracted two hours after the death of the mother. Jackson, Deleau, Huguier, Jolly, Duparcque, Laaverjat, Reicke, Green, Blundell, &c. have extracted living infants, by the caesarean operation, a few minutes after the deaths of the mothers. Dr. Blundell succeeded in fifteen minutes after the death of the woman. – (See former edition, 1831, of this work). Van Swieten, Baudelocque, and many other celebrated authors, cite cases of women supposed to be dead, who were roused from their lethargic, cataleptic, or hysterical state, by the caesarean section. Peu states, that in a case in which he commenced the first incision, the body trembled, the woman moved her lips and ground her teeth, to his great horror. Trinchinetti relates a case nearly similar. Rigaudeaux was summoned to a woman at Douay, whom he supposed to have been dead for two hours. Before he proceeded to open the abdomen, he judiciously examined the pelvis, found it natural, brought down the infant by the feet, and succeeded in restoring it to life in two hours. The limbs of the mother were still supple, and he properly advised that the body should not be buried while they remained so. The woman finally awoke from her lethargy, and, four years afterwards, told her surgeon she was not dead as yet. I have also recorded many singular cases of premature dissection and burial, well worthy of perusal, in another work, to which I refer the reader, (*Manual qf Medical Jurisprudence, 1836*). Art. *Inhumation – Burial of the Dead,* p. 485).

When called to a woman who has suddenly died in labour, or in the last month of pregnancy, we should always examine the state of the pelvis, and when sufficiently capacious, dilate the mouth of the womb, if possible, and bring down the feet of the infant when there is the slightest chance of saving its life. I have, however, failed in such attempts, in two cases in the last month of pregnancy, and the caesarean section would not be allowed. The operation is much easier made than effected.

If the Caesarian operation is tolerated, it should be performed according to the rules hereafter described, and with the same care and caution as during life.

[269] The [Caesarian] operation was performed at Saltzburgh, but delayed a day, because it was doubtful whether the infant was alive or not; a decision which made all the difference between the caesarean section and embryotomy. The motions of the infant were perceived next day, when the former operation was determined upon. The infant was extracted alive, but died after half an hour, and the woman recovered. A reviewer remarks on this case, "we consider the above procedure on the part of the surgeon as well deserving of condign punishment, whether the woman survived or not. To perform the caesarean operation, in preference to embryotomy (where the latter is practicable) is most unwarrantable, and evinces a lamentable, not to say a culpable want of judgment, as to the proper estimate of the value of human life." (Vide *ante*, p. 245). The French are deserving "of this condign punishment," as well as the Germans; for M. Duges, a late writer, and many others, assert that the crotchet aigu, which is the same as the perforator, should not be employed until after the death of the infant: "nous avons dit, qu'on ne pouvait l'appliquer qu'apres la mort certaine du fœtus; ne peut être applique que sur un enfant indubitablement mort." He likewise asserts that the caesarean section is the only resource in the excessive deformities of the pelvis. The Americans also deserve this condign punishment. My distinguished corre-

spondent, Professor Dewees, late of Philadelphia, well observes, "from an attentive consideration of both operations, the crotchet and caesarean section, we are free to confess ourselves in favour of the latter, and for the following reasons: First, because the infant must be destroyed by the crotchet; second, because the risks are often very great to the woman; third, because there are cases in which it is impossible to deliver with the crotchet; fourth, because where this instrument is employed, there is a great risk to the [269/270] mother, without a chance of benefit to the infant. These remarks refer to cases in which it is ascertained, or presumed, that the infant is living; if it is dead, then the crotchet may be used under a sufficient diameter of the pelvis. But if the infant is dead, and the delivery impossible by the crotchet, the caesarean operation should be proposed," (p. 594). He further observes, "for what reprehension, what punishment would be sufficiently severe for that practitioner, who after having destroyed the infant, should find it impossible to deliver it; and then, for its accomplishment, subject the poor woman to the caesarean section? He would scarcely merit the plea of *quo animo* in his favour." Par. 1479. (See Craniotomy and Symphyseotomy, as well as Gastrohysterotomy, in this work).

When we contrast the number of unsuccessful and successful authenticated cases in the eighteenth and nineteenth centuries, we shall find the results as follow: 147 deaths, 118 recoveries.

The operation is only performed in the British dominions in extreme cases of deformity, the sacro-pubic diameter being one ich and a half in. and upwards.

In eighty cases the following were the admeasurements in sixty-two:–one inch in 1 case; one inch and a half in 8; same and two lines in 23; two inches and a half and two lines in 25; two inches and a half and two three-quarters in 5=62.

From 1821 to 1830 the operation was performed in sixty-one cases, and only twenty-eight from 1810 to 1820.

In thirty-six operations in lying-in hospitals, eleven were fortunate and twenty-five unfortunate. In one practice, thirty-one succeeded out of sixty.

When the operation was performed before or immediately after the discharge of the amniotic fluid or water, the infants were extracted alive. In such cases, the proportion of women saved to that of those lost, is as 4:3.

Total of infants living 67, dead 29. The general proportion of favourable to unfavourable cases of the caesarean operation is 3:4.

According to M. Velpeau, from whom I quote, the causes of death were the following:

Peritonitis and enteritis, 13; gangrene, 8; hæmorrhage, 7; effusion into the abdomen, 3; meteorism or tympanites, 3; prostration or sinking, 3; shock of operation, 2; convulsions, 2; and colliquative diarrhœa, 1.

Days on which death occurred.— 1st day eight died; 2—six; 3—ten; 4—five; 5—twelve; 6—four; 7—one; 8—three; 18, 20, 27, 30, 45 —one on each day.

Time of recoveries.— In 3 weeks three recovered ; 4 in three ; 5--five; 6—five; 7—three; 8—three; and 10 in two weeks.

It is utterly impossible, in my opinion, in the present state of science, to arrive at any positive conclusion, either as to the mortality of women, caused by the caesarean operation, because no two cases, constitutions, or circumstances are alike, or can be determined.

It is not quite certain, perhaps, indeed it is very improbable, that all the unfavourable cases have been recorded, but there can be no doubt [270/271] as to successful ones. M. Velpeau observes, I think very properly, that up to the present time, the caesarean operation has been fatal at least in one in three, if not, in one in two cases. The great majority of authors are likewise of this opinion, and a great majority agree with the sentiment of Sir Fielding Ould, of Dublin, "that to practise it is a proof of detestable, illegal, and barbarous inhumanity." – I should say, if it can possibly be avoided.

The preceding details must convince obstetricians of the great danger of the operation, and that they should never have recourse to it without absolute necessity.

As the English law now stands, the destruction of the infant is a felony, even to save the life of the mother; yet the contrary opinion is the prevailing one in this country. "In this country and in France, however painful it may be to destroy the life of the child, the mischief is considered less serious in its consequences, than the destruction (or we ought, perhaps, rather to say) the probable destruction of the mother." (*Lancet*, 1828, p. 328). Dr. Kind asserts, that in Germany, the expediency, as in this country, of destroying the child, is always left to the judgment of the medical attendant. (*Op. Cit.* p. 415. *Vide ante*, p. 245). The first quotation is not correct.

The validity of this conclusion, will be further acknowledged, when we consider the number of important parts, divided by the incisions in gastro-hysterotomy – the parietes of the abdomen; the double division of the peritoneum; the incision of the enlarged arteries, veins, and nerves of the gravid or pregnant womb; and of the substance of that organ itself, all of which may be followed by inflammation, gangrene, or haemorrhage; especially when the powers of life are greatly weakened, or almost destroyed. The escape of the amniotic fluid, by tedious labour, as in the cases in this country, and of blood into the cavity of the abdomen, may, unless removed, induce fatal peritonitis.

It is, however, to be borne in mind, that the substance of the womb is not very sensitive, so that it has been often wounded or ruptured, without the supervention of inflammation.

The experiments of Blundell, the operations of gastrotomy performed by Lizars, and the numerous cases of penetrating wounds of the abdomen, even the intestines being transfixed, as recorded in the works on military surgery, and in Professor Cooper's valuable Dictionary of Surgery, in which recoveries took place, though there was effusion of blood and faeces into the cavity of the abdomen, clearly show, that when the powers of life are not weakened, as in the inferior animals, in women of good constitution, who are operated on in France and Germany, but rarely in this country: or in soldiers who are well fed, that gastrotomy is by no means so fatal, as gastro-hysterotomy in women of the worst constitutions, suffering from osteomalaxia, who are greatly exhausted by long continued labour-pains, and who are too often, indeed generally sinking before the operation is commenced, which has almost been hitherto the case in this kingdom, except in Mr. Barlow's solitary example, or perhaps the case of Mary Dunnally, already noticed.[6]

[277] The fatal results are so great, that some modern obstetricians of eminence have gravely proposed to excise a portion of either Fallopian or uterine tube, or the whole uterus, when the abdomen is laid open, and before inserting the sutures, for the purpose of inducing sterility. This proposal requires to be rigidly examined, on moral, social, and medical grounds. (see p. 261).

Two obstetric authors of eminence have proposed operations for the prevention of future conceptions when gastrohysterotomy has been performed. These proposals are, in my opinion, untenable in a moral point of view, and have never been hitherto adopted. Michaelis proposed to prevent future conceptions and the necessity of the caesarean operation, and to avoid a very great reaction, by which, I presume, he means conjugal dissension and infidelity, to extirpate the uterus (*Kilian, Op. Cit.*); and Dr. Blundell not only sanctions this operation, but also proposes a substitute, viz., the removal of a line of the Fallopian tube, right or left, so as to obliterate its calibre – the larger blood-vessels being avoided: mere division of the tube might be sufficient to produce sterility; but the further removal of a portion of the tube appears to be the surer practice. I recommend this precaution, therefore, as an improvement of the operation. (Work, by Lee and Rogers, 1840, already quoted, p. 359).

I cannot admit the correctness of this physiology, because it is on record and well authenticated, that a woman who had but one ovary and uterine tube, and who was found to possess no more on actual dissection, not only was not sterile, but was the mother of infants of the different sexes. No less an authority than M. Velpeau asserts this fact from actual dissection.

It therefore clearly follows, that obliteration, or even absence of one fallopian or uterine tube, or even the absence of one tube and ovary, neither might nor can produce sterility. It is equally well known, that a man who has one testicle destroyed by disease or removed by castra- [277/278] tion, may beget offspring of both sexes, proofs of which will be found in all the standard treatises on Medical Jurisprudence of modern times.

I am perfectly at a loss to perceive upon what moral or physical grounds, women who may or may not be deformed, are to be rendered sterile by surgical operations, or other means, and men exempted, though really the sources from which our race is perpetuated. This is not according to the laws of nature. I shall not dwell further upon this subject, as its various bearings would lead me into a long digression, and more particularly as I have discussed it elsewhere (see *Philosophy of Marriage*), and shall only add, that perhaps at some future period, during the rapid march of intelligence and knowledge, it may be gravely proposed to emasculate or render sterile all husbands and others, whose wives or mistresses may have been subjected or liable to, the performance of the caesarean operation.

The removal of the womb, if justifiable, which I strongly deny, though much more dangerous, and so fatal, that even when the organ is incurably diseased, has been totally abandoned of late years, would be a more effectual preventive; but even this remains to be proved, as every one knows there are extra-uterine conceptions. I cannot here enter upon physiological discussions, which would be out of place, but I refer the interested reader to the account of generation in a preceding chapter.

I am well aware that the womb, when diseased, has been cut away, and also after delivery absolutely torn away by ignorant practitioners, and even separated during the action of the bowels or bladder; but such cases are not in point, and do not relate to the question under consideration. (See p. 184).

That question is this, is it right or moral to use any means or operation to prevent procreation? I boldly answer in the negative – because any such means or operation prove a doubt in the minds of those who employ them, of the omnipotence of Divine Providence, and of his power to preserve the human race. The induction of abortion, to destroy the life of the foetus, is a felony according to the laws of this country, and is considered homicide by the most eminent of the medical faculty in all countries.

The following proposals by Dr. Blundell have not been as yet tried, so far as I know, nor are they ever likely to be sanctioned by the profession, on account of their danger.

"Now, is there any other mode in which, when the obstruction of the pelvis is insuperable, the formation of a foetus may be prevented? In my opinion, there is. If a woman were in that condition, in which delivery could not take place by the natural passage, – provided she distrusted the circumstance in which she was placed, – I would advise an incision (of 1 in. in length) in the linea alba, above the symphysis pubis. I would advise further, that the Fallopian tube (on either side) should be drawn up to this aperture; and, lastly, that a portion of the tube should be removed; – an operation easily performed; when the woman would, for ever afterwards, be sterile. All this may be done after due consideration; – circumstances not forbidding. 'But the abdominal inci- [278/279] sion; – that is bad.' True; but the caesarean incision; – that is worse! Is not that true also?

"*Destruction of the Ovum.* – If a woman, in the earlier months of pregnancy, be known to have a pelvis contracted in a high degree, is there nothing which you may then do to prevent an ultimate need of the caesarean operation? Abortive medicines might, in this case, be thought of; or, these failing or rejected, – if you could feel the os uteri, – you might introduce a female sound, or any other instrument of that kind; and, passing this sound into the uterine cavity, you might completely break up the structure of the ovum, so as to prevent the progress of generation. In doing this, there would always be a risk of haemorrhage; but where you are endeavouring to avoid the necessity of the caesarean incisions, this risk would be justifiable. The substitution of the smaller evil for the greater, is frequently the principle of the healing art. But what if the os uteri be inaccessible? Is there, in such a case, any other expedient to which we may have recourse? In a case like this, were my opinion consulted, I should be inclined to reply, – 'As a substitute for the caesarean operation, let an incision be made as before, above the symphysis pubis; then let some instrument (such as a trocar, or canula) be carried into the cavity of the uterus; let this instrument be sufficiently stiff to enter the cavity, and retain its form there under pressure; and then let it be resolutely moved about in the uterus, so as to break up completely the texture of the ovum. The whole instrument need not be much thicker than a bell-wire. The process is allied to that of acupuncture. The point of the trocar, on entering the uterus, should be withdrawn with the canula; a finger should be carefully placed on the uterus, so as to guide the instrument, and guard against injury of the intestines or the bladder.' Scribblers had better content themselves with sneering at the operation; – surgeons had better perform it! – *Artem quisque suam!* To produce future sterility, the Fallopian tubes might be rendered impervious." – *Op. Cit.*

All these proposals are liable to the most serious objections. The recommendation of destroying the ovum or fetus in the womb by so high an authority, may so influence many junior members of the profession, as to lead them to attempt to accomplish it in cases of illegitimate pregnancies, and render themselves liable to criminal proceedings for felony, punishable by long imprisonment, flogging, transportation for different periods, and in some cases, by an ignominious death. (See Author's *Manual of Medical Jurisprudence,* second edition, 1836; see also pp. 241–243, 254, 255, 261, 262).

The induction of premature labour is effected for the purpose of saving the life of the infant, preventing the necessity of craniotomy and the caesarean operation, and thus saving the lives of both mother and infant, is a laudable and humane proceeding, the most contrary to that under consideration, which is, in my opinion, properly considered, most reprehensible and penal. (See p. 240).

The mode of inducing premature labour I described in a dead language, to prevent, as far as possible, its being practised for criminal purposes. (See p 240).[7] As to the above quotation, he who runs may read; but, fortunately, the description is so general and vague, that few, [279/280] if any, could succeed in breaking up the ovum without more or less injuring the mother, and most probably destroying her life, as well as that of her innocent and unoffending offspring.

The orifice of the womb is inaccessible in the early months of pregnancy (see p. 155), and no instrument can be passed into it. It is for this reason that we are advised to resort to other operations, nearly as dangerous as the caesarean section, which, though tried on inferior animals, have never been attempted on the human female, nor never can be, by any one who properly values the rights of humanity, and his professional reputation. They may appear very ingenious and bold in a lecture-room, to novices who are incapable of forming a correct opinion on the subject, or to practitioners unacquainted with the histories of gastrotomy and gastrohysterotomy, but certainly not to learned and experienced obstetricians, who will and must avoid them. Here I must add, in conclusion, the old adage, – "Amicus Socrates, amicus Plato, sed magis amica veritas." Socrates is a friend, Plato is a friend, but truth is a greater friend.

Notes

1. *Editors' note*: Dr. William Potts Dewees (1768–1841) was professor of midwifery at the University of Pennsylvania, author of many works on midwifery, and a favorite authority of Ryan's.

2. *Editors' note*: Ryan repeats most of this material in his 1841 edition, but intersperses it with a good deal more technical information on the diagnosis and management of labor and its complications; *Manual of Midwifery* (1841): 166–179.

3. *Editors' note*: Thomas Denman (1733–1815) was the author of an introduction to midwifery which went though numerous editions, and several volumes of "aphorisms" on various topics. Ryan edited a volume of his aphorisms posthumously.

4. *Editors' note*: The "opponent" Ryan here refers to would apparently be Dr. Blundell, whose ideas are discussed below, p. 277 of Ryan's text. This was presumably Dr. James Blundell, of Guy's Hospital (1790–1878), author of several obstetric works.

5. *Editors' note:* Francesco Emmanuele Cangiamila, 1702–1763.

6. *Editors' note:* Ryan had earlier, in some detail, relayed a 1793 case in which Mr. Barlow successfully performed a Caesarian section on a patient who then survived until 1826, although the infant was dead at the time of the operation; in this case the severely contracted pelvis made embryotomy impossible. Ryan also cited the case of Mary Dunnally, an Irish midwife, who extracted a dead infant with a razor in 1738, sewed the wound with tailor's thread, and applied egg whites. The patient recovered in 27 days. This was apparently the first recorded successful caesarian section in Britain. Ryan, *Manual of Midwifery* (1841), 256–260.

7. *Editors' note*: Ryan described the specific features of the "operation" in Latin, pp. 242–243 of this volume.

Bibliography

I Writings by Michael Ryan

Ryan, Michael. *Tentamentum Medico-Physicum Inaugurale, de Genere Humano ejusque Varietatibus...* Edinburgh: P. Neill, 1821.

Ryan, Michael. "Remarks on the Use of Hydrocyanic Acid in Angina Pectoris and other Diseases of the Heart, and in Bronchitis, Phthisis, and Dyspepsia," *London Medical and Physical Journal* 51 (1824): 369.

Ryan, Michael. *A Treatise on the Most Celebrated Mineral Waters of Ireland containing an Account of the Waters of Ballyspellan, Castleconnel, Ballynahinch, Mallow, Lucan, Swadlinbar, Goldenbridgre, Kilmainham, &c., &c.: and of the spa lately discovered at Brownstown, near Kilkenny, with Plain Directions during the Use of Mineral Waters and an Account of some of those Diseases in which they are Most Useful.* Kilkenny: Printed and published by J. Reynolds, for Hodges, M'Arthur, and J. Cummings, Dublin, 1824.

Ryan, Michael. "An Essay on the Natural, Chemical, and Medical History of Water, in its simple and Combined States; including an Account of the Chemical composition, and Medical Effects, of the principal Mineral Waters in the United Kingdom of Great Britain and Ireland, and also the Continent of Europe," *London Medical and Physical Journal* 54 (1825): 442–61.

Ryan, Michael. "Delirium Tremens, Treated as a Nervous Disease," *Lancet* 2 (1827–28): 791–93.

Ryan, Michael. "Introductory Lecture to the Theory and Practice of Midwifery," *Lancet* 2 (1827–28): 394–400.

Ryan, Michael. *A Manual on Midwifery; or a Summary of the Science and Art of Obstetric Medicine, including the Anatomy, Physiology, Pathology, and Therapeutics, Peculiar to Females; Treatment of Parturition, Puerperal, and Infantile Diseases; and an Exposition of Obstetrical-Legal Medicine.* London: Longmans, 1828.

Ryan, Michael. *Remarks on the supply of water to the metropolis: with an account of the natural history of water in its simple and combined states, and of the chemical composition and medical uses of all the known mineral waters being a guide to foreign and British watering places.* London: Longmans & Co., Anderson, and Messrs. Underwood and Highley, 1828.

Ryan, Michael. "To the Editor of the *London Medical Gazette* [reply to report of meeting of the Medical Society of London]," *London Medical Gazette* 3 (1828–29): 684.

Ryan, Michael. "Drs. Ryan and Gordon Smith [letter]," *Lancet* 1 (1830–31): 106–108.

Ryan, Michael. *Lectures on population, marriage, and divorce, as questions of state medicine, comprising an account of the causes and treatment of impotence and sterility, and of the morbid and curative effects of marriage; forming a part of an extended course on medical jurisprudence, delivered at the Medical Theatre, Hatton-Garden.* London: Renshaw and Rush, 1831.

Ryan, Michael. *Manual of Medical Jurisprudence*, 1st ed. London: Renshaw and Rush, 1831. *Note:* Material in this volume was initially published in Ryan's journal, as follows:

Hippocratic ethics, *London Medical and Surgical Journal* 3 (1829): 485–95.

Importance of the study of state medicine, *London Medical and Surgical Journal* 4 (1830): 217 ff.

[Ancient medical ethics], *London Medical and Surgical Journal* 4 (1830): 300 ff.

[Medical ethics of the Middle Ages], *London Medical and Surgical Journal* 4 (1830): 400 ff.

[Medical ethics of the present period], *London Medical and Surgical Journal* 4 (1830): 492 ff.

Ryan, Michael. "To the Editor of *The Lancet*" [in reply to *Lancet* review of *Manual of Medical Jurisprudence*]. *Lancet* 1 (1831–32): 222–24.

Ryan, Michael. "Tweedie v. Ramadge [with Ryan's addendum]," *London Medical and Surgical Journal* 7 (1831); reprinted, *Lancet* 2 (1831–32): 408–9.

Ryan, Michael. "Reason of the Verdict against Ryan and Co." [reply to *Lancet* editorial]. *Lancet* 2 (1831–32): 496–97.

Edwards, Milne and Vavasseur, P. *A New Practical Formulary of Hospitals of England, Scotland, Ireland, France, Germany, Italy, Spain, Portugal, &c., &c.,* [Translated from the French and augmented by Michael Ryan]. Henderson, London, July 1835.

Ryan, Michael. *A Manual of Medical Jurisprudence and State Medicine, Compiled from the Latest Legal and Medical Works, of Beck, Paris, Christison, Fodere, Ofila, etc.* 2nd ed. London: Sherwood, Gilbert, and Piper, 1836. [Available at http://books.google.com/books?id = vToEAAAAQAAJ&printsec = frontcover&dq = michael + ryan&lr =, accessed December 14, 2007.]

Ryan, Michael. "Hyoscyamus and Its Preparations," [letter] *London Medical Gazette* 18 (1836): 609–10.

Hooper, Robert. *Physician's Vademecum,.* Michael Ryan ed. London: H. Renshaw, 1837.

Ryan, Michael. *The Philosophy of Marriage in its Social, Moral, and Physical Relations; with an Account of the Diseases of the Genito-urinary Organs, Which Impair or Destroy the Reproductive Function, and Induce a Variety of Complaints; with the Physiology of Generation in the Vegetable and Animal Kingdoms; Being Part of a Course of Obstetric Lectures Delivered at the North London School of Medicine, Charlotte Street, Bloomsbury, Bedford Square.* London: John Churchill, 1837. [Available at http://books.google.com/books?id = x0kEAAAAQAAJ&printsec = frontcover&dq = michael + ryan&lr =, accessed December 14, 2007.]

Ryan, Michael. *The Medico-Chirurgical Pharmacopœia; or, a Conspectus of the Best Prescriptions in Medicine, Surgery, Obstetricy, and Infantile Medicine; with a Table of the Doses of All Medicines in Use; the Additions in the London Pharmacopoeia, 1836; M. Magendie's Formulary; Aphorisms on the Treatment of Poisoning; Reduction of Dislocations and Fractures; and on Natural and Difficult Parturitions, with Puerpal Diseases,* 2nd ed. London: John Churchill, 1838.

Ryan, Michael. "Dr. Ryan's Reply to the Notice of Mr. Houston's Manual on Diseases of the Eye," *Lancet* 1 (1838–39): 705–6.

Ryan, Michael. *Prostitution in London with a Comparative View of That of Paris and New York, as Illustrative of the Capitals and Large Towns of All Countries; and Proving Moral Depravation to Be the Most Fertile Source of Crime, and of Personal and Social Misery; with an Account of the Nature and Treatment of Various Diseases, Caused by the Abuses of the Reproductive Function.* London: H. Bailliere, 1839. [Available at http://books.google.com/books?id = zHsEAAAAQAAJ&printsec = frontcover&dq = michael + ryan + prostitution + in + london&lr =, accessed December 14, 2007.]

Ryan, Michael. *A Manual of Midwifery and Diseases of Women and Children...,* 4th ed. London: (self-published), 1841.

II Secondary Sources and Other Works

Aikin, John. *Thoughts on Hospitals/with a Letter to the Author, by Thomas Percival*. London: Joseph Johnson, 1771.

[Anonymous]. "Death of Dr. Ryan," *Provincial Medical and Surgical Journal*, 1 (1840–41): 206–7.

[Anonymous]. "Dr. Ramadge and St. John Long," *London Medical Gazette* 8 (1831): 117–20.

[Anonymous]. "[Review of] *A Manual of Medical Jurisprudence ...* by Michael Ryan..." *Lancet* 1 (1831–32): 137–43.

[Anonymous]. [Review of *Medical Ethics...*, by Thomas Percival], *Lancet* 12 (series 2) (1826–27): 696.

[Anonymous]. [Review of Ryan, *A Manual of Midwifery*, 3rd ed.] *London Medical Gazette* 9 (1831–32): 52.

[Anonymous]. [Review of Ryan, *A Manual on Midwifery*,] *London Medical Gazette* 2 (1828): 714–15.

[Anonymous]. [Review of Ryan, *Prostitution in London*], *British and Foreign Medical Review* 7 (1839): 540.

[Anonymous], *Review of Works on Medical Ethics*. Edinburgh: Murray & Gibb, 1850.

[Anonymous]. [Review of S. Little's *A Manual of Diseases of the Eye; or, Treatise on Ophthalmology*,] *Lancet* 1 (1838–39): 494.

[Anonymous]. Subscription for Dr. Ryan, *Lancet* 1 (1832–33): 351–352.

[Anonymous]. "Subscription for the Widow and Children of the late Dr. Ryan," *Lancet* 1 (1840–41): 903.

Bacon, Francis. *The Advancement of Learning*, ed. Michael Kiernan. Oxford, UK: Clarendon Press, 2000.

Baker, Robert B., Arthur L. Caplan, Linda L. Emanuel, and Stephen R. Latham (eds.) *The American Medical Ethics Revolution*. Baltimore, MD: Johns Hopkins University Press, 1999.

Baker, Robert B. "The Discourses of Practitioners in Nineteenth- and Twentieth-Century Britain and America." In *Cambridge World History of Medical Ethics*, eds. Baker, Robert B. and Laurence B. McCullough. Cambridge, UK: Cambridge University Press, 2009:446–464.

Baker, Robert B., Arthur L. Caplan, Linda L. Emanuel, and Stephen R. Latham, "Introduction," In *The American Medical Ethics Revolution: How the AMA's Code of Ethics Has Transformed Physicians' Relationships to Patients, Professionals, and Society*, eds. Baker, Robert B., Arthur L. Caplan, Linda L. Emanuel, and Stephen R. Latham. Baltimore, MD: Johns Hopkins University Press, 1999.

Baker, Robert B. and Laurence B. McCullough. "What Is the History of Medical Ethics?" In: *Cambridge World History of Medical Ethics*, eds. Robert B. Baker and Laurence B. McCullough. Cambridge, UK: Cambridge University Press, 2009:3-15.

Bartrip, Peter. "An Introduction to Jukes Styrap's *A Code of Medical Ethics* (1878)," in *The Codification of Medical Morality*, ed. Robert Baker. Boston, MA: Kluwer, 1995: 145–148.

Beecher, Henry K. *Research and the Individual: Human Studies*, Boston, MA: Little, Brown & Co, 1970.

Berlant, Jeffrey. *Profession and Monopoly: A Study of Medicine in the United States and Great Britain*, Berkeley, CA: University of California Press, 1975.

Burns, Chester R. "Reciprocity in the Development of Anglo-American Medical Ethics," in *The Codification of Medical Morality*, ed. Robert Baker, Boston: Kluwer, 1995: 135–143.

Caldwell, John C. "Malthus and the Less Developed World: The Pivotal Role of India," *Population and Development Review* 1998; 24: 675–96.

Castro, Roderic a (Rodrigo de Castro), *Medicus-politicus: sive de officiis medico-politicis tractatus, quatuor distinctus libris: in quibus non solum bonorum medicorum mores ac virtutes exprimuntur, malorum vero fraudes et imposturae deteguntur ...* (The Politic Physician: or a Treatise on Medico-Political Duties ...), Hamburg, [no publisher identified] 1614.

Celsus, Aulus Cornelius. *De Medicina*. Florence: Nicolaus Laurentii, Alamanus, 1478.

Clarke, James F. *Autobiographical Recollections of the Medical Profession*, London: J. & A. Churchill, 1874.

Cope, Zachary. "The Private Medical Schools of London, 1746–1914." In F. N. L. Poynter, ed. *The Evolution of Medical Education in Britain*. Baltimore, MD: Williams & Wilkins, 1966: 89–109.

Corden, Thompson. *A Letter to the Public on the Necessity of Anatomical Pursuits; with Reference to Popular Prejudices, and to the Principles on Which Legislative Interference in These Matters Ought to Proceed*. London: Taylor, 1830.

D., H.W. "Obituary: John Gordon Smith, M.D., F.R.S.L.," *Gentleman's Magazine*, September 1833: 278–79.

Denman, Thomas. *The Obstetrician's Vademecum; or Aphorisms on Natural and Difficult Parturition; the Application and Use of Instruments in Preternatural Labours; on Labours Complicated with Hemorrhage, Convulsions, etc* (edited and augmented by Michael Ryan). London: Cox, 1836.

Dermott, George Darby. *A Discussion on the Organic Materiality of the Mind, the Immateriality of the Soul, and the Non-Identity of the Two*. London: Callow & Wilson, 1830.

Deslandes, Leopold. *De l'Onanisme et des autres Abus Veneriens*. Paris: A. Lelarge, 1835.

Desmond, Adrian. *The Politics of Evolution: Morphology, Medicine, and Reform in Radical London*. Chicago, IL: University of Chicago Press, 1989.

Epps, Elizabeth, ed. *Diary of the Late John Epps*. London: Kent & Co., 1875.

Epps, John. *Internal Evidences of Christianity Induced from Phrenology*, Edinburgh: J. Anderson, 1827.

Fraser, William. *Queries in Medical Ethics* [read before the Medico-Chirurgical Society of Aberdeen, 5th April 1849]. *London Medical Gazette*, Vol. 2, 1849: 181–7, 227–32.

Godman, John D. *Addresses Delivered on Various Public Occasions*. Philadelphia, PA: Carey, Lea & Carey, 1829.

Godman, John D. *Professional Reputation: An Oration before the Philadelphia Medical Society* [on February 8, 1826]. Philadelphia, PA: Benjamin & Thomas Kill. 1826.

Grant, Alexander. *The Story of the University of Edinburgh During Its First Three Hundred Years*. London: Longmans, Green &Co., 1884.

Gregory, James. *Conspectus Medicinæ Theoreticæ ad Usum Academicum*. Edinburgh: A. Constable, 1818.

[Gregory, John.] *A Comparative View of the State and Faculties of Man with those of the Animal World*, 2nd ed. London: J. Dodsley, 1766.

Gregory, John. *A Father's Legacy to his Daughters*. Edinburgh: A. Strahan and W. Creech, 1788.

Gregory, John. *Elements of the Practice of Physic: For the Use of Students*. 2nd ed. London: W. Strahan and T. Cadell, 1774.

Gregory, John. *Lectures on the Duties and Qualifications of a Physician*. London: W. Strahan and T. Cadell, 1772.

Gregory, John. *On the Method of Prosecuting Enquiries in Philosophy*, 1770.

Haakonssen, Lisbeth. *Medicine and Morals in the Enlightenment: John Gregory, Thomas Percival, and Benjamin Rush*. Atlanta: Rodopi, 1997.

Harris, James. *Hermes: or, a Philosophical Inquiry Concerning Language and Universal Grammar*. London: J. Nourse and P. Vaillan, 1771.

Haughton, Brian. "Feral Children: The Unsolved Mystery of Kaspar Hauser – Wild Child of Europe." Downloaded on August 25, 2004, from http://www.mysteriouspeople.com/Hauser1.htm

Hoffmann, Friedrich. *Medicus Politicus sive regulae prudentiae secundum quas medicus juvenis studia sua et vitae rationem dirigere debet...* Lugduni Batavorum: Apud Philippum Bonk [etc.], 1746.

Knips Macoppe, Alessandro, 1662–1744. *La politica del medico nell'esercizio dell'arte sua esposta in cento aforismi .../Traduzione italiana con note del Dottore Ignazio Lomeni*. Milan: G. Pirotta, 1826.

Leake, Chauncey D. Preface to Percival's *Medical Ethics*, Baltimore, MD: Williams & Wilkins, 1927.

Levine, Robert. "Clarifying the concepts of research ethics," *Hastings Center Report* 9 (3) (1979): 21–26.

Malthus, T. Robert. *An Essay on the Principle of Population as It Affects the Future Improvement of Society with Remarks on the Speculations of Mr. Godwin, M. Condorcet, and Other Writers.* 1798. [accessed from http://www.faculty.rsu.edu/~felwell/Theorists/Malthus/Essay.htm#35 on 16 November 2007]

McCullough, Laurence B. *John Gregory and the Invention of Professional Medical Ethics and the Profession of Medicine,* Hingham, MA: Kluwer, 1998.

McCullough, Laurence B. (ed.), *John Gregory's Writings on Medical Ethics and Philosophy of Medicine,* Boston, MA: Kluwer, 1998.

McCullough, Laurence B. "Virtues, Etiquette, and Anglo-American Medical Ethics in the Eighteenth and Nineteenth Centuries," in *Virtues and Medicine: Explorations in the Character of Medicine,* ed. Earl E. Shelp, Boston, MA: D. Reidel, 1985.

Moore, Theophilus. *Marriage Customs and Modes of Courtship of the Various Nations of the Universe. Remarks on the Condition of Women, Penn's Maxims, and Counsel to the Single and Married, &c. &c.* 2nd ed. London: [printed for John Bumpus, Holborn] Hamblin, 1820.

O'Brien, Eoin (ed.) *The Charitable Infirmary, Jervis Street, 1718–1987: A Farewell Tribute,* Dublin: Anniversary Press, 1987.

O' Meara, Barry Edward. *Napoleon in Exile, or, A Voice from St. Helena: The Opinions and Reflections of Napoleon on the Most Important Events in His Life and Government, in His Own Words.* London: Printed for W. Simpkin and R. Marshall, 1822.

Parent-Duchatelet, Alexandre Jean Baptiste. *De la Prostitution dans la ville de Paris, considérée sous le rapport de l'hygiène publique, de la morale et de l'administration; ouvrage appuyé de documents statistiques puisés dans les Archives de la Préfecture de Police; et précédé d'une Notice sur la vie et les ouvrages de l'auteur* par Fr. Leuret. Paris: J.-B. Baillière, 1837.

Paris, J.A. *Pharmacologia, Corrected and Extended, in Accordance with the London Pharmacopoeia of 1824, and with the Generally Advanced State of Chemical Science.* London: W. Phillips, 1825.

Percival, Thomas. *Essays Medical, Philosophical, and Experimental.* London: Printed by W. Eyres for J. Johnson, London, 1788–1789.

Percival, Thomas. *Medical Ethics; or, a Code of Institutes and Precepts adapted to the Professional Conduct of Physicians and Surgeons…* Manchester: Russell, 1803 (modern reprint by the Classics of Medicine Library, Birmingham, AL: 1985).

Percival, Thomas. *Medical Ethics; or, a Code of Institutes and Precepts, Adapted to the Professional Conduct of Physicians and Surgeons…by the late Thomas Percival…,* London: W. Jackson, 1827.

Pernick, Martin. The Patient's Role in Medical Decisionmaking: A Social History of Informed Consent in Medical Therapy," in: President's Commission for the Study of Ethical Problems in Medicine and Biomedical and Behavioral Research, *Making Health Care Decisions: The Ethical and Legal Implications of Informed Consent in the Patient-Practitioner Relationship; Volume Three: Appendices: Studies on the Foundations of Informed Consent,* Washington, DC: U.S. Government Printing Office, 1982: 1–35.

Phelan, Denis. *A Statistical Inquiry into the Present State of the Medical Charities of Ireland: With Suggestions for a Medical Poor Law, by Which They May be Rendered Much More Extensively Efficient.* Dublin: Hodges and Smith, 1835.

Pickstone, J.V. and Butler, S.V.F. "The Politics of Medicine in Manchester, 1788–1792: Hospital Reform and Public Health Services in the Early Industrial City," *Medical History* 28 (1984): 227–49.

Place, Francis. *Illustrations and Proofs of the Principle of Population: Being the First Work on Population in the English Language Recommending Birth Control* (1822); London: Routledge/ Thoemmes, 1994.

Pliny the Elder. *The Natural History* (translated with notes and illustrations by John Bostock and H.T. Riley). London: Taylor & Francis, 1855.

Porter, Roy. *The Greatest Benefit to Mankind,* New York: W.W. Norton, 1997.

Ramadge, Francis H. "London Medical Society: Expulsion of Dr. Ramadge," *Lancet* 2 (1830–31): 251–252; *London Medical and Surgical Journal* 7 (1831).

Ramadge, Francis H. "[John] Long the Quack" *Lancet* 2 (1830–31); No. 398 (16 April 1831): 90–93; No. 400 (30 April 1831): 154–156.

Rivett, Geoffrey. *The Development of the London Hospital System, 1823–1982*. London: King's Fund, 1986. [Section: Voluntary Hospitals" accessed http://www.nhshistory.net/voluntary_ hospitals.htm (accessed January 23, 2005.)]

Schleiner, Winfried. *Medical Ethics in the Renaissance*. Washington, DC: Georgetown University Press, 1995.

Smith, John Gordon. *An Analysis of Medical Evidence Comprising Directions for Practitioners in the View of Becoming Witnesses in Courts of Justice, and an appendix of Professional Testimony*. London: T. & G. Underwood, 1825.

Smith, Gordon. "Drs. Gordon Smith, Ryan, and A. Thomson," *Lancet* 1 (1830–31): 72–73.

Styrap, Jukes. *A Code of Medical Ethics*. London: J. & A. Churchill, 1878.

U.S. National Commission for the Protection of Human Subjects of Biomedical and Behavioral Research. *The Belmont Report: Ethical Principles and Guidelines for the Protections of Human Subjects*. Washington, DC: U.S. National Commission, 1978.

Veatch, Robert M. *Disrupted Dialogue: Medical Ethics and the Collapse of Physician-Humanist Communication (1770–1980)*. New York: Oxford University Press, 2005.

Velpeau, Alfred Armand L.M. *Traité complet de l'art des accouchemens*. Paris: Baillière, 1835.

Vinten-Johansen, Peter, Brody, Howard, Paneth, Nigel, Rachman, Stephen and Rip, Michael. *Cholera, Chloroform, and the Science of Medicine: A Life of John Snow*, New York: Oxford University Press, 2003.

[Wakley, Thomas.] [Editorial]. *Lancet* 1 (1838–39): 706.

[Wakley, Thomas.] [Editorial]. *Lancet* 2 (1831–32): 92, 21 April 1832.

[Wakley, Thomas.] "Letter from Dr. Ryan," *Lancet* 1 (1831–32): 222–26.

[Wakley, Thomas.] "Ramadge v. Wakley," *Lancet* 2 (1831–32): 408–409.

[Wakley, Thomas.] "Ramadge v. Ryan et al.," *Lancet* 2 (1831–32): 408–409.

[Wakley, Thomas.] "Tweedie v. Ramadge," *Lancet* 2 (1830–31), no. 419 (10 Sept. 1831).

Watson, Richard. *An Apology for the Bible in A Series of Lectures Addressed to Thomas Paine*. Boston, MA: James White, 1796.

Widdess, J.D.H. *The Royal College of Surgeons in Ireland and its Medical School, 1784–1984*, Dublin: Royal College of Surgeons in Ireland, 1984.

Index